RED DEER

WILDLIFE BEHAVIOR AND ECOLOGY
George B. Schaller, Editor

RED DEER

*Behavior and Ecology
of Two Sexes*

T. H. Clutton-Brock
F. E. Guinness
S. D. Albon

With original drawings by
Priscilla Barrett

EDINBURGH UNIVERSITY PRESS

T. H. Clutton-Brock, F. E. Guinness, and S. D. Albon are members of the Large Animal Research Group, Department of Zoology, University of Cambridge.

British Library Cataloguing in Publication Data
Clutton-Brock, T. H.
Red deer.
1. Red deer—Behavior
I. Title   II. Guinness, F. E.   III. Albon, S. D.
599.73′57   QL737.U5
ISBN   0   85224   446   0   hardback
ISBN   0   85224   447   9   paperback
Printed in
the United States of America

Contents

FOREWORD   xiii
PREFACE   xv
ACKNOWLEDGMENTS   xix

1   NATURAL SELECTION IN MALES AND FEMALES   1
    1.1   Natural Selection and Reproduction   2
    1.2   Selection Pressures in Males and Females   4
    1.3   Constraints on Sexual Dimorphism   6
    1.4   Summary   7

2   RED DEER AND THEIR HABITAT   9
    2.1   Introduction   10
    2.2   The Evolution of Red Deer   10
    2.3   Red Deer in Scotland   12
    2.4   The Red Deer of Rhum   13
    2.5   Topography, Climate, and Vegetation of Rhum   20
          Topography   20
          Climate   20
          Plant Communities   21
          Seasonal Changes in Food Availability   23
    2.6   The Study Area and Its Deer Population   26
    2.7   Summary   31

3   METHODS, SAMPLES, AND DEFINITIONS   33
    3.1   Introduction   34
    3.2   Recognition of Individuals   34
          Identification   34
          Calf Catching   36

Marking Adult Deer 39
Mortality 39
3.3 Data Collection 39
Census Data 39
Rut Censuses 41
Mother/Offspring Surveys 42
Calving Surveys 42
Continuous Watches of Individuals and Groups 42
Continuous Observation of Rutting Groups 44
Incidental Records 44
3.4 Definitions 44
Reproduction in Hinds 44
Reproduction in Stags 46
Social Behavior 47
Habitat Use and Feeding Behavior 48
Population Dynamics 48
3.5 Statistical Analysis 49
Selection of Tests 49
Results of Tests 50
3.6 Summary 50

4 THE BREEDING BIOLOGY OF HINDS 51
4.1 Introduction 52
4.2 Mating 52
4.3 Harem Membership and Mate Selection 56
4.4 Maternal Investment 61
Gestation 61
Birth 62
Early Care 64
Lactation and Milk Yield 68
Suckling Frequency 70
Duration of Suckling Bouts 72
Weaning 73
Early Growth 74
4.5 The Costs of Reproduction 74
Body Condition 74
Subsequent Reproductive Performance 76
Mortality 77
4.6 Summary 78

5   REPRODUCTIVE SUCCESS IN HINDS   80
   5.1   Introduction   81
   5.2   Differences in Lifetime Reproductive Success among
       Hinds   81
   5.3   The Proximate Causes of Variation in Lifetime
       Reproductive Success   82
   5.4   Factors Affecting Fecundity   83
   5.5   Factors Affecting Calf Mortality   84
       Frequency and Timing of Calf Mortality   84
       Immediate Causes of Calf Mortality   85
       Factors Affecting the Survival of Calves through the
         Summer   87
       Factors Affecting the Survival of Calves through the
         Winter   89
   5.6   The Ultimate Causes of Variation in Lifetime
       Reproductive Success   91
   5.7   Maternal Age, Parental Investment, and Calf Survival
       through the Winter   93
   5.8   Summary   98

6   THE RUTTING BEHAVIOR OF STAGS   104
   6.1   Introduction   105
   6.2   Rutting Activities   105
       The Early Rut   105
       Displays and Interactions   107
       Rutting Activities and Age   117
       Activity Budgets of Harem-Holders   121
       Quantitative Aspects of Displays and Interactions   123
   6.3   Interactions between Harem-Holders and Hinds   126
   6.4   Interactions between Harem-Holders and Young
       Stags   127
   6.5   Interactions between Mature Stags   128
       The Form of Contests   128
       Fighting Success   131
       The Costs of Fighting   132
       Adaptive Aspects of Fighting   135
   6.6   Assessment   136
   6.7   Summary   139

7   REPRODUCTIVE SUCCESS IN STAGS   143
    7.1   Introduction   144
    7.2   Harem Size   144
    7.3   Variation in Reproductive Success within
          Seasons   144
    7.4   Lifetime Reproductive Success   146
    7.5   The Proximate Causes of Variation in Lifetime
          Reproductive Success   149
    7.6   Factors Influencing Fighting Success and
          Reproductive Success   150
    7.7   Reproductive Success in Stags and Hinds   152
          Variance in Lifetime Reproductive Success   152
          Reproductive Value   154
          Factors Affecting Reproductive Success   154
    7.8   Sexual Dimorphism and the Factors Affecting
          Reproductive Success   156
    7.9   Summary   157

8   PARENTAL INVESTMENT IN MALE AND FEMALE OFFSPRING   161
    8.1   Introduction   162
    8.2   Adaptive Variation in the Sex Ratio?   162
    8.3   Comparative Investment in Individual Sons
          and Daughters   164
    8.4   Total Investment in Male and Female Progeny   168
    8.5   Summary   173

9   THE STRUCTURE OF SOCIAL GROUPS IN HINDS AND
    STAGS   176
    9.1   Introduction   177
    9.2   Party Size and Composition   178
          Segregation between the Sexes   178
          Variation in Party Size   178
    9.3   Association between Mothers and Calves   182
          Changes Related to Calf Age   182
          *Variation with Mother's Subsequent Reproductive
           Status*   182
          Sex Differences   183
    9.4   Association between Hinds   184
          Daughters and Mothers   184

Individual Differences   186
Association with Other Relatives   186
9.5   Matrilineal Groups among Hinds   187
9.6   Association among Stags   190
Dispersal   190
Stag Groups   191
9.7   Functional Considerations   192
Why Do Red Deer Aggregate?   192
Why Do the Sexes Segregate?   193
Why Do Hinds Associate with Relatives?   194
*Why Do Stags Disperse from Their Mothers' Home
    Ranges?*   196
9.8   Summary   197

10   SOCIAL INTERACTIONS AMONG HINDS AND STAGS   201
10.1   Introduction   202
10.2   Threat Types   202
10.3   Threat Frequency   204
Environmental Factors Affecting Threat Frequency   204
Sex Differences   207
10.4   The Distribution of Threats between Hinds   210
10.5   The Distribution of Threats between Stags   212
10.6   Affiliative Behavior   214
10.7   Functional Considerations   214
10.8   Summary   217

11   FEEDING BEHAVIOR AND HABITAT USE   219
11.1   Introduction   220
11.2   Ingestion and Rumination   222
11.3   Activity Budgets   222
Grazing Bouts   222
Diurnal Variation in Grazing   223
Daytime Grazing Budgets   224
Nocturnal Grazing Budgets   225
Twenty-four-Hour Grazing Budgets   226
11.4   Use of Different Altitudes   227
11.5   Use of Different Plant Communities   229
Seasonal Changes   229
Differences between Years   235

Individual Variation   238
Differences between Milk and Yeld Hinds   238
Sex Differences   239
11.6   Ranging Behavior   242
Day Range Length   242
Home Range Size and Core Area Size   242
11.7   Use of Shelter   246
11.8   Origins of Sex Differences in Feeding and
        Ranging Behavior   247
11.9   Summary   249

12   POPULATION DYNAMICS   258
12.1   Introduction   259
12.2   Changes in the Reproductive Performance of
        Hinds   260
Calf Birth Weight   260
Conception Date   262
Fecundity   262
Calf Mortality in Winter   263
Calf/Hind Ratios   267
Coat Change   267
Mortality and Emigration   268
Habitat Use   269
12.3   Changes in Growth and Reproductive Performance
        among Stags   270
Antler Growth   270
Mortality   270
Adult Sex Ratio   272
Rutting Behavior   273
12.4   Density Effects in Other Ungulates   275
Effects among Females   275
Population Density and Sexual Dimorphism   275
Population Density and Male Mortality   278
12.5   Summary   280

13   THE EVOLUTIONARY ECOLOGY OF MALES AND FEMALES   286
13.1   Sex Differences in Red Deer   287
13.2   Interspecific Comparisons   288
Weight Dimorphism   289
Antlers   289

Testis Size   292
Growth   293
Mortality   294
Feeding Behavior and Dispersal   296
13.3   Implications of Sex Differences   296
For Population Dynamics   296
For Management   297
For Studies of Evolution   299
13.4   Variation in Breeding Systems among Cervids   299
13.5   Summary   304

APPENDIXES   305
   1   Taxonomy of Contemporary Cervidae   307
   2   Red Deer Population of Rhum, 1957–80   309
   3   Climatological Summary for Rhum   310
   4   Characteristics of Soils Connected with Different Plant
         Associations on Rhum   311
   5   Digestibilities of Common Upland Plants   312
   6   Number of Hinds and Stags One Year Old or Older
         Resident in the Study Area in Different Years   313
   7   Altitude and Vegetation in the Study Area   314
   8   Estimates of Lifetime Reproductive Success
         in Hinds   315
   9   Reproductive Value   316
  10   Fighting Success in Stags   317
  11   Estimates of Lifetime Reproductive Success
         in Stags   318
  12   Definition of Party Size   319
  13   Range and Core Area Size   321
  14   Grazing Bouts   324
  15   Milk Composition   326
  16   Suckling Bout Duration   327
  17   Frequency of Rutting Activities   328
  18   Grouping Criteria   329
  19   Habitat Use by Month in Hinds and Stags   330

REFERENCES   331
AUTHOR INDEX   363
SUBJECT INDEX   369

Foreword

My association with red deer started when I was a boy on Exmoor, where the deer were, and indeed still are, hunted with hounds. This presented me with a conflict between the enjoyment of riding a horse fast over the moor and my growing awareness of the cruelty being imposed on the deer. Perhaps fortunately, the conflict was resolved for me by the outbreak of the war and by the fact that I have never subsequently had the time or money to hunt. What has remained is an abiding excitement at the sight of red deer in the wild.

It is therefore particularly satisfying to me that with the publication of this book the red deer becomes the best known wild mammal. The authors' aim has been not merely to describe the ecology of the deer, but to interpret that ecology in terms of modern evolutionary theory and to use their study to illuminate evolution theory in general, as David Lack did for an earlier generation in his *Life of the Robin.* They have been driven, by the nature of the animal they were studying, to concentrate on the differentiation of the sexes and on the operation of sexual selection. These topics, too long neglected since they were introduced by Darwin, are now again in the center of the stage.

The study is likely to be remembered for four principal results. The first is the demonstration that variance in lifetime reproductive success is greater among males than among females. Although this has been predicted by a variety of studies since Darwin's *Descent of Man,* lifetime reproductive success has, to my knowledge, never before been measured in both sexes in any wild animal. While the direction of the difference is not surprising, the fact that it is possible to put numbers to it and to assess the importance of the factors contributing to variation in reproductive success will be crucial in helping us understand the operation of sexual selection. Second is the finding that hinds invest more heavily in their sons than in their daughters before weaning but reverse the situation afterward. As the authors argue, this situation has presumably come about because the reproductive success of males is more strongly influenced by their size and hence by their early growth rates, while that of females is more strongly affected by resource access in adulthood. Third is the evidence that the calves of old hinds are more likely to survive starvation over the winter months—apparently because the reproductive effort that mothers allocate to each breeding attempt increases as they age. And last is the suggestion that, as a result of differences in the energetic costs of reproduction to males and females, both the proximate and the ultimate factors limiting population density may differ between the sexes.

All four results provide good examples of how evolutionary theory can provide a basis for predicting and interpreting ecological and behavioral processes. All of them could have been achieved only by a long-term study of a species in which it is possible to follow individuals throughout their life spans. The moral for future studies is clear: as research in evolutionary ecology increases in sophistication, long-term studies of easily accessible species will come to play a progressively more important role.

This long-term study of a single species, in which individuals can be recognized in the field and lifetime reproductive success measured, will be crucial in helping us understand sexual selection. The persistence and ingenuity needed to collect and analyze the data presented here are impressive. The answers that emerge make the effort worthwhile.

JOHN MAYNARD SMITH

Preface

In October 1838, that is, fifteen months after I had begun my systematic enquiry, I happened to read for amusement "Malthus on Population," and being well prepared to appreciate the struggle for existence which everywhere goes on from long-continued observation of the habits of animals and plants, it at once struck me that under these circumstances favourable variations would tend to be preserved, and unfavourable ones to be destroyed.

Charles Darwin (1876)

Charles Darwin died one hundred years ago. Although today he is almost universally recognized as the founder of modern evolutionary biology, it is less frequently remembered that he probably contributed more than any other biologist to founding modern ecology. Throughout his work, the twin themes of ecology and evolution are inextricably interwoven: he himself attributed his realization of the basic principle of natural selection to a fortuitous reading of Malthus, and the theoretical framework provided by his theory of evolution enriched and extended his grasp of ecological processes—of competition between and within species, of geographical distribution and colonization, and of population fluctuation and regulation.

However, during most of the century following Darwin's death, ecology and evolutionary biology followed divergent paths. The ecologist's emphasis on populations and interspecific relationships, on community structure and energy flow, directed students away from Darwin's focus on the behavior and reproduction of individuals. As a result, explanations for ecological phenomena that Darwin himself would have summarily dismissed became common in biological literature (see Williams 1966): disease was explained as a mechanism adapted to limiting population growth; male-biased mortality as an adaptation for removing individuals surplus to the requirements of the population; dominance hierarchies as mechanisms for ensuring that only the best adapted males bred. Perhaps no single example illustrates the divergence between ecology and evolutionary biology so well as the persistent difficulties some vertebrate ecologists experienced in explaining why most species produce approximately equal numbers of males and females when only a small proportion of males are necessary to propagate the species—a phenomenon that was explained in outline by Darwin (1871) and in detail by Fisher (1930) (see chap. 8).

During the past two decades, ecology and evolutionary biology have again converged (Lack 1954; Mayr 1963; Ford 1964; Wilson and Bossert 1971; Geist 1971b; MacArthur 1972; Maynard Smith 1974; Hutchinson 1978; Krebs and Davies 1978), and the importance of setting explanations of ecological phenomena within the framework provided by natural selection is now generally accepted. What is less often appreciated—and is the burden of this book—is that Darwin's theory of sexual selection has direct relevance to ecology, too. As Darwin realized, the factors influencing reproductive success commonly differ between males and females, and different adaptations have consequently evolved in the two sexes: for example, male and female mammals commonly differ in size, shape, growth rate, metabolic rate, life span, and many other aspects of physiology and biochemistry (Glucksman 1974, 1978). Pronounced sex differences also exist in many behavioral characteristics: in particular, in social behavior, dispersal, and food choice (Eisenberg 1966; Alexander 1974; Clutton-Brock 1977). And, as a result, environmental changes can have different effects on males and females—for example, recent studies of vertebrates have shown that changes in food availability, climate, or population density can affect the two sexes differently (Latham 1947; Jackes 1973; Klein 1968; Redfield, Taitt, and Krebs 1978a,b).

Although we know that ecological and behavioral differences are common between male and female mammals, and though the theory of sexual selection provides a framework for explaining these differences and predicting their distribution (chap. 1), there is much that is not yet clear. In particular, we know relatively little about the extent to which reproductive success differs among males and females or about the principal causes of variation in reproductive success in each sex, and as a result we often have no firm idea of the selective forces responsible for particular sex differences. In addition, relatively few studies have compared the reactions of males and females to environmental changes, and our understanding of the ecological consequences of sex differences in morphology, physiology, and behavior is rudimentary.

Since it is principally through differences in reproductive success between individuals that selection operates to produce adaptations in members of both sexes (see chap. 1), the most direct way to understand the adaptive significance of differences between males and females is to examine the factors affecting reproductive success in each sex. In this book we examine the evolutionary causes and ecological consequences of sex differences in red deer (*Cervus elaphus*). The early chapters (4–7) describe the extent to which reproductive success varies among males and females and examine the proximate and ultimate causes of these differences. Contrasts in the factors affecting reproductive success in males and females provide a basis for interpreting the adaptive significance of sex differences in body size and weapon development (chap. 7), maternal investment (chap. 8), social organization and dispersal (chap. 9), and social relationships (chap. 10). These differences, in turn, help explain why habitat use differs between stags and hinds and why the sexes react differently to changes in population density (chaps. 11 and 12). If, as we suggest, many of the differences between stags and hinds are a consequence of the fact that red deer are a highly polygynous species, the distribution of sex differences across other ungulate species should be closely associated with the distribution of polygyny. In the last chapter (13), we describe comparative evidence indicating that this is the case and speculate about the ecological factors controlling the distribution of polygyny itself. Since, throughout the book, we draw extensively on current evolutionary theory, we have outlined the principal theories involved in chapter 1.

We selected red deer as a suitable species in which to study the ecology of reproduction partly because their physiology and ecology had already been well studied and partly because they provided an unusual opportunity for monitoring the reproductive success of individuals on account of their large body size, short life spans, and open habitat. Though we did not originally intend to

concentrate our research on the comparative reproductive strategies of the two sexes, the contrasts between stags and hinds grew
out of our data as surely as if we had been studying two different
species. In retrospect, it is clear that this was an inevitable outcome
of examining the ecology of reproduction at the individual level.

T. H. CLUTTON-BROCK

Acknowledgments

Many people have helped us with this project in many capacities. Our greatest single debt is to the Nature Conservancy Council, which owns and maintains Rhum and whose members have generously provided us with access to their facilities on the island over the past ten years. Without their help and the support of their staff in Edinburgh, in Inverness, and on Rhum, the project would have been impossible. In particular it is a pleasure to thank Morton Boyd, Jim McCarthy, and Niall Campbell. Martin Ball, the officer in direct charge of Rhum, has been a generous ally, and his infectious enthusiasm for the project has provided a continuous source of encouragement.

To the Nature Conservancy Wardens on Rhum and to their wives we owe thanks for many forms of practical help as well as for encouragement and support—in particular, we are grateful to Peter and Jessie Wormell, George and Fay MacNaughton, Peter and Margaret Corkhill, John and Meg Bacon, Brian and Sarah Lightfoot, and Bob and Barbara Sutton. The Nature Conservancy's stalkers on Rhum—Geordie Sturton and Louis Macrae—have been generous with their time and expertise and have carefully minimized conflicts of interest between our studies and the annual cull.

Life on Rhum would have been much poorer without the help and friendship of members of the Rhum community, who have contributed to our work in a host of ways. they have brought us coal and mail, sewn ear flashes, taken messages, repaired our vehicles, and recharged our batteries, both literally and metaphorically. In particular we are grateful to Walter and Lois Bee, Ian and Linda Black, Sam and Georgina Glenn, John Love, Angus and Pam MacKintosh, Ian and Kathy Simpson, and Nan and Alec Taylor.

More than thirty persons have helped in some way to collect

data, catch calves, or organize our lives at Kilmory. They include Rhian Powell, Paul Greenwood, Sue Wells, Jane Shrivnan, Chris Spray, John Murray, Anna Baird, Tony Martin, Richard Davis, Ben Osborne, Tim Johnson, Mark Avery, Pamela Moncur, Peter Dratch, Norman Owen-Smith, John Odling Smee, Linda Partridge, Mark Stanley Price, John Hambrey, Belinda Grant, Vivvy Brun, Mary Holt, Tessa Scott, Sally Temple, Callan Duck, Rebecca Short, Mike Rands, Finn and Kieran Guinness, Becky Meitlis, Gail Simpson, Jane Groves, Alexandra Gejmuchasian, and David Mount. We owe particular thanks to Gerald Lincoln, John Fletcher, and Janet Brooke, who have helped us catch deer of all sizes and enriched our records and our evenings by their presence.

We are also grateful to previous research workers on Rhum—to Pat Lowe, Gerald Lincoln, Brian Mitchell, Brian Staines, Nigel Charles, Roger Short, and John Fletcher, whose earlier work on Rhum provided the foundation for this study and who have all been generous with advice, criticism, and access both to published results and unpublished data. Five postgraduate students—Marion Hall, Robert Gibson, Rosemary Cockerill, Mike Appleby, and Mike Reiss—have completed or are completing doctorates on the Rhum deer population, and we are grateful to all of them for their contributions to the project and for permission to quote some of their unpublished work. In addition, four diploma students from the Department of Applied Mathematics and Statistics (Jane Allin, Howard Gilbert, Robert Lee, and Andrew Watts) completed dissertations using parts of the Rhum data and have permitted us to use some of their results. We would also like to thank the staff of the former Nature Conservancy and of the Red Deer Commission, whose annual counts provide records of population density and performance in the Rhum deer population since 1958.

For help with computing we are deeply grateful to the staff of the Cambridge Computer Laboratory, who have extended our shares and pursued our bugs with persistent patience. They include Judy Bailey, Margaret Oakley, Richard Stibbs, and Roger Stratford, while Anna Harvey and Tony Martin developed some of the programs used in analysis of the census data. Val and John Bray have punched many thousands of computer cards for us with apparently tireless accuracy, and John Hambrey, Mark Avery, Glen Iason, Ben Osborne, David Mount, Tessa Scott, Pam

Moncur, and Anna Baird have all helped with data analysis. Paul Harvey, John Harwood, Dan Rubenstein, Elizabeth Thompson, and Charles Free gave us the benefit of their statistical wisdom. The workshop staffs at Sussex and Cambridge have designed, made, and mended equipment from crossbows to calf nets. In particular, Andrew Smith of Sussex University and Dennis Unwin at Cambridge provided important technical assistance, and Don Manning, Collin Atherton, and Les Barden have taken, developed or printed countless photographs for us. The BBC and Wharfedale, Ltd., kindly lent us equipment, and Imperial Chemical Industries provided us with Darvic for calf collars.

We owe a particular debt of gratitude to Roger Short. Without his encouragement and guidance, this project would never have started, and we have benefitted in many ways from his help and advice. Academic colleagues at Oxford, Sussex, Banchory, and Cambridge have stimulated us with their suggestions, criticisms, or ideas—including Robert Hinde, Paul Harvey, John Maynard Smith, Richard Wrangham, Pat Bateson, Dan Rubenstein, Brian Bertram, Robin Dunbar, Peter Jarman, Geoff Parker, Adam Lomnicki, Peter Slater, Richard Andrews, Nick Humphrey, Richard Dawkins, Tim Halliday, Nigel Leader-Williams, Dafila Scott, and Hans Kruuk. We are also grateful to Paul Harvey, Brian Staines, Martin Ball, Brian Mitchell, Robin Kay, and Jim Suttie for their comments on an earlier version of this manuscript.

Four academic institutions have provided a base for the project: The Department of Zoology at Oxford, the School of Biological Sciences at Sussex, King's College Research Centre (Cambridge), and finally the Department of Zoology at Cambridge. We are grateful to all these bodies, and to their staff, and to our colleagues who have tolerated our vagaries when present and our absences when needed. In particular we should like to thank Richard Andrew, Peter Slater, Donald Parry, and Gabriel Horn for their support. Hazel Clarke and Phyllis Osbourn typed parts of the manuscript, though the main load fell on Sarah Cartwright, who has persevered patiently through multiple drafts of appalling handwriting.

For permission to reproduce material, we are grateful to: the Nature Conservancy Council (Appendixes 3 and 4), Robin Kay and *Nutrition Abstracts and Reviews* (Appendix 5), Nigel Charles (table 2.2, figs. 2.6, 11.11), the Royal Aircraft Establishment (fig. 2.2), the *Journal of Animal Ecology* (fig. 5.2*a*,*b*), Marion Hall (fig. 4.8*a*,*b*), Sir Kenneth Blaxter and Her Majesty's Stationery Office (fig. 4.17), *Behaviour* (fig. 6.27), *Animal Behaviour* (figs. 6.19, 6.26, 6.28, 6.29, 6.30), *Nature* (fig. 8.2), Brian Mitchell and the *Journal of Zoology* (figs. 4.18, 6.22). Jim Gammie drew figure 2.3, and Priscilla Barrett drew the illustrations at the beginning of each chapter. We are grateful to her for their use and for the care and precision of her draftsmanship.

Finally, we would like to thank the Natural Environment Research Council (NERC), the Science Research Council (SRC), the Royal Society, the provost and fellows of King's College (Cambridge), and the Leverhulme Trustees, who have provided the funds necessary for the project. We are especially grateful to Michael Schultz of NERC and to Ronald Tress of the Leverhulme Trust for their guidance in the difficult business of trying to carry out a long-term project on short-term funding.

1 Natural Selection in Males and Females

One can, in effect, treat the sexes as if they were different species, the opposite sex being a resource relevant to producing maximum surviving offspring. Put this way, female "species" usually differ from male species in that females compete among themselves for such resources as food but not for members of the opposite sex, whereas males ultimately compete only for members of the opposite sex, all other forms of competition being important only in so far as they affect this ultimate competition.

R. L. Trivers (1972)

1.1 NATURAL SELECTION AND REPRODUCTION

The well-designed fit between animals and the environmental pressures that impinge on them provides ample evidence of the precision with which animal species are adapted to their environments (Williams 1966). But what are they adapted to do? The short answer is to reproduce. The genetic material of an individual that fails to reproduce is lost from the population, and successive failures by individuals carrying the same genes will lead eventually to the disappearance of those genes from the gene pool. Conversely, genes that increase the reproductive success of their carriers will spread through successive generations until they occur in all members of the species.

But does natural selection operate principally through differences in reproductive success between species? Or individuals? Or genes? The answer to this question is of central importance to our thinking, for on it depends the level at which we can expect organisms to be adapted to their environments. For example, if we believe that natural selection operates principally through differences in reproductive success between species, we should not expect to find individuals behaving in a way that increases their own reproductive success at the expense of the success of the average individual. Conversely, if we believe that natural selection operates through differences between individuals, we should not expect to find individuals behaving in a way that reduces their own reproductive success, even if it increases the average success of their population or species.

Most evolutionary theorists are now agreed that selection between populations or species is unimportant compared with selection between individuals or genes (Williams 1966, 1971; Maynard Smith 1976) and that traits that reduce the reproductive success of their carriers are unlikely to be favored even if they increase the average success of all members of the population (Hamilton 1971; Parker 1978; Hrdy 1977). Briefly, the reason for this is that individuals bearing genetic instructions to restrict their own reproductive success in order to increase that of other members of the population would decline in number compared with individuals that did not bear such genes, and the "unselfish" genes would quickly disappear. In theory, this might not occur if some populations consisted entirely of "unselfish" individuals and the rate at which "selfish" populations became extinct was very fast

(Maynard Smith 1964), but such conditions are likely to be rare, especially in longer-lived organisms. Consequently, we should be skeptical of arguments that suggest that particular individuals should relinquish access to food resources or restrict their reproductive rate or reduce their survival for the good of the species (e.g., Wynne-Edwards 1962).

There is still disagreement about whether selection operates principally at the level of individuals or at the level of genes (Ford 1975; Dawkins 1978). Ultimately, it is genes that endure, that spread or become extinct. And, in a variety of species, selection has apparently favored genes that ensure their own replication at the expense of the reproductive success of the individuals that carry them. For example, in some species of social Hymenoptera, a large proportion of individuals forgo reproduction, spending their efforts in helping relatives to reproduce instead (Hamilton 1964a,b; Dawkins 1978). However, there are problems with regarding genes as the unit of selection. Genes cannot propagate themselves, and changes in their frequency can occur only through differences in the reproductive success of individuals (Dawkins 1982). Moreover, adaptation is evident only in the phenotype, so that questions about the functional significance of particular traits must be confined to this level. For those concerned principally with understanding adaptations, the most useful view is to consider individuals as being adapted to propagate their genes.

In animals like red deer where cooperation between relatives is limited, it is reasonable to assume that the number of surviving offspring an individual produces during its lifetime provides a measure of its success in passing on its genetic material to the next generation. As Darwin realized, it is through individual differences in breeding success that natural selection acts to spread and maintain advantageous traits, leading to the evolution of contrasting adaptations in species occupying contrasting environments: "for those individuals which generated or nourished their offspring best would leave, coeteris paribus, the greatest number to inherit their superiority; whilst those which generated or nourished their offspring badly would leave but few to inherit their weaker powers."

1.2   SELECTION PRESSURES IN MALES AND FEMALES

While Darwin's theory of natural selection provided an explanation for differences between species, it did not account for variation between the sexes. In most species, males and females live in the same habitat and are subject to similar environmental pressures, so why should they differ so widely?

The Descent of Man (Darwin 1871) offered a partial answer to this problem. Darwin realized that males often compete intensely with each other for access to breeding females, and that many traits that are more highly developed in males have evolved because they either enhance an individual's fighting ability or improve his capacity to attract mates. He called the evolutionary process giving rise to these traits "sexual selection" and distinguished it from natural selection on the grounds that it "depends on the success of certain individuals over others of the same sex, whilst natural selection depends on the success of both sexes, at all ages, in relation to the general conditions of life" (Darwin 1871).

Darwin was not specific as to why males should compete more strongly than females for access to breeding partners, and it was left to biologists of this century to provide the answer (Fisher 1930; Bateman 1948; Trivers 1972). The reason why males usually compete more intensely than females is most easily understood by considering the costs of reproduction to members of each sex. In most mammals the energetic costs of fertilization are minimal, and copulation requires no more than a few minutes. In contrast, it takes a female many weeks or months to produce and rear her offspring, since the energy costs of gestation and lactation are high. The general principle is more accurately stated by saying that a female allocates a greater proportion of her total reproductive effort to each offspring than does a male (see Fisher 1930). However, since "effort" is a word that sits uneasily on the tongues of biologists, it has been replaced by the term "parental investment," defined as a process or action by a parent that increases the reproductive value of its offspring but reduces its own ability to invest further in the future (see Trivers 1972).

Because males are capable of fathering more progeny than females can bear and rear, a male's reproductive success is usually limited by the number of breeding females to which he can gain access. Where successful males can monopolize breeding access to large numbers of females, direct or "interference" competition is likely to be intense, aggressive interactions will be frequent, and the selective advantages of possessing traits that affect success in interactions will be high (see Bateman 1948; Sherman 1976; Payne 1979; Howard 1979). Several studies of mammals have shown that the reproductive success of males is closely related to their fighting ability and depends on their body size, strength, or weapon development (Geist 1971b; Packer 1979a,b; LeBoeuf 1974; Clutton-Brock et al. 1979), though as yet there is little firm evidence that female preferences for particular males make an important contribution to differences in reproductive success.

In contrast, the reproductive success of a female will depend not so much on the number of males she can mate with as on her ability to rear offspring. Since the energetic costs of lactation are high (Pond 1977), her reproductive success is likely to be determined by her ability to acquire nutritional resources and to transfer them to her progeny. Although in some species food acquisition in times of shortage may depend on fighting ability, where resources are widely dispersed and the energetic value of individual food items is low (as is the case for most herbivores), it is likely to depend to a greater extent on the rate and efficiency of food collection and processing, and selection will favor traits that increase a female's capacity for indirect or "scramble" competition for resources (see Miller 1967; Wilson 1975). Of course this comparison between males and females is an oversimplification, since the efficient collection and use of food is also important to males and since direct competition between females will sometimes occur, even in herbivores. However, the generalization that direct competition is of greater importance to males than to females whereas indirect competition may be more important to females than to males is central to understanding the costs and benefits of sex differences among herbivores.

As Darwin realized, differences between the sexes in the intensity of direct competition for breeding partners provide an explanation for sex differences in many traits associated with fighting ability or display, including horn size, skin thickness, and body size (Geist 1966). Darwin termed such traits "secondary" sexual characters and distinguished them both from "primary" sexual characters, such as the structure of the sexual organs, and from other traits that differed between the sexes but were not immediately associated with fighting or display. As he pointed out, secondary sexual characters should be less developed in the males of monogamous species than in those of strongly polygynous ones, where differences in reproductive success among males are likely to be large and where particularly intense competition for access to females will occur (see Trivers 1972). Quantitative comparisons of several secondary sexual characters, including dimorphism in body size and in the development of weapons, have confirmed that they are reduced or absent in monogamous species compared with polygynous ones and provide a clear indication of their functional significance (Gill 1871; Crook 1972, 1973; Brown 1975; Gautier-Hion 1975; Clutton-Brock, Harvey, and Rudder 1977; Leutenegger and Kelly 1977; Harvey, Kavanagh, and Clutton-Brock 1978; Spassov 1979; Alexander et al. 1979).

However, competition between members of the same sex can affect many traits that Darwin would not have regarded as "secondary" sexual characters. For example, the degree of direct competition between males can influence the form and size of the testes (Harcourt et al. 1981), and it seems likely that the faster growth rates, higher hemoglobin levels, and larger relative lung sizes of many male animals (Glucksman 1974, 1978) may be a consequence of selection favoring fighting ability in males. Intermale competition may also affect female characteristics: if large body size confers major advantages on males and large mothers produce large sons, intense competition between males may select for increased body size in females (see Maynard Smith 1978). Moreover, since competition for limited resources is typically most intense between members of the same sex, many traits that are more developed in females and have arisen through selection favoring resource acquisition or lactation capabilities probably also fall within Darwin's definition of sexual selection. Consequently it is usually more useful to attempt to relate traits that are more developed in one sex to differences in the selection pressures operating on males and females than to try to distinguish between differences that are a consequence of natural selection and those that have been produced by sexual selection.

1.3   CONSTRAINTS ON SEXUAL DIMORPHISM

Of course, not all sex differences are adaptive. The development of any trait involves costs as well as benefits, and we should expect to find that many differences are nonadaptive side effects of selection for functional traits—the shorter life spans, higher rates of heat loss, and greater susceptibility to disease of males are likely examples.

Indeed, it is presumably these costs that limit the extent of adaptive sex differences. In particular, many of the characteristics that improve a male's success in direct competition for females (such as large body size) have obvious energetic costs. Selection operating through starvation presumably constrains the development of such traits at a point where their benefits in terms of reproduction are balanced by their energy costs. In redwing blackbirds *(Agelaius phoeniceus)* there is evidence that, although larger males are more successful competitors, they lose their fat reserves faster than smaller ones (Searcy 1979). In the great-tailed grackle *(Quiscalus mexicanus),* males' large body size makes them less efficient at foraging than females and exposes them to a greater risk of starvation (Selander 1965, 1972).

This argument has important implications for ecology. In species where access to females depends principally on direct interactions while resource acquisition depends on success in in-direct competition, males may be less well adapted to competing for food than females. Consequently, when males are forced to compete for limited resources either with females or with members of other species, they may be less likely to survive. Such effects may be pronounced, for, if the benefits of traits that improve fighting ability are substantial, they may be favored by selection even if they have heavy costs.

1.4 SUMMARY

1. In many animals, natural selection operates principally through differences in the reproductive success of individuals. However, the factors affecting reproductive success commonly differ between the sexes: since the energetic costs of reproduction are usually higher in females than males, access to females usually limits the reproductive success of males, while access to resources usually limits that of females.
2. Where access to resources depends mainly on success in in-direct competition but access to females depends principally on direct competition, traits promoting fighting ability will be more strongly favored in males than females, while those pro-

moting feeding efficiency will be more strongly favored in
females.

3. Traits that favor fighting ability in males can evolve even if they
   have heavy energetic costs, particularly in strongly polygynous
   species. Consequently, in species where food acquisition depends primarily on indirect "scramble" competition while access to females depends principally on fighting ability, males
   may be more likely to starve in times of food shortage.

2 Red Deer and Their Habitat

The ruminants are eminently polygamous, and they present sexual differences more frequently than almost any other group of mammals; this holds good especially in their weapons, but also in other characters.

Charles Darwin (1871)

2.1 INTRODUCTION

To understand the significance of differences in behavior and ecology between red deer hinds and stags, it is necessary to appreciate the kind of habitat they occupy and how this changes throughout the year. Sections 2.2 to 2.4 briefly outline the phylogeny of red deer and their history in Scotland and on Rhum. Section 2.5 describes the topography, climate, and vegetation of Rhum, and section 2.6 describes our study area in the northeastern quarter of the island.

2.2 THE EVOLUTION OF RED DEER

Though they extend into the tropics, the Cervidae have always been principally confined to temperate lands. Deerlike animals first became distinct from giraffelike lineages in the Oligocene. The most ancient forms were of small or medium size, and males had long incisors (Flerov 1952). The first true Cervidae first appeared in the early Miocene of Eurasia about twenty million years ago and resembled modern musk deer. Forms with deciduous antlers are found by the middle Miocene, when muntjaclike deer (e.g., *Dicrocerus*) occurred widely in Eurasia (Viret 1961). It was probably during this period that deer first extended their range out of marshy thickets into drier forest zones (Flerov 1952).

By the Upper Miocene, Eurasian cervines had begun to invade open habitats and to feed extensively on grassy plants, and, by the late Miocene or early Pliocene, identifiable ancestors of the genera *Cervus* and *Axis* had developed. The marked sexual dimorphism characteristic of many modern deer probably dates from this period. In the late Pliocene, members of the genus *Cervus* show a progressive complexity of antler forms as a result of ramification of the upper tines, and by the Cromerian a *Cervus* species closely resembling modern red deer, carrying antlers with a simple top fork, is found in Eurasia (Beninde 1937; Flerov 1952). The taxonomy of contemporary cervids is shown in Appendix 1.

It is now usual to regard contemporary populations of red deer (*Cervus elaphus* Linnaeus, 1758), which extend from western Europe to central Asia, as conspecific with wapiti (*Cervus canadensis* Erxleben, 1777) (Ellerman and Morrison-Scott 1951; Flerov 1952; Corbet 1978), which extend from central Asia westward through North America (see fig. 2.1). Since both vocalizations and color patterning differ strikingly between the two populations, we

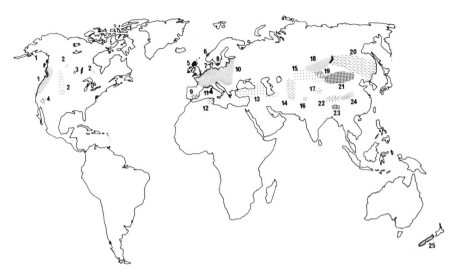

FIG. 2.1. The distribution of red deer and elk (from Whitehead 1972). 1, *Cervus canadensis roosevelti;* 2, *C. c. nelsoni;* 3, *C. c. manitobensis;* 4, *C. c. nannodes;* 5, *Cervus elaphus scoticus;* 6, *C. e. atlanticus;* 7, *C. e. hippelaphus;* 8, *C. e. elaphus;* 9, *C. e. hispanicus;* 10, *C. e. hippelaphus;* 11, *C. e. corsicanus;* 12, *C. e. barbarus;* 13, *C. e. maral;* 14, *C. e. bactrianus;* 15, *Cervus canadensis songaricus;* 16, *Cervus elaphus hanglu;* 17, *C. e. yarkandensis;* 18, *Cervus canadensis asiaticus;* 19, *C. c. wachei;* 20, *C. c. xanthopygus;* 21, *C. c. alashanicus;* 22, *C. c. macneilli;* 23, *Cervus elaphus wallichi;* 24, *Cervus canadensis kansuensis;* 25, introduced *Cervus canadensis* from North America and *Cervus elaphus* from Great Britain.

believe that, until the Asian populations have been studied, any firm decision is premature, and, principally for the sake of convenience, we have retained the specific distinction in this book. It is unfortunate that "elk" is used in Europe to refer to *Alces alces* while in America and Canada it refers to *Cervus canadensis:* throughout the book we use it only in the latter sense.

Late Pleistocene red deer in Europe were very much larger than members of the same species from post-Pleistocene and Recent times (Beninde 1937; Walvius 1961). Body size apparently reached its peak during the last glaciation (Delpech and Suire 1974) and has gradually declined until the present day (Cameron 1923), though contemporary red deer on the Continent are substantially larger than those in Scotland and have relatively large antlers (Huxley 1931, 1932). The decline in body size and relative antler size in Scottish deer appears to be a result of stunting in the harsh and mineral-poor environment of the Scottish uplands rather than of genetic changes (Huxley 1932; Lowe 1961): not only were the recent ancestors of Scottish red deer similar in size to contemporary eastern European forms (Ritchie 1920), but Scottish red deer imported to New Zealand attain weights and antler sizes comparable to those of eastern European deer (Huxley 1932).

Though many contemporary red deer populations exist principally by grazing, comparison of their digestive anatomy and incisor breadth with those of other ungulates indicates that they are adapted to feeding on a mixed diet of browse and graze (Hofmann 1976; Hofmann, Geiger, and König 1976; van de Veen 1979).

2.3   RED DEER IN SCOTLAND

Red deer populations have existed in Scotland since the end of the Pleistocene. During most of this time they have lived in an environment that was heavily forested but probably always included substantial tracts of open country. Ritchie (1920) suggests that, about seven thousand years ago, about 54% of Scotland's land surface may have been covered by woodlands. Birch, willow, and alder were common, and Scotch pine forests covered substantial proportions of the country, reaching their maximum spread about this time.

From the earliest colonization of Scotland by man, red deer were hunted for food and sport (Lowe 1961). Records of deer drives in the sixteenth century in which several hundred were killed show that red deer were numerous in the Highlands. However, in the latter half of the eighteenth century sheep farming began to spread into the eastern Highlands and quickly became the most profitable form of land use. By the end of the eighteenth century, deer populations were at their lowest ebb owing to exclosure, competition with sheep, and removal of cover by burning and felling (Evans 1890; Lowe 1961): in 1790 only nine deer "forests" (large tracts of hill land, commonly with few trees, where deer populations were preserved for sport) were left (Lowe 1961). Subsequently, a number of the larger landowners, wishing to prevent the decline of deer stocks, cleared their ground of sheep, and by 1838 the number of Scottish deer forests had increased to forty-five (Scrope 1897). In the latter part of the eighteenth century, increasing demands for sport and a decline in the economic advantages of sheep farming led to further increases in the number of deer forests, and by 1912 there were 213 forests, covering about 1,450,850 ha (Lowe 1961).

After 1945, when meat was scarce and laws limiting poaching were ineffective, several attempts were made to introduce legislation enforcing legal close seasons and increasing penalties for poaching. These culminated in Fraser Darling's appointment in 1952 to survey Scottish deer populations, in the formation of the Red Deer Commission to provide a permanent basis for monitoring Scottish deer populations, and in the Deer (Scotland) Act of 1959, which gave legal support to close seasons and prohibited shooting at night. The Red Deer Commission has extended Darling's original survey and, using teams of eight to ten experienced stalkers, has now counted more than 85% of the 2,850,000 ha of ground used by red deer in Scotland at least once, while many areas have been counted several times (Red Deer Commission annual reports, 1960–79; see also Stewart 1976). Estimates of present deer populations indicate that there are about 260,000 red deer in Scotland, at densities of between one deer to over 120 ha and one deer to 10 ha, and that numbers are still rising, despite progressive enclosure of their wintering grounds for forestry and, more recently, increases in sheep stocks (Miller 1979).

2.4   THE RED DEER OF RHUM

The island of Rhum (57°0′N, 6°20′W) lies south of Skye and north of Mull, some 19 km off the Scottish mainland (see fig. 2.2). It measures nearly 14 km from north to south and, at its widest point, approximately 13 km from east to west. There is no record of whether red deer occur naturally on Rhum or whether they were originally introduced by man. Their numbers on the island have been closely related to the size of the human population. In the sixteenth century they were evidently common: in 1549 Dean Monro noted "an abundance of little deire" on the island, and they were still numerous in 1703 (Martin 1703). However, in the eighteenth century the human population of the island grew rapidly: by 1764 Rhum had 304 inhabitants, and by 1796, 445. Felling, burning, and overgrazing by sheep cleared the extensive woodlands that previously grew on the lower ground. No accurate information on the decline of the woodland is available, but it was already well advanced by the middle of the eighteenth century, and by the beginning of the nineteenth woodland had totally disappeared (*Old Statistical Account* 1796). As the human population grew, deer numbers fell: by 1772 only 80 deer were left (Pennant

1774), and the population became extinct about 1787 (Otter 1824). The *Old Statistical Account* (1796) links their disappearance with the removal of woodland: "There were formerly great numbers of deer: there was also a copse of wood that afforded cover . . . while the wood throve, the deer also throve; now that the wood is totally destroyed the deer are extirpated." However, the ability of subsequent populations to live without woodland

FIG. 2.2. The west coast of Scotland showing Skye and the Inner Hebrides. Rhum lies to the south of Skye at 57° N, 6° 20′ W.

suggests that other factors were responsible for their disappearance.

During the first quarter of the nineteenth century, the human
population of the island declined as a result of poor living conditions, overcrowding, and famine until, in 1827, the entire
population emigrated, mostly to Canada. About 1845 the island
was again restocked with red deer, and by 1895 their numbers
had increased to 800. During the latter part of the nineteenth
century, the island was developed as a sporting estate first by the
Marquis of Salisbury and later by the Bullough family. By 1939
there were at least 1,200 and possibly as many as 1,700 deer on the
island. During this period the deer shared the resources of the
island with a substantial population of sheep (Miller 1967). In
1886 there were about 6,000 sheep on Rhum when Harvie-Brown
judged it to be "over-stocked and carrying at least 1,000 head too
much." By 1895 the numbers of sheep had dropped to 3,000, and,
after a brief period between the wars when sheep were removed,
it increased again to 6,750 in 1943, falling to 3,300 in 1945 and to
1,750 by 1957.

In 1957 the Nature Conservancy purchased the island from the
Bullough family and removed the entire sheep population. The
first systematic deer count on Rhum in 1957 gave a total of 1,584
animals, and after this the traditional cull of about 40 stags and 40
hinds per year was raised to about 100 stags and 140 hinds until
1960, then stabilized at about one-sixth of the adult population.
Culling at this level led to a decline in the size of the population in
the late sixties (see Appendix 2), and in the seventies culls were
reduced to allow the population to increase to about 1,400 head.
For culling purposes, the island is divided into five blocks (see fig.
2.3) that are culled in proportion to the number of deer counted
in them during the spring counts. Following traditional practice,
stags are culled mostly in August and September, hinds in
November, December, and January.

Both stags and hinds are widely distributed across the island
and spend most of their time on low ground (Lowe 1966). Studies
of recognizable individuals have shown that hinds have well-
defined home ranges and that stag calves are likely to disperse
farther from their birthplaces than hinds. A suggestion that deer
do not disperse freely throughout the island is provided by evidence that the frequencies of some serum proteins appear to be

higher among deer shot in the south and west of the island than among those shot in the north and east (McDougall and Lowe 1968).

RHUM

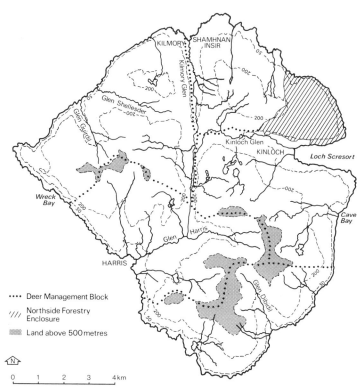

FIG. 2.3. Rhum, showing the outlines of the five culling blocks used by the Nature Conservancy Council.

Though the densities of both sexes (about 15 deer/km²) are higher than in many mainland deer forests, they are not exceptional, and, on most measures of growth and reproductive performance, Rhum deer fall close to the average for Scottish populations (see table 2.1). Calves average about 6.5 kg at birth and grow rapidly until they are three years old in the case of hinds or six in the case of stags (see fig. 2.4). The mean larder weight (weight of the entire animal less alimentary tract and bleedable blood) of milk hinds (individuals with a surviving calf) shot between November and January lies about 51 kg, and yeld hinds

TABLE 2.1   Measures of Reproduction and Performance in Different Scottish Deer Forests

Measure	Glen Dye[4]	Glen Fiddich[1]	Glenfeshie[1,3]	Rhum	Invermark[1]	Scarba[2]
Density (km²)	1.6	8.0	13.1	13.9	15.3	34.4
Adult sex ratio (hinds per 100 stags)	198	137	140	116	556	157
Calf/hind ratio (calves per 100 hinds)	45.8	30.7	34.6	38.1	41.5	39.2
Percentage of yearlings pregnant	64	20	0	0	43	0
Yearling hind larder weight (kg)	42.6	45.3	38.0	37.0	40.4	37.2
Percentage of milk hinds pregnant	94	92	44	51	83	39
Milk hind larder weight (kg)[b]	54.7	55.8	46.5	51.0	51.7	58.9
Stag larder weight (kg)[c]	78.6	105.8[a]	83.8	88.7	92.1	93.7

Sources: (1) Mitchell (1967); (2) Mitchell and Crisp (1981); (3) Mitchell, Staines, and Welch (1977); (4) Staines (1978).

[a] Stags fed in winter.
[b] Milk hind larder weight 60% of live weight.
[c] Stag larder weight 73% of live weight.

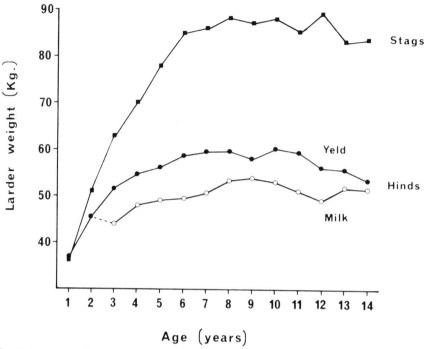

FIG. 2.4. Age-specific larder weights for stags, yeld hinds, and milk hinds shot on Rhum (1958–76 inclusive).

(mature females that do not have a calf, either because they failed to conceive or because their calf died shortly after birth) average about 60 kg. Mature stags weigh about 89 kg larder weight when they are shot in late summer and early autumn. The annual rut occurs in October, and calving occurs eight months later in late May and early June (Fletcher 1974). During the study, no hinds produced more than one calf per year. As in most other Scottish populations, hinds usually conceive for the first time in their third or fourth year of life, and a substantial proportion fail to breed following the years when they have reared a calf successfully (Lowe 1969; Mitchell 1973; Mitchell, Staines, and Welch 1977). After the first year of life, mortality is low until the eighth year, when it increases in both sexes (see fig. 2.5). Between the ages of two and seven, hinds show a higher frequency of mortality, which may occur partly because stag culls are restricted to mature animals while hind culls are not.

Sheep have not been reintroduced to the island since they were removed in 1957. A free-ranging herd of about twenty-five

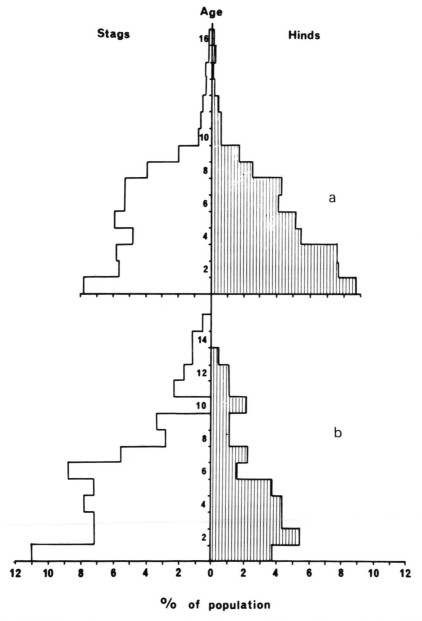

Age

Stags   Hinds

16

14

12

10

8

6   a

4

2

b

12 10 8 6 4 2 0 2 4 6 8 10 12

% of population

FIG. 2.5. (a) Year class percentages in the red deer population on Rhum on 1 June 1957 reconstructed from subsequent deaths and estimated survivors up to May 1966 (yearling represent the survivors from the calves born in 1956) (redrawn from Lowe 1969). (b) Year class percentages in the North Block in 1971.

Highland ponies uses both Kilmory and Harris glens (Clutton-Brock, Greenwood, and Powell 1976), and approximately two hundred feral goats live on the steep, grassy slopes of the southern half of the island (Boyd, 1981). In 1970 the Nature Conservancy Council introduced a herd of twenty highland cows at Glen Harris, and this has now increased to over one hundred head. Plantations were established under the Bulloughs and, more recently, by the Nature Conservancy, but these are fenced and the deer are excluded.

Apart from man, the deer have no mammalian predators. Two to four pairs of golden eagles usually breed on the island and take a number of calves each year (see chap. 5) and both black-backed gulls and hooded crows may occasionally attack sickly calves.

2.5 Topography, Climate, and Vegetation of Rhum
Topography

Rhum's 10,600 ha are divided into three parts by glens radiating north, east, and south from a point in the center of the island near the Long Loch (fig. 2.3). The mountain mass in the southeast that occupies almost half of the island includes four hills over 600 m and consists largely of basic and ultrabasic rocks with some Lewisian gneiss. The southwestern portion of the island contains two major glens (Guirdil and Shellesder), surrounded by hills of basalt and Torridonian sandstone, and contains much of the island's richest grazing on the slopes of Fionchra (Eggeling 1964). The northeast segment, or North Block, includes two gently rounded east-west ridges of Torridonian sandstone rising to nearly 300 m at the highest point on Mullach Môr. On the west, Kilmory Glen is separated from Shellesder by two other sandstone hills, Minishal (320 m) and Sgaorishal (274 m).

Climate

The island has a mild, wet, oceanic climate. January is the coldest month, followed by February and December, and August is the warmest, followed by July. Estimates of average minimum and maximum temperatures at Kinloch in different months of the year are shown in Appendix 3.

Rainfall varies widely between areas of the island. From 1971 to 1980 average rainfall at the mouth of Kilmory Glen was 1,913 mm, whereas at Kinloch it was 2,511 mm. April is usually the driest month and November or January the wettest, though, since 1958, every month has at some time been recorded as the wettest of the year. Snow is common on high ground in winter, but, in contrast to many areas of the central Highlands, it rarely lies long.

Plant Communities

The vegetation of Rhum is typical of west-coast Scotland, though not all the twenty-one plant associations described by McVean and Ratcliffe (1962) are represented, and it is convenient to reduce this list to nine plant communities (Ferreira 1970). In the descriptions below we show both the original name of each community and, in parentheses, the name and abbreviation used throughout this book.

1. *Calluna heath: Calluna* dominant with a cover of 80% or more. It is often associated with hypnoid mosses as well as with *Deschampsia, Potentilla,* and *Vaccinium* species. The vegetation type is usually associated with well-drained podzols that have low base status, and pH varies (see Appendix 4).

2. *Wet heath* (heather moorland, CT): Two forms of wet heath occur on Rhum, one where *Calluna, Trichophorum,* and *Molinia* are codominant, with the abundance of *Trichophorum* increasing in wetter areas, and one dominated by *Rhacomitrum* and *Calluna.*

3. *Blanket bog* (EC): Blanket bog, dominated by *Eriophorum,* is widespread in the valley bottoms and other badly drained, peaty areas (except where boggy areas overlie ultrabasic rocks where *Schoenus* fen occurs). In the commonest form of blanket bog, *Eriophorum vaginatum, Calluna vulgaris,* and *Sphagnum* are dominant, and *Erica tetralix* and *Trichophorum* also occur. However, in some areas, *Eriophorum angustifolium* replaces *E. vaginatum* and is usually associated with *Carex* spp.

4. *Nardus heath:* Grassland dominated by *Nardus* is found in sheltered areas of the acid, igneous hills, grading at lower altitudes into *Calluna* heath.

5. *Schoenus fen.* This is a tall, tussocky community with *Schoenus nigricans* and *Molinia* as dominants, though *Erica, Trichophorum,* and *Potentilla* also occur. This community is found on the lower ground in the southeast, where groundwater arises from ultrabasic rocks.

6. *Molinia flush* (*Molinia* grassland, MG): In many areas of the island, *Molinia caerulea* achieves almost complete dominance.

Molinia flushes are abundant in acid areas wherever there has been slight enrichment of groundwater. Where pH is relatively high, *Molinia* is mixed with *Agrostis tenuis, Anthoxanthum odoratum,* and *Festuca vivipara* and grades into *Agrostis/Festuca* grassland. *Molinia/Agrostis* mixtures are common along the edges of streams and on land that has previously been ridged and cultivated ("lazy" beds).

7. *Herb-rich heath* (short greens, G1): On Ca-rich soils this community consists of *Calluna, Erica cinerea,* and a wide variety of grasses (including *Agrostis* spp., *Anthoxanthum odoratum, Festuca vivipara, F. rubra,* and *Koeleria cristata*) and herbs (including *Euphrasia officinalis, Linum catharticum, Lotus corniculatus, Potentilla erecta, Thymus drucei,* and *Viola riviniana*). Sites are typically high in pH and well drained. In Mg-rich areas, *Carex, Molinia, Anthyllus vulneraria,* and *Primula vulgaris* are common.

8. *Juncus marsh* (MR): The commonest form of freshwater marsh on the island is dominated by *Juncus acutiflorus* associated with grasses, herbs, and sedges including *Festuca rubra, Holcus lanatus, Cirsium palustre, Potentilla erecta, Prunella vulgaris, Ranunculus acris, R. flammula,* and *Succisa pratensis.* In some areas *Juncus effusus* replaces *J. acutiflorus* and is typically associated with *Sphagnum recurvum* and *Carex nigra.* Marshes occur mostly in the valley bottoms and along streambeds. On the hillsides they are associated with flushed peats overlying a gleyed mineral horizon. The pH and Ca, N, and P concentrations are typically high.

9. *Agrostis/Festuca grassland* (short greens, G1; long greens, G2): Two distinct forms of *Agrostis/Festuca* grassland occur on Rhum. One is species-rich, with no single dominant species but including *Agrostis tenuis, Festuca rubra, F. vivipara, Sielingia decumbens, Bellis perennis, Euphrasia* spp., *Galium verum, Lotus corniculatus, Plantago lanceolata, P. maritama, Prunella vulgaris, Ranunculus acris, Thymus drucei,* and *Viola riviniana.* It occurs on well-drained soils with high pH and N values as well as on the calcareous dunes and the sandy soils or "machair" lying behind them. It is typically closely cropped except during the early summer and difficult to distinguish from herb-rich heath. In our analysis we combine these two communities and refer to them both as "short greens."

The second form of *Agrostis/Festuca* grassland is species-poor and is typically dominated by *Agrostis canina, A. tenuis, Anthoxanthum odoratum*, and *Festuca vivipara*. It occurs on acid soils, base-poor podzols, or podzolic loams showing lower pH values than the species-rich form. In sandstone areas it is largely confined to the valley bottoms, though on the igneous hills it occurs at all levels and is often mixed with *Rhacomitrum* and *Vaccinium*. Though heavily used by the deer, tussocky stands are not uncommon, and standing crops are almost always substantially greater than in species-rich areas: we refer to this form of *Agrostis/Festuca* grassland as "long greens."

Seasonal Changes in Food Availability

As elsewhere in the Highlands (McVean and Ratcliffe 1962; Miller 1971; Nicholson and Robertson 1958; Poore and McVean 1957), production in different plant communities on Rhum is closely related to seasonal changes in climate and day length. Grasses, herbs, sedges, and dwarf shrubs all show a period of rapid growth between April and June, followed by lower production during the summer and (in some species) a secondary growth peak in the autumn before the cessation of growth during the winter months.

Though production on the short greens is higher during the summer months, they are heavily used by the deer, and standing crops are never large (see fig. 2.6). By early winter almost all the available food has been removed, and estimates indicate that little vegetation is removed during the latter part of the winter (see table 2.2). On the long greens, *Juncus* marsh, and *Molinia* grasslands, standing crops are higher, but at all seasons these include a substantial proportion of dead grass, making it harder for herbivores to feed selectively. In addition, the digestibility of these swards is typically lower than of the short greens as a result of increased cellulose and lignin levels. Intake from the long greens and *Molinia* grasslands increases later in the year as the abundance of food on the short greens falls. Standing crops on heather moorland and blanket bog are large throughout the year, though, as in the case of the long grasslands, they contain a high proportion of dead material during most months and the digestibility of plant material is typically low.

From a herbivore's viewpoint, the quality of most of the main food plants is closely related to their growth stage and is typically

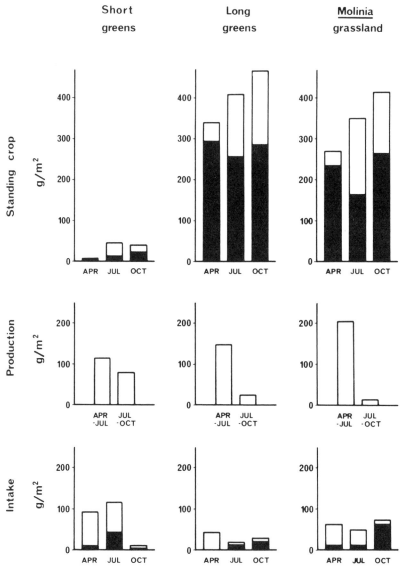

FIG. 2.6. Standing crop, production, and intake from individual plots of low-lying short (herb-rich) greens, long greens, and *Molinia* grassland in the southwest of Rhum (from N. Charles, unpublished data). Blocked columns: dead matter.

highest in spring and lowest in winter (see Appendix 5). *Molinia*, in particular, is relatively digestible in its early growth stages, but it quickly becomes coarse and rank and its digestibility falls rapidly. *Eriophorum, Trichophorum,* and many other sedges change almost

TABLE 2.2　Mean Standing Crop, Production, and Intake of Forage from Grasslands in the Southwest Block of Rhum (g dry weight per m²)

	Short Greens[a]	Long *Agrostis/Festuca* and *Molinia*[b]
Standing crop		
April	20 ± 10	325 ± 32
July	58 ± 12	402 ± 35
October	52 ± 13	461 ± 38
Production		
April–October	208 ± 48	173 ± 102
Intake		
April–July	83 ± 29	16 ± 86
July–October	102 ± 42	31 ± 58
October–April	17 ± 12	49 ± 54

Source: N. Charles, unpublished data.

Note: Totals from 1964–66, together with estimates of half-width of 95% confidence intervals. Estimates of production and intake based on comparisons of exclosures and grazed plots.

[a] Six groups of plots.
[b] Four groups of plots.

as rapidly, while *Agrostis, Festuca, Holcus,* and *Deschampsia* show renewed growth in autumn and a higher digestibility in winter than most other upland plants. The digestibility of heather does not vary as widely as that of the grasses and sedges: it is typically highest in spring and declines as photosynthesis increases throughout the summer, though protein levels increase again in winter (Kay and Staines 1981; Grace and Woolhouse 1970).

Altitude and grazing pressure have far-reaching effects on the quality of hill vegetation. Spring growth is delayed at higher altitudes, and grasses that develop a high fiber content when mature consequently show higher digestibility later in the season at higher altitudes (see Mitchell, Staines, and Welch 1977). At higher altitudes, heather shows a higher leaf/stem ratio, increased digestibility, and higher sugar levels (Grant and Hunter 1962; Nicholson and Robertson 1958; Mathews 1972). This effect is particularly marked in late winter, possibly because snow cover inhibits growth and maintains sugars at a high level, which may account for deer's tendency to browse heather immediately below retreating snow lines (Mathews 1972).

Grazing pressure can stimulate growth in many grasses (see

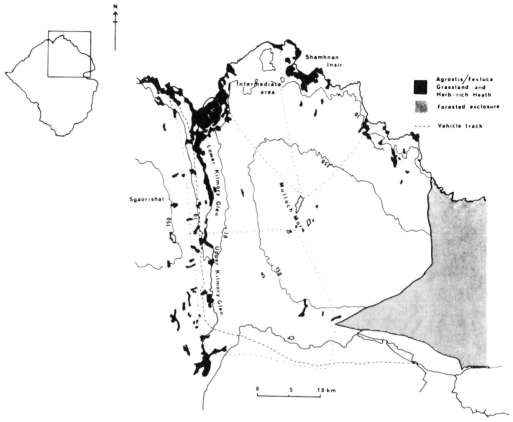

FIG. 2.7. The study area, showing the four parts and distribution of greens (*Agrostis/Festuca* grassland). Hinds allocated to one part of the study area typically ranged over other parts of the study area but concentrated their time in the part to which they were allocated.

McNaughton 1976) as well as in dwarf shrubs, including heather (Grant and Hunter 1962, 1966), and it can lead to the production of swards with high leaf/stem ratios and low fiber levels. On Rhum these effects on grazing are particularly marked on the short greens, where grasses are seldom allowed to reach maturity and leaf/stem ratios are high, as well as in particular patches of heavily grazed heather.

2.6 THE STUDY AREA AND ITS DEER POPULATION

Most of the work described in this book was carried out on the deer population of the North Block of Rhum (see fig. 2.7). Between 1960 and 1965 the North Block supported approximately 90 hinds over one year old and 67 stags. Subsequently, research on the reproductive biology of stags (Lincoln 1971a,b, 1972; Lincoln, Guinness, and Short 1972) required close observation of individuals, and stags using Kilmory Glen were fed with concentrates once a day on the flats in front of the Kilmory laundry (Lincoln, Youngson, and Short 1970). This apparently attracted stags into the North Block, and stag numbers rose to 171 in 1970, when regular feeding ceased. Since 1973, culling of both hinds and stags in the North Block has been completely suspended in connection with our research, and the number of individuals resident in the study area has risen from 70 hinds and 129 stags over a year old in 1972 to 149 hinds and 135 stags in 1979 (see Appendix 6). An individual was defined as resident in the study area if it was seen in at least 10% of censuses in at least four months of the year.

The study area consists principally of a single gently sloping hill (Mullach Môr) and the glens surrounding it, and nearly 70% of its area lies below 120 m (see Appendix 7). Only five of the vegetation communities found on Rhum are common in the North Block: wet heath, blanket bog, *Molinia* flush, *Juncus* marsh, and *Agrostis/Festuca* grassland, though *Calluna* heath and herb-rich heath do occur. For simplicity's sake, we have combined *Calluna* heath with heather moorland and herb-rich heath with short greens. In addition, the published vegetation map of the island (Ferreira 1970) does not distinguish between short and long greens, and, where we wished to use this as a basis for calculations of the areas covered by different plant communities, we combined these two categories.

The greater part of the study area is covered by heather moorland and blanket bog, and smaller areas support *Molinia* grassland, *Agrostis/Festuca* greens and *Juncus* marsh (fig. 2.8). The distribution of the different communities is closely related to altitude: *Agrostis/Festuca* greens and *Juncus* marsh occur mostly at the lowest altitudes, blanket bog and *Molinia* grassland occur at all levels, and heather moorland increases in importance at higher altitudes (Appendix 7).

The quality of the vegetation varies between different parts of the study area. The deer are fenced out of the large areas of alluvial soil at the mouth of Kinloch Glen, and the western end of the glen is poor and stony. The southern end of Kilmory Glen is also poor and contains little *Agrostis/Festuca* grassland (see fig. 2.9),

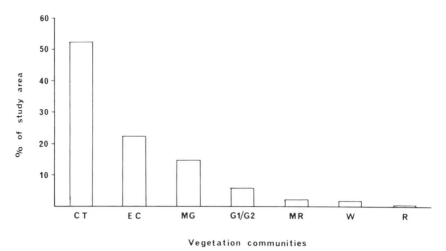

FIG. 2.8. The abundance of different plant communities in the study area: *CT*, heather moorland; *EC*, blanket bog; *MG*, *Molinia* grassland; *G1/G2*, short and long *Agrostis/Festuca* greens; *MR*, *Juncus* marsh; *W*, water; *R*, rock (calculated from Ferreira 1970).

whereas the northern half of the glen has a broad bottom with sizable areas of fertile alluvial soils and extensive *Agrostis/Festuca* greens (fig. 2.10*a,b*). Eastward from Kilmory, the slopes of Mullach Môr grade into peaty flats that terminate in low sandstone cliffs broken by many small coves, while due east of Kilmory lies another sandy bay (Samhnan Insir), sheltered by sandstone headlands with sand dunes lying behind the main bay and smaller areas of alluvial soil and *Agrostis/Festuca* greens lying behind these.

On the basis of their ranging behavior, hinds could be allocated to four topographical divisions within the study area: the Upper Glen (the western end of Kinloch Glen and the southern half of Kilmory Glen), Kilmory (the northern half of Kilmory Glen), the Intermediate Area between Kilmory and Samhnan Insir, and Samhnan Insir itself (see fig. 2.7). Comparisons between animals using Kilmory and Samhnan Insir were of particular interest, for, though the topography and vegetation of the two areas was similar, the density of hinds was considerably higher at Samhnan Insir than at Kilmory (see Appendix 6). Probably as a consequence of this, standing crops on the short greens were lower at Samhnan Insir (see table 2.3).

Although most aspects of the reproductive performance of deer on Rhum fall close to the average for Highland populations, three important differences between Rhum and many areas of the Scottish mainland should be borne in mind. First, the topography on Rhum is on a small scale, and local shelter is almost always

FIG. 2.9. View across the southern half of Kilmory Glen to Mullach Môr, showing the long greens and *Juncus* marsh beside the river with *Eriophorum* bog in the foreground.

available within a relatively small area. Second, snow seldom lies for long, and grasses are usually available to the deer throughout the winter. And third, the relatively low level of human activity allows the deer to use low ground during the daytime with little disturbance. Consequently it seems likely that individuals may have smaller home ranges, feed less on dwarf shrubs, and be less affected by climatic factors on Rhum than on the mainland. In addition, diurnal patterns of activity and movement may differ from those in areas of the mainland where human disturbance is common.

Fig. 2.10. (a) Kilmory Bay, looking northeast toward Skye. The greens behind the beach are visible on the right. Old "lazy beds," covered by *Molinia* grassland, show up as parallel lines. (b) Kilmory Glen, looking south. The remains of Kilmory village are visible in the foreground and beyond them are the flats covered with long *Agrostis/Festuca* greens, *Juncus* marsh, and *Molinia* grassland.

TABLE 2.3   Standing Crop on the Short Greens at Kilmory and Samhnan
Insir in April 1978

Measure (dry weight, kg/ha)	Kilmory	Samhnan Insir
Live grass	82.50 ± 22.75	52.80 ± 21.25
Dead grass	193.00 ± 101.25	118.50 ± 78.75
Moss	81.00 ± 41.75	54.25 ± 78.75
Total	370.00 ± 156.00	225.50 ± 173.00

Note: Measures based on eighteen samples collected from separate patches of greens
in each area. Each sample consisted of as much vegetation as could be plucked with
fingers from a randomly chosen 20 cm quadrat in 60 sec.
   The two areas differed significantly on all three measures ($p < .01$).

2.7   SUMMARY

1. Deerlike animals first evolved from giraffelike lineages in the
   Oligocene, and animals ancestral to the present red deer were
   present in Eurasia by the Cromerian. Body size reached its peak
   during the last glaciation and has declined to the present day,
   though Scottish red deer are substantially smaller than Continental populations.
2. As a result of clearance and exclosure, Scottish red deer have
   been progressively confined to open moorland. The total Scottish population is currently more than 250,000.
3. On Rhum, original deer stocks were extinct by the end of the
   eighteenth century, but the island was restocked about 1845,
   and the population rose to about 1,600 by 1958. Between 1959
   and 1980, numbers have ranged from 1,175 to 1,842.
4. On most measures of size, growth, and reproductive performance, the Rhum red deer population falls close to the average
   for Scottish populations. Mortality is heavy during the first year
   of life and subsequently is low until the eighth year, when it
   increases in both sexes. The majority of hinds and stags die
   before the age of twelve, though maximum life span is longer in
   hinds.
5. Rhum is largely covered by *Calluna-* and *Molinia*-dominated
   moorland. Small areas of *Agrostis/Festuca* grassland are widely
   distributed and heavily used by the deer.
6. The bulk of this study was carried out in the North Block of
   Rhum. The population of our study area has not been culled
   since 1973 and has grown from 70 hinds and 129 stags over a

year old in 1972 to 149 hinds and 135 stags in 1979. Hinds using the study area can be allocated to four home range areas: Upper Glen, Kilmory, the Intermediate Area, and Samhnan Insir. Both Kilmory and Samhnan Insir offer sizable areas of *Agrostis/Festuca* grassland, but population density is higher at Samhnan Insir and food availability is lower.

3 Methods, Samples, and Definitions

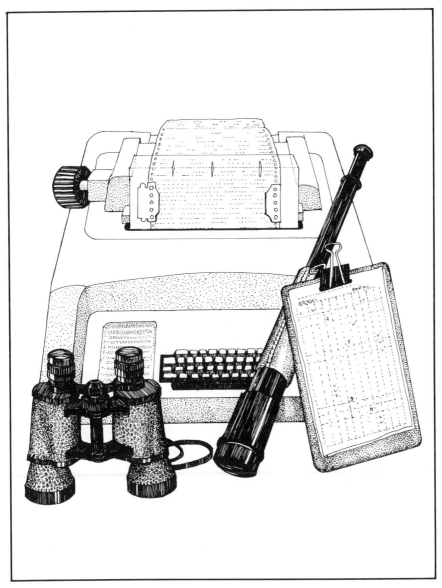

Quantification as such has no merit except insofar as it helps to solve problems.

P. B. Medawar (1979)

3.1   Introduction

In this chapter we outline the techniques of recording and analysis used in the rest of the book: section 3.2 describes how we recognized individuals, section 3.3 covers the various methods of data collection and the samples used in the analyses, and section 3.4 defines the variables we used. In the last section we describe the techniques of statistical analysis we employed.

3.2   Recognition of Individuals
Identification

Much of the study depended on our ability to recognize individual deer. Stags are comparatively easily recognized from individual differences in antler form, which are remarkably stable from year to year (see fig. 3.1). Hinds, and stags without antlers, can be identified from facial idiosyncrasies, from body shape, and from coat coloration (see fig. 3.2). As in other vertebrates (Bateson, Lotwick, and Scott 1980), relatives often look alike (see fig. 3.3).

To check our ability to recognize individuals, we carried out a number of interobserver reliability tests. In 1974, Clutton-Brock and Guinness censused the whole study area, separately noting the identity of each hind in turn. Analysis of these data showed that, of sixty-one hinds seen, only one had been misidentified. Subsequent tests revealed a similarly high degree of interobserver reliability and indicated that experienced observers failed to recognize animals only when they identified them without a clear view of their faces. On several occasions we have been surprised to find that our ability to recognize individual deer is regarded as unusual. Recognition of individual animals by natural markings is now widely used in field studies (Goodall 1968; Douglas-Hamilton 1972; Bertram 1976), and tests of accuracy on species whose individual differences are considerably more discrete than those of red deer have confirmed their reliability (see, for example, Bateson 1977; Scott 1979). In fact, in long-term field studies, individual recognition by natural idiosyncrasies is often more reliable than recognition based on artificial marks, since the latter can fall off or become unrecognizable with age. Had we been unable to recognize stags or hinds when they lost their collars or ear flashes, our sample of recognizable individuals would have diminished rapidly with time.

Fɪɢ. 3.1. CLYD (a) as a two-year-old in November 1972; (b) as a five-year-old in September 1975; (c) as an eight-year-old in February 1978; and (d) as a ten-year-old in December 1980.

FIG. 3.2. Close observation of hinds reveals many facial idiosyncrasies. The shape and patterning of the ears are particularly useful. BIGH *(a)* has long, thin ears with dark markings only on the lower edge, whereas MOM5 *(b)* has shorter ears with rounded lower sides. In addition, the shape and coloration of rump patches and the breadth of the "eel" strip on back and neck aid recognition.

To aid recognition in poor weather, we marked as many calves as possible with expanding collars made from 1.2 mm thick Darvic (see fig. 3.4) and also with multicolored polyvinyl chloride ear flashes. By 1979, over 50% of the deer regularly using the study area carried some artificial mark.

As the study progressed, individuals became accustomed to the presence of observers. After 1971, most animals tolerated an observer in full view within 50 m, provided he did nothing suspicious, though, when accompanied by young calves, hinds were considerably less tame (see chap. 4). Both hinds and stags were tamest during the rut.

Calf-Catching

During the birth season, one observer made a daily check of as many hinds as possible, recording udder size and any visible abdominal distension. These, and changes in the hind's behavior after calving, allowed us to identify when individuals had calved. When a hind that had recently calved was located, an observer was set to watch her from far enough away to be undetected, and we waited until the hind moved to her calf to allow it to suck. When the mother left the calf again, and after it had settled, we approached the calf and caught it, sometimes by hand, sometimes

Fig. 3.3. Related hinds often look alike: *(a)* BLTL *(left)*, a thirteen-year-old hind, with her six-year-old daughter *(right)*. The two hinds closely resemble each other in ear shape and patterning as well as in the general conformation of their faces. *(b)* MOMH *(third from left)* has a pronounced black mark at the base of her tail, as do her calf *(far left)* and her two daughters *(second left and far right)*. In contrast, her grandson *(center)* has a particularly light-colored rump patch.

FIG. 3.4. A month-old calf wearing an expanding Darvic collar. Expanding collars were fitted only to female calves, since we found it impossible to make them large enough to allow for the neck development of stags in late summer.

using a long-handled net. The calf was then sexed, weighed, marked, and inspected for abnormalities. Desertion resulting from our interference undoubtedly occurred occasionally. However, from observations following marking, we believe that no more than 5% of all calves marked between 1974 and 1979 died as a result of desertion.

Marking Adult Deer

To reduce the risk that they would be shot when they ranged outside the study area, we marked as many as time allowed of the adult stags and of the adult hinds with home ranges on the periphery of the study area. Animals were immobilized using Etorphine injected by a projectile syringe fired from a crossbow and were subsequently marked with Dalton cow collars (see fig. 3.1*b*). We took measurements and weights at the same time, as well as tooth impressions of individuals whose age was uncertain.

Mortality

Before deer died, their body condition usually declined visibly. During the last few days of life, they seldom moved far, and when individuals were consistently missing in censuses we usually knew where to search for their carcasses. In addition, each spring the study area was searched for dead animals by the Nature Conservancy's stalkers. As a result, we found the carcasses of over 70% of all mature animals that died in the study area and were able to collect details of skeletal size and tooth wear (see fig. 3.5).

3.3 DATA COLLECTION
Census Data

Many of the results described in this book are based on data collected in censuses of our study population. A satisfactory system of censusing was developed in 1973: one observer searched the study area, recording the identity, position, activity, and location of each animal seen. Weather data, the plant community (see p. 27) on which each individual was standing, and group composition were also noted, as well as the distance between each mature hind and her offspring when they were in the same group. Different parts of the study area were searched in different order in successive censuses: the census either started at the Kinloch end of the study area and proceeded along the road to Kilmory Bay

FIG. 3.5. Last rites: measuring the antlers of stags that have died in early spring.

before covering the Intermediate Area, Samhnan Insir, the northern slopes, and the top of Mullach Môr, or it started at Samhnan Insir, covered the Intermediate Area and Mullach Môr, and finally reached Kilmory Glen and the southern part of the study area. Censuses began as soon as it was light enough to see clearly, and in winter they typically took all day. Censuses in 1973 and 1974 showed that the deer seldom used the rocky plateau at the top of Mullach Môr, and after 1974 this area was omitted from censuses between November and February, since the days were short and as much time as possible was needed to search the rest of the area thoroughly.

Not all deer resident in the study area were seen in every census. For hinds, the mean proportion of censuses in which the average individual was seen was 84%, while for stags the same figure was 67%, though some animals were seen in nearly all censuses. Some of the animals not recorded must have been overlooked, and a proportion probably were temporarily outside the study area. The lower percentage of stags recorded was probably because the two main areas where stags were found (the Laundry Greens and the Kilmory Fank [sheepfold]) were on the periphery of the study area and the stags were consequently more likely to move outside than hinds.

During 1973, fifty-five censuses spanning the different seasons were carried out by one observer (Clutton-Brock). Between 1974 and 1980, the study area was usually censused five times each month by Guinness, providing a total of 60 censuses per year. Data were recorded onto check sheets in the field or, when the weather was wet, dictated into a pocket tape recorder and later transferred to paper. They were subsequently punched on cards and analyzed using the Statistical Package for the Social Sciences (SPSS) computer package (Nie et al. 1975).

Not all our analyses used the entire sample of census data. In a number of cases it would have been impossibly time-consuming to have used the whole sample, and analysis was restricted to particular years. In others we describe analyses we completed several years ago, which there was little point in extending.

Rut Censuses

Between 12 September and 1 November, the system of full censuses was supplemented by daily censuses of all ground below 120 m, with the aim of monitoring the harem size of each rutting stag: virtually all rutting groups in the study area formed within this area. The same information was collected for each individual seen, supplemented with records of the condition and activity of each stag.

Mother/Offspring Surveys

Outside the rut, the system of full censuses was supplemented by approximately fifteen surveys per month of ground below 120 m. These had the aim of locating as many as possible of the breeding hinds as well as their offspring, and they were used to measure changes in the frequency with which offspring associated with their mothers. In addition, the linear distance between each mother and each of her offspring was also estimated.

Calving Surveys

As we have already described, the study population was surveyed each day during the calving season (20 May–30 June). These surveys did not provide systematic coverage of the study area, since their aim was to locate hinds that had recently calved, and they were not included in the analyses of the routine census data.

Continuous Watches of Individuals and Groups

To measure activity patterns, we watched particular individuals continuously for a day at a time. When an animal was selected for observation, its activity was recorded at the end of every minute during the daytime or at the end of every five minutes at night (using infrared viewing equipment) by teams of two to four observers taking four-hour shifts. Day watches began at dawn or soon afterward (approximately 0400 GMT in July and August, and 0800 in February and March) and lasted till dusk (2200 in July and August, 1800 in February and March), and night watches began at dusk and lasted till dawn. Every effort was made to avoid allowing subjects to move out of sight, but, partly because it often took some time to locate a suitable subject at the beginning of the day and partly because individuals inevitably spent part of the day out of sight, all watches included a portion of time when the animal could not be seen. This fraction was excluded from the analysis of the proportion of time spent in different activities. The sample of day records was restricted in summer to days where

more than 720 minute records were collected and, in winter, to
days when more than 480 records were collected, causing us to
reject about 10% of all watches started.

For hinds, eight different samples of daytime activity data were
collected. In August 1973 one milk hind (TALH) and the members of her family group were observed for four consecutive days,
and in July and August 1974 we followed the same hind (then
yeld) for a further eight days and followed two other milk hinds
(RGRH, COLL) for eight days each. In July and August 1975, we
observed nine different hinds for one day each, and in February
and March 1976 we observed eleven. To investigate the effects of
changes in food selection on activity, we followed one milk hind
(COLL) for twenty-four days in July and August 1976 and for
twenty days in February and March 1977. Finally, to investigate
the effects of population density on grazing behavior, in July and
August 1977 we watched five hinds from Upper Kilmory and
from the Intermediate Area and ten each from Lower Kilmory
and Samhnan Insir, repeating the latter two samples in March
and April 1978. Nighttime observation samples of hinds included
five nights spent watching one milk hind and her two mature
daughters in July and August 1973 and five 24-h periods spent
watching one milk hind (COLL) and the members of her family
group in March 1977. We again followed the same hind and her
family for five consecutive 24-h days in July 1977 when she was
yeld.

For stags, two different daytime samples were collected. These
included eighteen different animals of five to eleven years observed for one day each in July and August 1975 and eleven
watched in February and March 1976. Nighttime observations
included five stags followed in July and August 1975 and six
watched in March 1977 and 1979.

During the day watches, the frequency of social interactions
involving the target animal was recorded, as well as the animal's
distance from its two nearest neighbors and their identity. In addition, we collected data on the activity and composition of the
group they were in. At 15-min intervals we recorded the identity
of all animals in the group, their activities, and (in some watches)
their positions relative to each other, using a check sheet with a
series of concentric rings on which the position of all animals was
recorded.

Continuous Observation of Rutting Groups

During the rut, we also carried out a number of full day watches of stags and hinds. Samples included seven day-watches and five night-watches of mature stags (six years or older) when they were holding harems, seven day-watches of five- and six-year-old stags not holding harems, ten day-watches of anestrous hinds, and four day-watches of hinds in estrus. In addition, the activity of all hinds belonging to the harems of the stags that were watched at night were recorded at 15-min intervals, giving a measure of nighttime activity among hinds.

In connection with detailed studies of the rutting activities of stags (chap. 6), we also carried out shorter periods of continuous observation of individuals. Since in this case we were principally interested in the relative frequency of interactions, we watched only individuals that were standing. These watches varied in length between 10 min and 90 min. In analysis of these data, we combined samples for the same individual, and our sample size is the number of individuals sampled.

Incidental Records

During censuses and continuous watches, certain kinds of data were collected ad libitum. Whenever possible, suckling bouts were recorded and timed using a stopwatch; all threats and interactions were noted, whether or not they involved the target animal; and, during the rut, all approaches, parallel walks, fights, and copulations were recorded and timed. In analyses of fights and threats, we typically combined data from a wide variety of different individuals, since sample sizes for particular animals were too small to provide a reasonable basis for measurement. In analyses of suckling data, data for particular dyads were analyzed separately.

3.4 DEFINITIONS

This section lists definitions of variables used in the rest of the book that are not provided in the relevant chapters. Where development of a nonarbitrary rationale for a particular definition required considerable analysis, this is described in the relevant appendix.

Reproduction in Hinds

Harem: The group of hinds defended by a stag at a particular time on a given day during the rut.

Harem stability: The proportion of individual hinds seen in the harem of the same stag on successive days.

Turning point: The point at which cumulative sum analysis (see below) revealed an alteration in the rate of change of a process.

Mother's age: The age of a hind in years, assuming she was born on 1 May of her year of birth (i.e., a hind who was born in 1970 and conceived for the first time in 1972, giving birth in the spring of 1973, would be classified as a three-year-old mother). The ages of hinds born after 1968 were known accurately because individual recognition had been maintained since they were calves. Ages of hinds born before 1968 were known to within a year, from maintaining recognition of them since they were yearlings, from ear tags used in a previous study, or from examination of tooth wear.

Fecundity: A hind's fecundity was the proportion of years in which she gave birth to a calf. The fecundity of a category of hinds was the proportion of individuals that produced calves over a given period.

Mother's previous reproductive status: The reproductive status of a hind in the season that preceded the birth of her calf. Five categories of hinds were recognized: first breeders; true yelds (hinds that had given birth before but had not produced a calf in the previous year); summer yelds (hinds that had borne a calf and lost it before the end of September); winter yelds (hinds that had borne a calf and lost it between October and May); and milk hinds (hinds that had successfully reared a calf the previous year). In most comparisons, we combined true yelds and summer yelds (on the basis that, since most summer mortality occurred within a week of birth, summer yelds had not suffered the strains of lactation) and winter yelds and milk hinds (both these categories of mothers had supported calves through the winter, since most winter calf mortality occurs in March and April). We refer to these two categories as "yelds" and "milk hinds."

Mother's subsequent reproductive status: The reproductive status of a hind in the winter following the birth of her calf. Two categories were recognized: pregnant mothers (those that conceived again and gave birth the following year) and barren mothers (those that apparently failed to conceive again and did not give birth the following year).

Mother's home range area: The division of the study area in which the mother's core area was located (see fig. 2.7).

Kidney fat index (KFI): The total weight of kidneys plus peri-
nephric fat divided by the weight of the kidneys (see Mitchell,
McCowan, and Nicholson 1976).

Birth weight: The weight of the calf at birth calculated for calves
caught when fourteen days old or younger, assuming a weight
gain of 0.4 kg/day (see Guinness, Albon, and Clutton-Brock
1978).

Conception date: The date on which a calf was conceived, estimated
by backdating from its birth date by 234 days for female calves
and 236 for male calves.

Birth date: Date of calving assessed from daily observations of the
hind's udder size and flank distension, the behavior of the
mother, and sightings of the calf.

Suckling bout duration: Sucks not separated by more than 15 min
were considered a single bout, and the cumulative number of
seconds for which an animal sucked was calculated. Isolated
bouts of < 1 sec were scored as rejections.

Date of death: The date at which we judged a calf had died, based
on observations of its mother, on the shrinkage of her udder, or
on the location of the carcass (see Guinness, Clutton-Brock, and
Albon 1978).

Reproductive success: The number of calves a mother reared to one
year old over a specified period of time.

Lifetime reproductive success: The total number of calves a mother
reared to one year old during her lifetime (see Appendix 8).

Reproductive value: The total number of *female* calves an individual
of a given age could expect to produce during the rest of her
life (see Appendix 9).

Reproduction in Stags

Sparring: When two stags locked antlers and twisted and turned
their heads. Stags seldom pushed hard during sparring encounters. Sparring typically occurred between younger stags,
and it never involved harem-holders.

Harem size: The total number of hinds one year old or older seen
with a particular stag at one time on a given day.

Fighting success: The number of stags an individual beat plus the
number that they beat, divided by the number of stags he lost
to, plus the number that they lost to (Appendix 10).

Reproductive success: The estimated number of calves fathered by a
stag during a given period of time (see Appendix 11).

Lifetime reproductive success: The estimated number of calves *sur-viving to one year* fathered by a stag during its lifetime (see Appendix 11).

Reproductive value: The number of *female* calves a stag of a given age could expect to father in the future (see Appendix 9).

Peak rut: The period between the turning points of harem stability, defined by cumulative sum analysis (see p. 50). This was closely associated with the conception peak and changed from year to year.

Social Behavior

Party: An aggregation of deer where no individual was more than 50 m from any other animal in the same party (see Appendix 12). The term carries no implication that membership was stable (see Crook 1970).

Party size: The number of animals in a party at a given time (excluding calves).

Matrilineal group: The total number of a hind's known matrilineal relatives.

Association: Two different measures of association between animals were used, though both were based on the frequency of occurrence in the same party. The first of these was the number of occasions on which two animals were seen in the same party, divided by the number of occasions on which either or both were seen in separate parties (coefficient 1). This measure was used in analysis of dyadic association between mothers and their calves and yearlings (sec. 9.3; see Guinness, Hall, and Cockerill 1979). The second measure used (coefficient 2) was essentially the same as the first with the exception that the denominator was the number of times the pair were seen together plus the number of times *each* of the animals was seen separately (Everitt 1974) and consequently gave an absolutely lower figure. This coefficient was used in analyses of association between hinds and their mature daughters as well as between stags (chap. 9, analyses 6, 8, 9, 10, 11, 13; figs. 9.7, 9.8, 9.9).

Nearest neighbor distance: The linear distance between the head of an animal and the head of the animal nearest to it excluding its own calf.

Mother-offspring distance: The linear distance between a mother and her offspring when both were members of the same party.

Range (or core area) overlap: The number of hectare quadrats in an

individual's range (or core area) that also fell in the range (or
core area) of another animal, divided by the total number of
hectare quadrats in the first animal's range (or core area).

Habitat Use and Feeding Behavior

Season: Some comparisons of feeding behavior and habitat involved only the "central" months of summer and winter. For
summer these were May, June, and July; for winter, January,
February, and March.

Grazing: When an animal either was actively biting or was chewing
recently ingested food in a standing position. Over 95% of grazing records involved standing animals.

Ruminating: When an animal was either chewing or in the process
of passing a bolus up to the buccal cavity. Over 95% of ruminating records involved lying animals.

Moving: When an animal was actively locomoting at end of the
1-min period.

Rutting: Included roaring, herding, chasing, or fighting.

Inactive, standing: When an animal was not engaged in any of
previous activities and was standing.

Inactive, lying: As above, but lying.

Range: The area enclosed by a boundary set around the total
number of sightings of an individual during a specified period
(see Appendix 13).

Core area: The smallest number of contiguous hectare quadrats
within an animal's range that accounted for 65% of all sightings
of that individual (see Appendix 13).

Grazing bout: A period of feeding activity not interrupted by more
than 10 successive min of some other activity (Appendix 14).

Day range length: The total number of hectare quadrats an animal
entered during a day.

Shelter: Animals were recorded as sheltered if they were not exposed to the direct force of the prevailing wind.

Population Dynamics

Larder weight: The weight of the whole animal less alimentary tract
and bleedable blood. This approximates to 73% live weight in
stags shot in early autumn, and to 66% and 60% respectively in
yeld and milk hinds shot between November and January
(Mitchell, Staines, and Welch 1977; Mitchell and Crisp 1981).

Member of the study population: An animal that was seen in at least

10% of surveys of the study area in at least four months of the
  year.
Population size: The number of animals belonging to the study
  population (see above).
Cleaning date: The day on which the velvet of a stag's antlers began
  to fray.
Casting date: The day on which a stag cast its first antler.
Coat change: Animals were scored as changing into winter coats if
  more than half their summer coats had been shed.
Antler weight: The weight of a single dry antler measured at least
  six months after casting: only intact antlers were used.

3.5   STATISTICAL ANALYSIS
  Selection of Tests

Standard parametric and nonparametric tests were used (Sokal
and Rohlf 1969; Siegel 1956): where we used the former, we
checked to see that the data were approximately normally distributed. For frequency comparisons, we used G tests (Sokal and
Rohlf 1969) except where calculations of expected frequencies
involved weighting or where the analysis was extracted from earlier papers. An advantage of the G test is that, like the analysis of
variance, it permits investigation of whether the effects of two
independent variables interact (i.e., whether the relationship between two variables differs across the range of a third variable).
Where comparisons involved values for individual animals, we
used the Mann-Whitney U test or, where k > 2, the Kruskal-Wallis
one-way analysis of variance for independent samples, and the
sign test or the Wilcoxon matched-pairs signed-ranks test for related samples (Siegel 1956).

  Where we wished to investigate the effects of more than one
linear variable at the same time, we used multiple stepwise regression (Snedecor and Cochran 1967), which removes the effects of
the most strongly correlated variable, permitting one to examine
relationships between other variables and the residual variance.
In most cases we restricted such analyses to two independent variables. Where we needed to remove the effects of a nonlinear or
categorical variable, we calculated adjusted means using analysis
of variance and subsequently related these to other independent
variables. Values that were corrected in this way are referred to in
the text as "adjusted."

  In a number of cases it was necessary to identify the point at which the rate of a process changed—for example, to identify the peak rut, we needed to locate the date when the proportion of hinds changing harem began to decline. Cumulative sum testing (Woodward and Goldsmith 1964) is suitable for problems of this kind. Briefly, it treats the number of events, from first to last, as the outcome of a process whose rate can vary and calculates the cumulative sum of events that have occurred by each point in time. When the cumulative sum is plotted against position in the sequence, changes in process rate that might otherwise be obscured by short-term variability in the data appear as changes in slope. The locations of such changes (termed "turning points") were determined visually before testing whether changes in slope were significant (Woodward and Goldsmith 1964).

Results of Tests

The statistical tests used are indicated in the text by numbers in square brackets and are listed at the end of each chapter. All probability values are two-tailed unless otherwise specified. Where we state in the text that two samples differ, this indicates a significant difference at the .05 level or at a higher level of significance, while statements that two categories "tend" to differ indicate a nonsignificant result where there were additional grounds for believing that a difference existed. One-tailed probability values were used only when there were firm reasons for predicting the direction of an effect.

3.6 SUMMARY

1. Individual deer were recognized by ear tags, collars, and idiosyncrasies of coloring and shape. After 1971, the population of our study area in the North Block of Rhum increased from 60 hinds and 124 stags to 149 hinds and 135 stags in 1979.
2. The study area was censused five times each month, and the locality, activity, and associations of each animal seen were recorded. Less complete surveys of social relationships between mothers and their offspring were carried out on most other days.
3. Individual deer were also observed for a day (or, in some cases, a day and a night) at a time.
4. Definitions of variables used in the text are listed and our techniques of statistical analysis are described.

4 The Breeding Biology of Hinds

The Male and the Female individual may be compared in various ways with the spermatozoon and ovum. The Male is active and roaming, he hunts for his partner and is an expender of Energy; the Female is passive, sedentary, one who waits for her partner, and is a conserver of Energy.

W. Heape (1913)

4.1   INTRODUCTION

In order to reproduce, a female must select a fertile breeding partner. In practice this seldom presents difficulties, and in most species receptive females can rely on males to find them. It is not yet clear to what extent female mammals can enhance their reproductive success by selecting particular males. Theoretically, they ought to do so either by choosing partners who are able to invest heavily in their joint offspring or by choosing ones that possess superior genotypes (Halliday 1978; Partridge 1980). In red deer, where males play no part in rearing their offspring, the first kind of choice is redundant. Does the second kind of choice occur? In sections 4.2 and 4.3, we describe the mating behavior of red deer hinds and examine evidence for mate selection by hinds.

Since a hind's life span is relatively short, each calf she bears represents a large portion of her reproductive potential, and strong selection pressures will favor mothers that minimize calf mortality. To accomplish this, hinds must minimize the chance that their calves will be caught by predators, at the same time ensuring that they grow rapidly so that they are large and healthy by the onset of the winter. In section 4.4 we describe gestation, birth, and early care, and in the final section we describe the costs of maternal investment and the effects of rearing a calf through the summer on the mother's subsequent condition, reproductive performance, and survival.

4.2   MATING

The first change that was evident in the behavior of hinds at the onset of the rut was a restriction in the size of their ranges, which occurred from early September onward, several weeks before the stags joined the hinds on the rutting grounds. For example, eight hinds whose ranging behavior was monitored in 1975 reduced the total area they used by 45% between 6 September and 26 September, compared with 16 August to 5 September [1] (see fig. 4.1). This change was associated with an increase in the amount of time they spent on the short greens, where they collected in large parties. At approximately the same time, they began to give off a characteristic scent from the region of the vulva that grew in intensity during the following weeks (Guinness, Lincoln, and Short 1971).

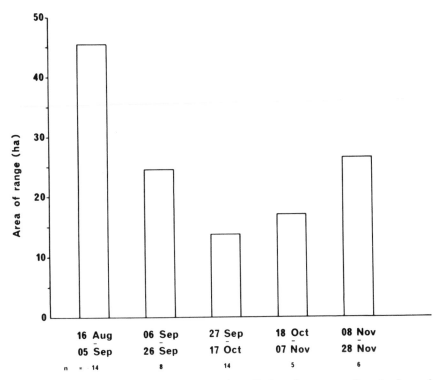

Fig. 4.1. Median range size in hinds during the 1975 breeding season. Samples for each three-week period were based on animals recorded in censuses at least five times during the three weeks. Sample size is shown below each histogram. Range size was determined by the method described in Appendix 13.

Mature stags began to leave their normal ranges in mid-September and usually joined the hinds on the short greens before 1 October, becoming progressively intolerant of each other's presence. At first they showed little interest in the hinds, but by the first few days of October they started to associate more regularly with parties of hinds and became increasingly intolerant of the presence of male yearlings and young stags in the vicinity. As far as possible, they controlled the movement of hinds, preventing individuals from leaving wherever possible by herding them back to the group and driving in extra hinds that approached. The proportion of hinds belonging to harems grew rapidly, and by 5 October only a small proportion of animals were seen outside harems (fig. 4.2). The day range length of hinds decreased from about 2 km per 12-h day to less than 1 km [2], and range size also declined further.

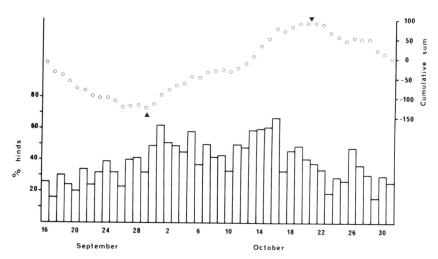

Fig. 4.2. Proportion of hinds associating with stags older than five years during the rut (1975 data). Cumulative sum plots *(open circles)* are also shown.

Conception was highly synchronized, and over 70% of all conceptions occurred during the second and third weeks of October (fig. 4.3). A variety of behavioral changes occurred during the 12–24 h the hind was in estrus. The proportion of time she spent feeding tended to decline [3], and the proportion of time she spent inactive standing or moving tended to increase (fig. 4.4). Similar changes have been described in other deer species (Morrison 1960; Ozoga and Verme 1975). The timing of breeding in hinds is probably under photoperiodic control and may be triggered by changes in day length relative to the solstices or equinoxes: within both hemispheres, red deer populations at different latitudes breed at approximately the same time, but breeding is displaced by exactly six months between hemispheres (Fletcher 1974). Variation in body condition is probably also involved (Mitchell and Lincoln 1973).

Stags took an intense interest in estrous hinds, standing close to them, licking their preorbital glands or vulvas, and intermittently chasing them through the harem. In the early stages of estrus, hinds ran when the stag moved toward them or tried to place his chin on their backs, and only as estrus progressed were they willing to stand and allow him to mount (see fig. 4.5). At this stage hinds also showed interest in the stag, licking him and sometimes mounting him. Stags usually mounted several times before

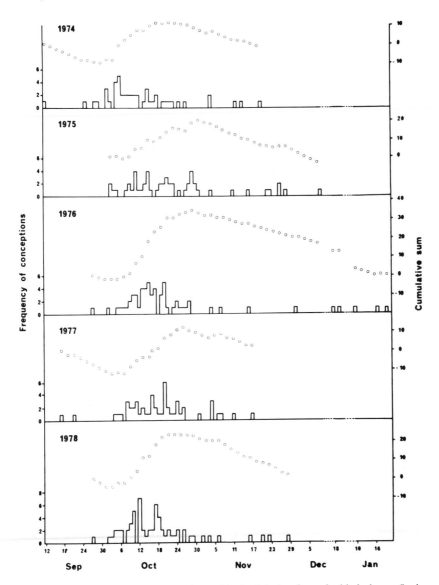

FIG. 4.3. Frequency of conceptions (estimated by backdating from the birth dates of calves using a gestation interval of 236 days for males and 234 days for females). Cumulative sum plots *(open circles)* and turning points *(solid triangles)* are also shown.

ejaculating, and after ejaculation their interest in the hind apparently waned rapidly, though several successive sequences of chasing, mounting, and ejaculation sometimes occurred. Although most hinds were successfully fertilized at their first estrus, some cycled a second and even a third time. Observations

of hinds held in enclosures without stags show that, if still un-
mated, they continued to cycle as late as April (Guinness, Lincoln,
and Short 1971; F. E. Guinness, unpublished data). Mean cycle
length in a sample of hinds kept in an enclosure on the island was
18.3 days ± 1.7 days (Guinness, Lincoln, and Short 1971).

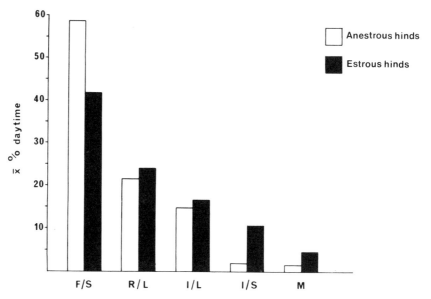

Fig. 4.4. Comparison of the proportion of daytime spent in different activities by estrous
versus anestrous hinds. F/S, feeding, standing; R/L, ruminating, lying; I/L, inactive, lying;
I/S, inactive, standing; M, moving.

4.3   Harem Membership and Mate Selection

The stability of harem membership gradually increased during
the early weeks of the rut, peaking at the same time as the frequency of conceptions among hinds (see figs. 4.3, 4.6). However,
even when harems were most stable, changes in harem membership were common, though they were often temporary.

The causes of harem changing were diverse. Hinds often left a
harem to feed elsewhere and were subsequently gathered up by a
neighboring stag. When family groups had been split up as a
result of conflicts between stags, individuals often left the harem
they were in to rejoin their relatives in another rutting group.
And, in cases where the harem-holding stag persistently
threatened a hind's calf or yearling, the mother often left the
harem. The activities of competing stags were also important in
influencing harem membership. Young stags that infiltrated

FIG. 4.5. An estrous hind stands with back slight arched, allowing the stag to press his chin on her back *(a)* and finally to mount her *(b)* and *(c)*.

harems often chased out hinds when the harem-holder was otherwise occupied, and fights between harem-holders were often associated with changes in harem membership among the hinds.

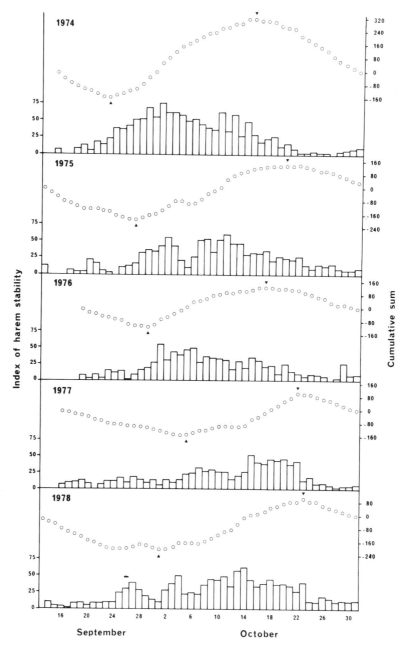

FIG. 4.6. Harem stability at different stages during the rut (1974–78). Our measure of stability was the proportion of hinds seen in rut censuses that were in the same harems on the following day. Solid lines show this figure for each day on which a rut census was carried out. Cumulative sum plots *(open circles)* and turning points *(solid triangles)* are also shown.

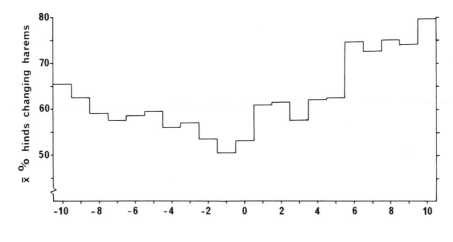

FIG. 4.7. Mean percentage of hinds changing harems between days in relation to the time of conception, estimated from the calf's birth date. Data for 1974–77 were combined, dates of conception were determined by backdating (see p. 46) for each individual in each year, and the mean percentage of individuals that changed was calculated across years *(solid line)*.

The frequency with which hinds changed harems was related to the estrous cycle. As ovulation approached, there was a decline in the frequency of harem changing (see fig. 4.7), and just before estrus the frequency of change was relatively low. Around estrus and afterward, harem changing increased [4] and hinds were seen with more different stags [5].

Hinds were often seen in the harems of the same stags in successive years. However, partly because of the short duration of the breeding life span of stags (see chap. 7) and partly because conception date varied between years, a relatively small proportion of hinds bred with the same male in two successive years: of thirty-two hinds whose breeding partner was identified by back dating in 1974 and 1975, only 15.6% bred with the same stag in both years. Similarly, of twenty-nine hinds whose mates were identified in 1975 and 1976, 24.1% bred with the same stag.

Related hinds tended to be found in the same harems, but, because their conception dates varied and the rutting periods of individual stags seldom spanned the whole conception peak, comparatively few relatives mated with the same stag. Of hinds over two years old that conceived in the same year as their mothers, our estimates of paternity indicated that only 15% bred with the same stag, while fewer than 10% of mature sisters that bred in the same year did so.

There was little firm evidence that hinds showed strong preferences for particular stags. Evidence that individuals sometimes bred with the same stag for several years did not necessarily mean they selected particular breeding partners, for both sexes typically spent the rut in the same area in successive years. Nor did evidence that hinds frequently escaped from harems necessarily indicate mate choice (see above). Investigation of the comparative frequency with which hinds moved to larger harems versus smaller ones, to the harems of older stags versus those of younger ones, and to the harems of stags with larger antlers versus those with smaller ones did not provide any evidence of consistent selection [6]. In a previous study, Gibson found that the number of estrous hinds mated by different stags could be predicted from their harem size, and he concluded that there was no evidence that hinds actively selected particular breeding partners (Gibson 1978; Gibson and Guinness 1980a,b).

In contrast, many observations suggested that hinds avoided mating with young stags. When particular harem-holders were fighting, stags less than five years old often chased hinds out of their harems. Hinds invariably ran away from young males, and we never saw a free-ranging hind mating with a stag of less than five years during the peak rut, though stags are fertile as yearlings (Lincoln 1971b) and though hinds confined in an enclosure on the island bred with a two-year-old stag when no older animal was available. However, hinds on the hill occasionally mated with younger stags late in the breeding season when competition between males was less intense.

Selection by females against breeding with immature males is common in other mammal species (e.g., Cox and LeBoeuf 1977; Eaton 1978). Although this behavior could be favored by selection because age provides some guarantee of fitness in males (Trivers 1972; Halliday 1978), there are several other reasons it may be advantageous. Young stags were rarely able to defend hinds effectively, and a hind held by an immature male was frequently disturbed by the attentions of other males. Moreover, young males were conspicuously inept in their attempts to mate with hinds, and this could have delayed fertilization or increased the risk of physical damage associated with mating. During the study, we observed two cases where females were served in the rectum by experienced stags: in one case the female appeared to be in considerable pain for several weeks afterward, and in the other case the hind died because of a rupture of the rectal wall. Presumably the risk of accidents of this kind is greater in matings involving inexperienced stags.

A similar reluctance to be mated by young males has been observed in elephant seals. Here females react to mating attempts by young or subordinate animals with loud vocalizations that commonly attract the attention of dominant males. Cox and LeBoeuf (1977) argue that this behavior ensures that the female is mated by a genetically superior male, and they predict that other forms of incitement of intermale competition will be widespread among mammals. The reduction of the hinds' ranging during the rut, the increase in the size of hind groups, the synchronization of conception, and the changes in hinds' behavior associated with estrus probably all helped to stimulate competition between stags, and all four traits could have evolved for this reason. However, it is also possible that the first two changes were caused by the autumn flush of the grasses growing on the short greens, that breeding synchrony was adapted to seasonal variation in food abundance, and that the estrous behavior of hinds principally served to ensure that their condition was noticed by the stags holding their harems.

One further form of mate selection might be expected. Several studies have shown that mating with close relatives reduces breeding success (Greenwood, Harvey, and Perrins 1978; Packer 1977; Willis and Wilson 1974). Opportunities for mother/son and father/daughter matings were not common, since stags rarely bred successfully before the age of six and had relatively short reproductive life spans, while the dispersal of a substantial proportion of young stags (see chap. 6) reduced the chance of brother/sister matings. However, we observed a few examples of brother/sister mating and several probable cases of father/daughter mating, though neither was common. Problems of sample size and difficulties in calculating a realistic random probability of such matings prevented us from investigating whether they occurred less frequently than would be expected by chance.

4.4 MATERNAL INVESTMENT
  Gestation

Conception was followed by a gestation period of about 34 wk: a sample of seventy cases where we observed the hind in estrus and knew the birth date of the calf gave a figure of 236.1 ± 4.75 days for gestation of male calves and 234.2 ± 5.04 for females. Other studies indicate that, although gestation length may be influenced by environmental variables, such effects are not pronounced and represent a small percentage of the total length (see Verme 1965; McEwan and Whitehead 1972; Clegg 1959; Alexander 1956).

Both resorption and fetal loss appear to be uncommon in red deer (Blaxter et al. 1974; Mitchell, Staines, and Welch 1977), and evidence for other cervids suggests that the rate of fetal loss usually lies between 10% and 15% (Robinette, Gashwiler, Jones, and Crane 1955; Teer, Thomas, and Walker 1968; Ransom 1967; Nellis 1968; Markgren 1969). In a small number of cases, hinds weaned a calf but failed to give birth the following season, and it seems likely that these animals may have aborted or resorbed the embryo. However, where hinds did not give birth, we believe they usually had failed to conceive in the previous autumn.

Birth

The first sign that a hind is about to calve is a marked swelling of the udder, usually 1–2 days before parturition (see Blaxter et al. 1974). A few hours before birth, the perineal area becomes red and slightly swollen (see Hall 1978). During labor, which usually lasts 30–120 min (Arman 1974; Hall 1978; Arman, Hamilton, and Sharman 1978), the hind becomes more and more restless, frequently nuzzling or grooming her flanks, udder, and perineal area. During the days before and after parturition, hinds sometimes bellow, raising their noses in the air and giving a loud deep moan similar to a stag's roar but softer in tone, a behavior that may be promoted by the high levels of estrogen characteristic of the perinatal period. In the later stages of labor hinds normally lie on their sides, straining as the fetus begins to emerge. Once it is partly exposed, they frequently stand, allowing it to fall out (see fig. 4.8a).

Immediately after the birth, the mother licks the calf clean, then eats the membranes, licking up the amniotic fluid from the ground (fig. 4.8b). Calves are usually able to stand within half an hour of birth, and the first suckling bout typically occurs within 40 min of the end of labor (Arman, Hamilton, and Sharman 1978). The placenta is expelled 1–1½ h after birth and is immediately eaten by the hind; the ground where it fell is cleaned and any

FIG. 4.8. (*a*) A hind, who has been lying during labor, stands as the fetus emerges, allowing it to fall out and (*b*) cleans her newly born calf. (Photographs by Marion Hall.)

stained grass is eaten (see Hall 1978). After cleaning their calves, hinds commonly rest close to them for a short period. Subsequently, another bout of licking and sucking occurs, and at the end of this the mother normally moves slowly away from the place of birth, encouraging the calf to follow her. At some stage during this movement (which seldom exceeds 300 m) the calf leaves the mother, adopting a characteristic hunched posture with head held low, and eventually lies down, typically curling up in a patch of deep grass or heather. During the next few hours the mother grazes or rests within sight of the calf but rarely approaches it closely. After a further 2–3 h she usually visits the calf, and a third bout of licking and sucking occurs—the mother eats the calf's feces and urine as the calf sucks (fig. 4.9). Subsequently the calf again lies down, and the mother moves farther away.

FIG. 4.9. A hind suckles her week-old calf. The mother is licking beneath the tail to stimulate the calf to defecate.

Early Care

Though many red deer populations live in areas where predators are uncommon, this is a comparatively recent circumstance, and studies of populations living where substantial numbers of mammalian predators still exist show that predation can have important effects on recruitment (e.g., McCullough 1969). Since each calf represents a large fraction of a hind's lifetime reproductive success, strong selection pressures should favor mothers that take care not to reveal the position of their calves.

Hinds about to calve typically move away from their matrilineal groups and their usual home ranges (Clutton-Brock and Guinness 1975) and keep their calves on high ground. Isolation of nursing mothers from their usual social groups and separation of the calf from the mother throughout most of the day probably help to make both of them inconspicuous to predators hunting by visual or olfactory cues (see Lent 1974; Geist 1982): across ungulate species, separation of mothers and offspring during the days following birth is associated with cryptic coloration of the young and with ingestion of their feces by the mother (Walther 1965, 1968, 1969).

When their calves were standing (and therefore visible), hinds were intensely vigilant. Before calving, most individuals alerted (raised the head and gazed around) during approximately 20% of minutes in which they were seen feeding (see fig. 4.10); after calving, the percentage increased to about 70% when they were accompanied by standing calves under twenty-one days old. If hinds with young calves were suddenly disturbed when the calf was visible, the hind commonly barked, and the calf quickly hid (see also Bubenik 1965). The hiding response was uncommon among calves over a week old, though evidence suggests that it may continue for up to three weeks in elk (Altmann 1963). On Rhum, hinds with calves over a week old quickly fled when disturbed. Their flight distance was considerably greater when they were accompanied by standing calves less than two weeks old than either before calving or after calving when the calf was lying, though this distance decreased rapidly with increasing calf age (see fig. 4.11).

As calves grew older, both their own behavior and their mother's behavior changed rapidly. During their first three weeks, calves spent over 80% of the time their mothers were with them within 3 m of the hind. However, they rarely lay down close to their mothers and usually moved at least 20 m away before doing so (Clutton-Brock and Guinness 1975). This appeared to represent a preference for lying away from the mother, since they frequently ignored lying places close to her that were apparently suitable. After the third week of life, though calves spent less time close to their mothers, they more often lay down closer to them.

Calves selected their lying sites carefully. They usually lay in long vegetation, in sites raised above the surrounding ground and sheltered from sight on at least one side (see fig. 4.12). In many cases sites were within a gully or a dip in the hill's face. This aspect of site selection appeared to be reinforced by the mother's behavior: mothers with standing calves tended to avoid moving out of dead ground, and they traveled swiftly when on the open face of the hill. As they grew older, calves more frequently lay on the short greens and in sites that were neither sheltered, raised, nor in gullies (fig. 4.13).

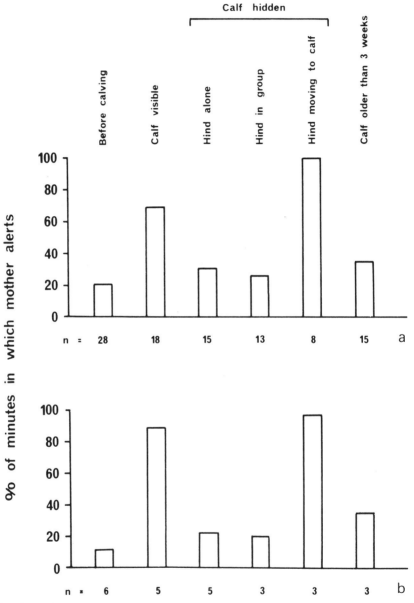

FIG. 4.10. Changes in vigilance among *(a)* multiparous hinds, *(b)* first breeders. Histograms show the median percentage of minutes in which mothers alerted. The number of different hinds tested is shown beneath each point (from Clutton-Brock and Guinness 1975).

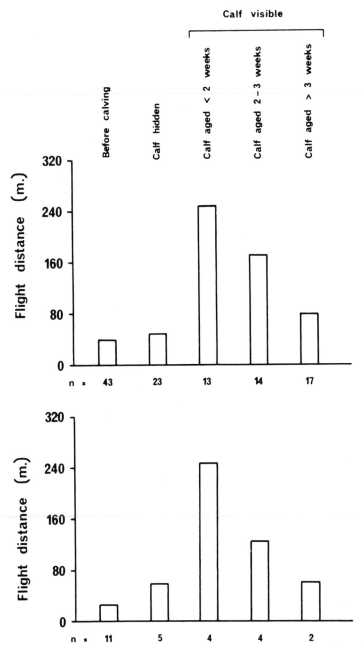

FIG. 4.11. Changes in median flight distance from an observer in breeding hinds. The number of different hinds tested is shown beneath each point (after Clutton-Brock and Guinness 1975). (*a*) Multiparous hinds; (*b*) first breeders.

FIG. 4.12. A young calf lying curled up, concealed in long vegetation, while its mother feeds.

Since a hind that had to learn by experience the behavior necessary to protect her calf from predators would be unlikely to realize her potential reproductive success, selection is likely to favor individuals in which the appropriate behavior occurs spontaneously the first time they give birth. In red deer this appears to be the case: the removal to high ground before calving, the avoidance of other deer, and the extreme wariness during the first two weeks of the calf's life all appeared in hinds calving for the first time (Clutton-Brock and Guinness 1975).

Lactation and Milk Yield

The period of lactation is of crucial importance to hinds. The growth rate of calves during the first six months of life is probably the principal determinant of the size at which they enter the winter and is likely to affect both their chances of survival and their body size as adults (see chap. 5). Lactation is associated with a rapid decline in the mother's body reserves, and a hind that invests too heavily during this period not only may be less likely to

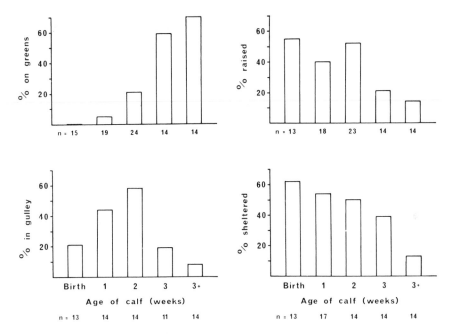

FIG. 4.13. Age changes in the lying sites used by calves of different ages. Sample size shown below each column (from Clutton-Brock and Guinness 1975).

conceive the following autumn but may also enter the winter in poor condition, risking both her own life and that of her calf.

Estimates of the milk yield of captive Scottish red deer hinds (fig. 4.14) show that both the volume and the total calorie value of the hind's milk yield peak during the first eighty days of lactation and subsequently decline. In this sample, calves took between 150 and 630 g of milk per feed until approximately their fifteenth week. The total milk yield of hinds peaked around the seventh to eighth week after birth, declining after the fifteenth week. Maximum yields ranged from 1,415 to 1,980 g per day, considerably lower than Bubenik's estimate of 3–4.5 l per day for Continental red deer (Bubenik 1965), which show both greater birth weights and faster growth rates.

In contrast to milk yield, the concentration of solids in the milk (including fat, protein, and minerals) and the calorie value per unit milk yield continue to increase with the age of the calf (see Appendix 15). Similar changes in the richness of milk occur in other cervids (e.g., Aschaffenburg et al. 1962).

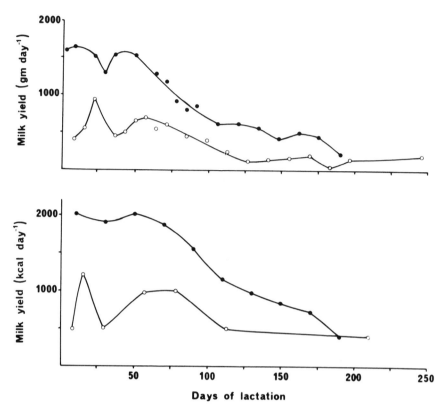

FIG. 4.14. Mean milk yield (estimated by weighing calves) of three hinds maintained on un-restricted food *(solid circles)* and one maintained on a food supply adequate for a nonlac-tating hind *(open circles)* (from Arman et al. 1974).

Suckling Frequency

During the first day or two after birth, calves sucked every 2–3 h while their mothers were with them (see Hall 1978). When they began to accompany their mothers to low ground after a week or so, suckling frequency was about eight times per 24 h, then it declined to four to five times per day by the time they were three months old (see fig. 4.15). Suckling frequency differed little be-tween day and night, though it peaked at dawn and dusk. Hinds rarely suckled calves that were not their own (see Geist 1982), and they often reacted aggressively to approaches by other calves. After declining in late summer, suckling frequency increased again at the beginning of the rut, apparently because the un-familiar circumstance of rutting groups led to increased proximity between mothers and their calves. Estimates of suckling frequency

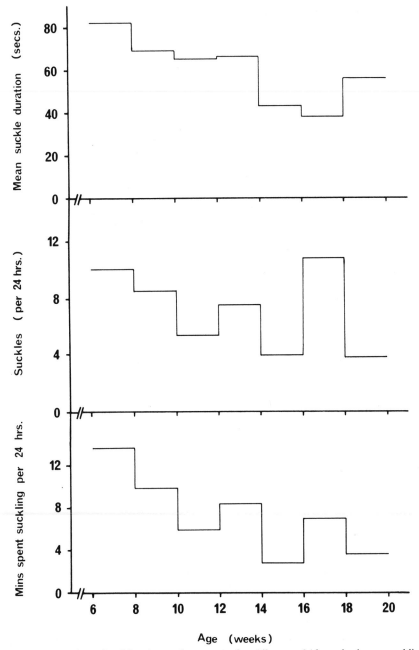

FIG. 4.15. Duration of suckling bouts, frequency of suckling per 24 h, and minutes suckling per 24 h by one male calf (COL6) in 1976. Estimates of suckling frequency were based on daytime observation only but are shown calculated for the 24-h period, since observation showed that neither frequency nor duration changed at night.

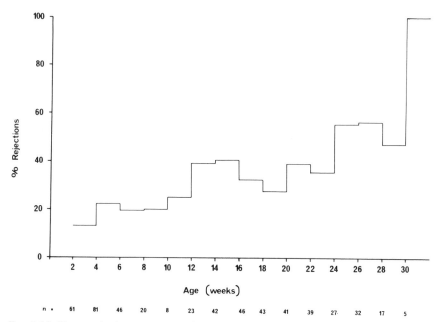

FIG. 4.16. Changes in the percentage of sucking attempts by the calf that were rejected by the mother. The figure shows data for the offspring of pregnant hinds only (see text) (from Cockerill, n.d.).

in red deer kept in enclosures have provided similar figures (Bubenik 1965; Arman et al. 1974; Hall 1978).

Duration of Suckling Bouts

During the first day or two after birth, calves sucked weakly and inefficiently, and suckling bouts were protracted. Bouts quickly became shorter as the calf sucked more strongly, and by the time calves were three weeks old they averaged about 100 sec (see Cockerill, n.d.; Bubenik 1965). After this, bout duration gradually declined to about 60 sec by the time the calf was ten weeks old and to about 30 sec by twenty-five weeks (see Appendix 16). Comparisons showed that there was a tendency for suckling bouts involving heavy-born calves to be shorter than those involving light-born calves, which was significant among females [7].

The duration of suckling bouts (allowing for variation in calf age) was related to the mother's previous breeding experience [8]: bouts involving first-breeders were longest, those involving yeld hinds averaged 18 sec shorter. However, no differences were evident between milk and yeld hinds, though suckling bout duration was related to maternal age (see sec. 5.6).

TABLE 4.1   Last Dates on Which Offspring of Pregnant versus Barren Mothers Were Observed to Suck

Year	Pregnant	Barren
1975	28 October 1975 (27)	16 July 1976 (4)
1976	1 November 1976 (22)	5 July 1977 (9)
1977	27 October 1977 (23)	13 June 1978 (23)

Note: See text for definitions. The table shows medians calculated across all calves that were seen to suck from their mothers after the age of three months. Sample sizes are shown in parentheses.

Weaning

As calves grew older, their mothers were progressively more likely to frustrate attempts to suck by walking forward as soon as they began to search for the nipple (Cockerill, n.d.). The proportion of sucking attempts that were rejected by the mother increased steadily over the first six months of the calf's life, from less than 15% during the first two weeks to over 50% by the time the calves were twenty-four weeks old (fig. 4.16). After calves were six months old, the proportion of rejections remained approximately constant, and the frequency of sucking attempts fell rapidly.

The timing of weaning differed between the offspring of mothers that became pregnant in the October following the calf's birth (pregnant mothers) and those of mothers that did not (barren mothers) [9]. Pregnant mothers often weaned their calves by the time they were five to seven months old—the last dates on which their calves were observed to suck typically fell before the end of November (see table 4.1). In contrast, barren mothers usually did not wean their calves till the following summer or autumn: although the frequency with which barren mothers allowed their calves to suck often dropped to lower levels in the winter and early spring, both the size of their udders and the frequency with which they permitted sucking usually increased again in May and June before declining in September and October.

The postponement of weaning by barren hinds may have been advantageous because it allowed them to increase the growth rate of their yearling offspring at little cost to their future reproductive potential. Even if the amount of milk they allowed the yearling was relatively small, this could have had an important effect on its growth by increasing protein levels in its rumen.

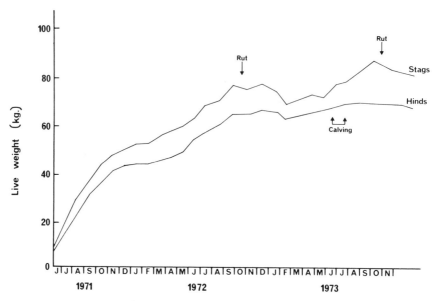

FIG. 4.17. Growth rates of male and female calves raised on a high plane of nutrition at the Glensaugh deer farm (from Blaxter et al. 1974).

Early Growth

The pattern of early growth in red deer resembles that of other cervids, as data for farm-reared deer show (fig. 4.17). Wild Scottish populations show slower growth rates, but detailed information on early growth is unavailable. Both in farm-reared animals and in wild populations, there is a slackening in the growth rate from the onset of the first winter to the following March or April (Blaxter et al. 1974; Mitchell, unpublished data).

4.5   THE COSTS OF REPRODUCTION
Body Condition

The energy costs of gestation and lactation have a marked effect on maternal condition during the summer months (fig. 4.18). Despite the abundance of food, milk hinds put on little fat during the summer, and by autumn they are usually in substantially poorer condition than yelds. In our study area this was associated with differences between yelds and milk hinds in the timing of the change into winter coats, which was later in milk hinds [10]. Though yeld hinds entered the winter in substantially better condition than milk hinds, they lost much of their body fat over the

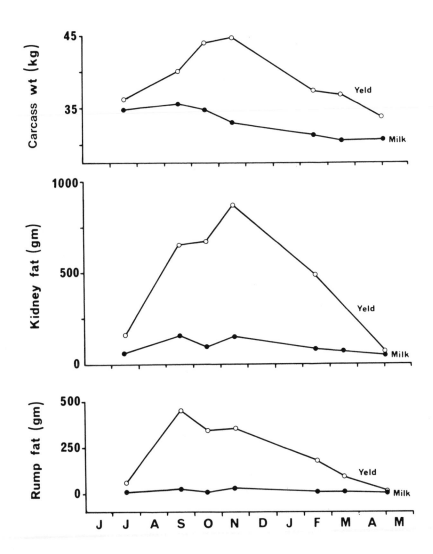

FIG. 4.18. Changes in carcass weight, kidney fat, and rump fat in milk and yeld hinds shot on Rhum throughout the year (from Mitchell, McCowan, and Nicholson 1976).

winter months, and by the following April or May they differed little in body condition from milk hinds.

The effects of increasing age on body condition are also important. Young hinds of three to seven years showed the best condition; after the age of eight condition began to decline (fig. 4.19), and it fell rapidly in old animals. Body weight shows similar

changes with age, though hinds of two to four years show relatively low weights and body weight does not peak till between
seven and ten (see fig. 2.4). The cumulative effects of both
breeding and tooth wear may contribute to this decline, and we
often found that individuals showing a sudden decline in condition had broken an incisor.

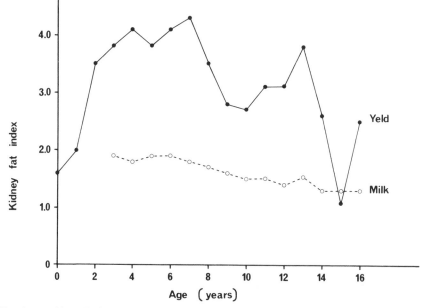

FIG. 4.19. Kidney fat index of milk and yeld hinds of different ages shot during the winter
at Glenfeshie (Invernesshire) (data from Mitchell, Staines, and Welch 1977).

Subsequent Reproductive Performance

Comparisons of the subsequent reproductive performance of
milk hinds with that of yelds provided additional evidence of the
costs of reproduction. Milk hinds showed substantially lower
fecundity than yelds (73.7% versus 90.1%) [11] and later calving
dates [12]. Both these effects intensified as population density rose
in the study area (see chap. 12, figs. 12.4, 12.5). However, the
calves of milk hinds and yelds did not differ either in birth weight
(see Guinness, Albon, and Clutton-Brock 1978) or in survival.
This contrast may have occurred because yelds were in better
condition than milk hinds during the mating season but lost much
of their fat during the winter months, so that by calving time they
differed comparatively little from milk hinds (see fig. 4.18). In

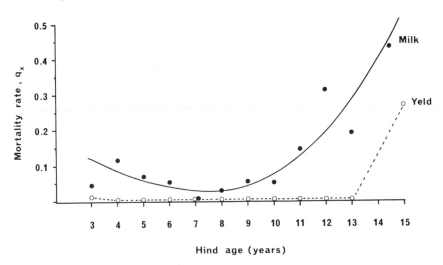

F<small>IG</small>. 4.20. Age-specific mortality rates for milk and yeld hinds. Points show the proportion of milk and yeld hinds dying at each age, the solid line the (smoothed) curve for milk hinds based on these figures. Data from 1974–80 combined.

addition, it is likely that a hind's access to food resources during the period of heaviest energy investment has a greater effect on her ability to invest in her calf than does her condition at the beginning of gestation: in blackface sheep, milk yields and lamb growth are more closely related to food availability during lactation than to body condition before parturition (Peart 1968).

Mortality

Reproduction also endangered the life of the hind. During the course of the study, three hinds died during the calving season as a consequence of birth complications. In addition, of thirty-seven parous hinds that died between 1974 and 1980 (mostly in late winter and early spring) only seven (18.9%) were yeld during the previous breeding season, and all twelve hinds that died when less than ten years old during late winter or spring (December–May) had raised a calf through the previous summer and in some cases were still lactating. Comparison of age-specific mortality rates for milk and yeld hinds shows that the effect of breeding on survival is slight throughout those of the lifespan but increases after the age of ten (see fig. 4.20).

4.6 SUMMARY

1. At the end of September, hinds gathered in traditional rutting
   areas, where they were joined by the stags.
2. Apart from avoiding mating with immature stags, there was no
   evidence that hinds consistently selected any particular category of stags as breeding partners.
3. Births were highly synchronized, and most occurred in early
   June. Hinds were careful to keep their calves concealed during
   the first few weeks of life.
4. The costs of reproduction to hinds were evidently high: compared with yeld hinds, milk hinds had lower fat reserves, were
   less likely to conceive in autumn, and were more likely to die in
   winter.

Statistical Tests

1. Comparison of range size in eight hinds between 16 August
   and 5 September versus 6 September and 26 September 1975.
   Wilcoxon matched-pairs test: $T = 3$, $N = 8$, $p < .05$.
2. Comparison of mean day range length (number of quadrats
   entered per day) in seven hinds between 2 October and 14
   October 1978 and mean day range length in (a) thirty-three
   hinds observed in winter and (b) twenty-eight observed in summer.
   (a) Student's t test: $t = 2.17$, d.f. $= 39$, $p < .05$.
   (b) Student's t test: $t = 2.09$, d.f. $= 34$, $p < .05$.
3. Comparison of proportion of daytime spent feeding in watches
   of estrous versus anestrous hinds in October 1976.
   Mann-Whitney U test: $U = 6$, $n_1 = 4$, $n_2 = 9$, $.1 > p > .05$.
4. Comparison of mean percentage of hinds leaving their previous harem per day during ten days before estrus versus ten
   days after (1974–77 data combined).
   Mann-Whitney U test: $U = 10$, $n_1 = 10$, $n_2 = 10$, $p < .002$.
5. Comparison of the number of different stags that hinds were
   seen with in the five days before estrus versus five days after
   (1974–77 data combined).
   Sign test: $x = 46$, $N = 130$, $z = 3.42$, $p < .001$.
6. Comparison of frequency with which hinds that changed
   harems during the five days before estrus (calculated by backdating from birth dates) moved to: (a) stags with larger harems

than the harems they were in previously; *(b)* stags that were older than their previous harem-holders; *(c)* stags with larger antlers (judged on point number) than their previous harem-holders.

(a) Sign test: $x = 20, N = 41, z = 0$, n.s.

(b) Sign test: $x = 22, N = 45, z = 0$, n.s.

(c) Sign test: $x = 20, N = 45, z = -.596$, n.s.

7. Correlation between the mean length of suckling bouts involving female calves and their weight at birth.

Pearson product-moment correlation: $r = -.493, t = 3.45$, d.f. $= 37, p < .01$.

8. Comparison of duration of suckling bouts between first-breeders, yeld hinds, and milk hinds (data collected 1975–78).

Analysis of variance: $F_{1.101} = 5.564, p < .02$. (Cockerill, n.d.).

9. Comparison of last dates on which the calves of pregnant versus barren mothers were observed to suck.

Mann-Whitney U tests:

1975: $z = 3.1, n_1 = 27, n_2 = 4, p < .001$.

1976: $z = 3.4, n_1 = 22, n_2 = 9, p < .001$.

1977: $z = 4.2, n_1 = 23, n_2 = 23, p < .001$.

10. Comparison of the frequency of milk hinds versus yeld hinds that had changed into winter coat by 1 November 1978 (36% and 70% respectively). $N = 50, 40$.

G test: $G = 10.51$, d.f. $= 1, p < .01$.

11. Comparison of fecundity in milk hinds versus yelds (1971–79).

G test: $G = 18.7$, d.f. $= 1, p < .001$. $N = 357, 152$.

12. Comparison of the frequency of early-born calves to yeld versus milk hinds (1971–79), $N = 131, 212$.

G test: $G = 7.03$, d.f. $= 1, p < .01$.

5 Reproductive Success in Hinds

for those individuals which generated or nourished their offspring best would leave, coeteris paribus, the greatest number to inherit their superiority; whilst those which generated or nourished their offspring badly would leave but few to inherit their weaker powers.

Charles Darwin (1871)

5.1 INTRODUCTION

Although it is through differences in the lifetime reproductive success of individuals that natural selection operates to spread and maintain adaptations, so far few studies have been able to measure lifetime breeding success in wild animals. In section 5.2 we describe our estimates of lifetime reproductive success among hinds.

A hind's lifetime success is a product of her fecundity, the frequency of mortality among her calves, and the length of her breeding life span: in section 5.3 we describe the contributions of each of these three variables to differences in breeding success, and, in sections 5.4 and 5.5 we examine some of the ecological factors that affect them.

Our results show that a hind's age is related to virtually all aspects of her reproductive performance. Many age effects are probably a consequence of variation in weight and body condition resulting from changes in the ability to collect and process food, but adaptive strategies may also be involved. In the final section we examine some evidence for effects of this kind in red deer.

5.2 DIFFERENCES IN LIFETIME REPRODUCTIVE SUCCESS AMONG HINDS

An ideal estimate of lifetime reproductive success would be based on a large cohort of individuals, all of which had been followed till their deaths. Such data do not yet exist for other vertebrates, and, since red deer can live until they are twenty, our study has not lasted long enough to provide this information. Nevertheless, our data extend over enough years to cover the bulk of the reproductive life span of a considerable number of individuals and can be used to provide an approximate estimate of the distribution of lifetime reproductive success.

Among seventeen hinds that reached breeding age during the study and died before 1981, the total number of calves reared to one year old varied from zero to seven. However, this sample is biased toward animals that died young, so to compensate for this, we extracted data for all forty-three hinds alive in 1979 for which we had at least six years of breeding data. (Complete data for all variables were not available in each case so sample size varies to some extent between analyses using this data sample.) Since no animals in our study area calved before they were three and only

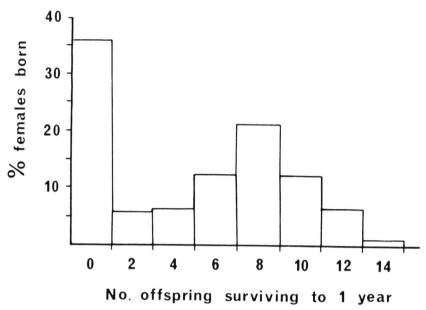

FIG. 5.1. Estimated distribution of lifetime reproductive success in red deer hinds in our study area.

10% of hinds reared a calf after age thirteen, this covered the bulk of their reproductive life spans. The numbers of calves these animals reared in the remaining years of their lives were extrapolated from their breeding record in the years for which data were available, taking into account age changes in reproductive success (see Appendix 8). These estimates were combined with our records of reproductive success for each individual to provide a frequency distribution of lifetime reproductive success that was smoothed by reiteration. This indicates a range of 0–13 calves reared per hind, with a median value of 4.5 (fig. 5.1). Most (but not all) individuals that failed to rear any calves during their lifetimes died before the age of four, and the majority of those that reared ≤ 2 offspring died before they were eight. Among animals that reared at least one calf, mean reproductive success was 7.38, with a variance of 6.05.

5.3   THE PROXIMATE CAUSES OF VARIATION IN LIFETIME REPRODUCTIVE SUCCESS

What were the immediate causes of variation in reproductive success? In the original sample of seventeen hinds, variation in lifetime reproductive success was correlated with differences in life span [1] and calf survival [2] but not with variation in fecundity between hinds. However, since this sample was biased toward individuals that died young (see above), these results probably overestimate the contribution of differences in life span.

To compare the effects of individual differences in fecundity and calf mortality, we used data for all hinds born between 1963 and 1971 for which we had records of breeding success in at least six seasons when they were over three years old. The total number of calves these hinds reared to one year old varied from one to six and was again best predicted by variation in calf survival, which accounted for 75.1% of the variance in reproductive success [3], while differences in fecundity accounted for 20.7% of the residual variance [4]. Individual differences in calf survival and fecundity were not significantly correlated with each other [5].

5.4   FACTORS AFFECTING FECUNDITY

What factors influenced a hind's fecundity? Fecundity changed with age and was low among three- and four-year-old hinds as well as among mothers aged thirteen and over (table 5.1; see also Lowe 1969; Mitchell 1973). Previous analyses have shown that differences in fecundity in wild red deer populations are closely related to the hind's weight (Mitchell and Brown 1974), and, in captive red deer of under seven years, weight differences fully account for age variation, though it is not yet clear whether this is true over the whole life span (Hamilton and Blaxter 1980).

Differences in weight between individuals were presumably a consequence of differences in skeletal size and body condition (Mitchell, Staines, and Welch 1977). Which of these variables accounted for the relationship between fecundity and body weight? There was little evidence that a hind's size influenced her fecundity: skeletal size (as measured by jaw length) was not correlated either with fecundity or with body condition in samples of red deer hinds shot on the Scottish mainland (Mitchell and Brown 1974; B. Mitchell, unpublished data). In contrast, fecundity was closely correlated with measures of body condition (Mitchell and Brown 1974).

What factors influence a hind's body condition in wild populations? Unfortunately neither studies based on carcass samples nor those based on observation alone can provide a satisfactory answer. However, several lines of evidence indicated that the quantity and quality of resources available in the mother's home range

TABLE 5.1   Fecundity in Hinds of Different Ages in Our Study Area on Rhum, 1971–79

Fecundity	3	4	5	6	7	8
Milk hinds	—	52.9	65.9	68.3	79.5	84.7
N	—	34	41	41	39	35
Yeld hinds	36.1	100	97.1	92.0	94.1	91.7
N	97	6	35	25	17	12

influenced her body condition and fecundity: hinds using Samhnan Insir showed lower fecundity than those using Kilmory, and as population density increased in the study area fecundity fell, especially in younger hinds (see chap. 12).

In species, such as red deer, where daughters adopt home ranges overlapping those of their mothers (see chap. 9), Clark (1978) has suggested that reproductive performance is likely to decline as the number of resident daughters increases, as a result of competition for food between relatives. There were several indications that this was the case in our population. In the sample of forty-three hinds for which we had at least six years breeding data (see sec. 5.2), members of large matrilineal groups (those above the median size for their part of the study area) gave birth to fewer calves per year (average 0.695 calves/year) than members of small groups (average 0.847 calves/year) [6], principally because the former were less likely to calve as three-year-olds (47.1% did so versus 78.6% of members of small groups, [7]) but also because their subsequent fecundity tended to be reduced (0.789 calves/year compared with 0.875 for members of small groups).

5.5   FACTORS AFFECTING CALF MORTALITY
Frequency and Timing of Calf Mortality

As we have already described, variation in the frequency of mortality among calves was the most important source of differences in reproductive success between hinds. On average, 20% of all calves died before the end of the September following their birth, and a further 11% died during the following winter. Nearly 80% of calves that died in their first summer did so within a week of birth, and winter deaths usually occurred toward the end of the season, in March or April (fig. 5.2). Summer and winter calf mor-

9	10	11	12	13	14+	Overall
67.6	92.0	84.0	90.5	80.0	50.0	73.7
37	25	25	21	15	14	327
75.0	86.7	66.7	85.7	100	81.3	90.1
4	15	9	7	6	16	152

tality explained much the same proportion of variance in reproductive success [8].

Immediate Causes of Calf Mortality

The immediate causes of calf mortality were diverse. Of the fifty-nine calves that died between 1974 and 1976 during either their first summer or the succeeding winter, we were able to locate the bodies of forty-two and to assess the probable cause of death of thirty-one, either through autopsy or because the calf had been observed during the days immediately before its disappearance. We judged that five were stillborn (from direct observation that there was no wear on the hooves and that the fur was still partly covered with mucus); three died because of sucking difficulties or because the mother had insufficient milk (from observation during the first few days of life or because there was a small amount of hoof wear but no wound on the dead calf); two died from drowning; two were involved in accidents during the summer (including being attacked by feral ponies); ten were deserted or killed by their mothers after being marked; and at least three were attacked by eagles (attacks actually observed). It is possible that eagle attacks may have been responsible for further deaths: of the eleven carcasses where cause of death was uncertain, all had been fed on by eagles, and in all cases these calves were apparently healthy and had died suddenly. Observations elsewhere in Scotland have shown that eagles can kill large calves: for example, there is a record of an immature bird's killing a 20.5 kg calf in Sutherland (Cooper 1969).

Postmortems of calves that died during their first winter indicated that all had suffered from malnutrition, though exposure or parasite infestation may have contributed to their deaths. All calves that died during this period previously had been observed

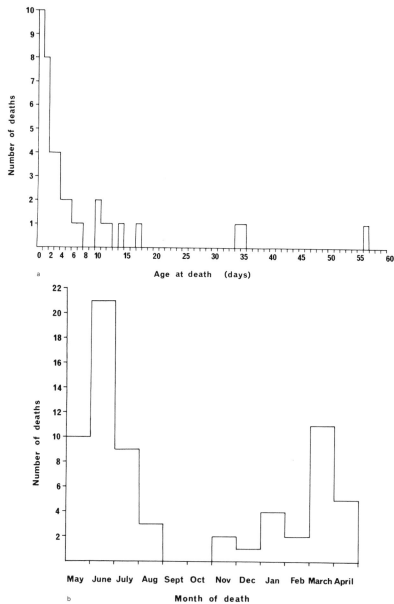

FIG. 5.2. (a) Ages of calves dying during their first summer. (b) Number of calf deaths occurring in each month of the year (from Guinness, Clutton-Brock, and Albon 1978).

to be in poor condition: the sample included three calves with deformities, one that had been in an accident, and two that had been suffering from an infection affecting the nervous system.

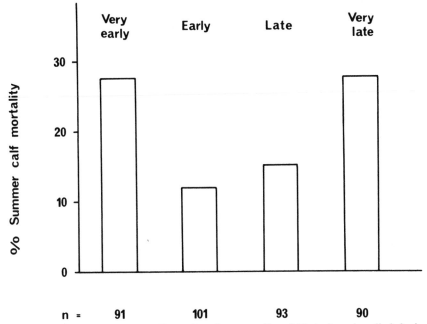

FIG. 5.3. Proportions of calves allocated to four quartiles of birth date that died during their first summer (May–September). The analysis took into account differences in the timing of calving between years and between areas. Figures beneath each histogram show the total number of calves in the sample, collected between 1971 and 1979.

Factors Affecting the Survival of Calves through the Summer

Though the immediate causes of death varied, the birth weight and date of calves evidently played an important part in determining whether they survived the summer. Calves dying before October had significantly lower birth weights (mean birth weight 6.04 kg) than those that survived to the winter (mean birth weight: 6.66 kg) [9], and similar results have been found in captive deer (Blaxter and Hamilton 1980). In addition, calves born very early and very late were more likely to die during their first summer than those born during the birth peak [10] (fig. 5.3), a result that agrees with studies of other seasonally breeding mammals (Reiter, Stinson, and LeBoeuf 1978; Nowosad 1975). Since there was no correlation between birth weight and birth date in our sample, both variables presumably had independent effects on summer survival. Low birth weight may have increased the calf's susceptibility to exposure during the days immediately following birth: higher mortality during the first months of life in light-born progeny has been recorded in other northern ungulates (e.g., Haukioja and Salovaara 1978). Early-born calves had the disadvantage that they were more likely to be born during periods of harsh weather or low food availability, and high mortality among late-born calves may have occurred because the mother's potential milk yield declined during the late summer: when hinds in an enclosure were prevented from conceiving until the latter part of the summer, only one out of six had sufficient milk to rear her calf (Guinness, Lincoln, and Short 1971).

What factors affected birth weight and date? Birth weight was probably influenced by the mother's weight and condition in spring: birth weights in our study area tended to be lower at Samhnan Insir than at Kilmory, and differences between years were related to spring weather conditions (see chap. 12). There was no significant difference in birth weight between the calves of milk and yeld hinds, though the calves of young hinds three and four years old were born about 600 g lighter than those of five- to ten-year-olds (see Guinness, Albon, and Clutton-Brock 1978), and birth weight began to decline in hinds over nine and fell rapidly after the age of eleven (see fig. 5.4). The probability that calves would die in the summer months also varied with maternal age: the calves of young and old hinds were less likely to survive than those of middle-aged hinds, and 50% of the variation in summer mortality between calves born to mothers of different ages was attributable to changes in birth weight with maternal age [11]. Evidence from captive populations supports the suggestion that these age changes were probably a consequence of the effects of age on maternal weight and condition (see Blaxter and Hamilton 1980).

Calf birth date was probably influenced by the mother's condition in autumn. A close relationship between maternal condition and conception date has been demonstrated in Scottish red deer populations (Mitchell and Lincoln 1973), and, as would be expected, milk hinds in our study area calved later than yelds. Birth dates also became later as population density increased and were later at Samhnan Insir than at Kilmory (chap. 12). Calving dates became gradually later after a hind was eleven years old (see Guinness, Albon, and Clutton-Brock 1978; Guinness, Gibson, and Clutton-Brock 1978), though age changes in birth date contributed relatively little to age-related differences in the frequency of

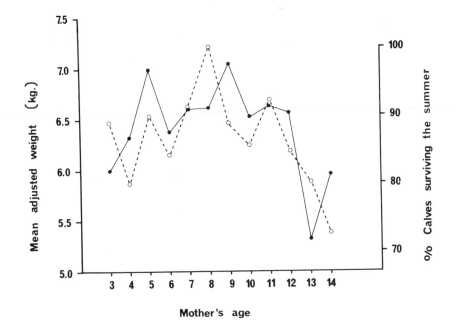

n = 27 30 29 25 23 27 18 21 13 13 10 11

FIG. 5.4. Birth weights (adjusted for sex of calf and year of birth) of calves born to mothers of different ages in the North Block of Rhum, 1971–79 *(solid line)*. The proportion of calves in each category that survived till October is also shown *(broken line)*.

calf mortality in summer: whereas changes in calf birth weight accounted for about 50% of variation in survival between the offspring of mothers of different ages, changes in birth date accounted for less than 14%.

Factors Affecting the Survival of Calves through the Winter

After calves were six months old, their birth weights had little effect on their chances of surviving. Birth weights of calves dying during their first winter (October–May) did not differ significantly from those of calves that survived (Guinness, Clutton-Brock, and Albon 1978). Presumably, the size and condition of calves at the onset of winter was the principal determinant of survival, and differences in growth obscured any effects of birth weight.

In contrast, birth date apparently had important effects on winter survival: 19.4% of calves born after the median birth date for their year died, whereas only 10.3% of calves born before the median date did so [12]. Late-born calves were probably lighter than average at the beginning of winter and may have been more likely to die for this reason (see Scrimshaw, Taylor, and Gordon 1968): even among calves reared with a plentiful food supply on deer farms, late-born calves are substantially lighter in early autumn than early-born calves, despite a tendency for them to be born heavier (Blaxter et al. 1974).

As we have already described, a calf's birth date was related to its mother's condition during the rut. Maternal condition probably had an important influence on the growth of calves, too: in a sample of seventy-nine mother/calf pairs shot throughout the year on Rhum, the body condition (kidney fat index), weight, and skeletal size (jaw length) of calves were correlated with the body condition, weight, and mammary-gland weight of their mothers, though not with their skeletal size (Mitchell, McCowan, and Nicholson 1976; B. Mitchell, pers. comm.). Analysis of samples of mother/calf pairs shot in winter at Glenfeshie [13], studies of captive red deer (Blaxter and Hamilton 1980), and studies of sheep (Geisler and Fenlon 1979) have produced similar results.

Other factors associated with winter calf mortality included the mother's home range area (calf mortality in winter was higher at Samhnan Insir than at Kilmory) and population size (see chap. 12). Mortality tended to be higher among calves born to mothers belonging to large matrilineal groups (30.9% during the first year of life) than among calves of mothers belonging to small matrilines (16.7%) [14]. It was interesting that significant differences in mortality between these two categories of calves occurred only among male calves [15] (see fig. 5.5) and that this represented a significant difference between the sexes [16]. Differences in the frequency of mortality among male calves born into small versus large matrilines were apparently not a consequence of variation in birth dates or birth weights, which did not differ between these two categories of calves, implying that they were a consequence of ecological factors affecting the calves after birth. Another line of evidence that supports this conclusion is that significant differences in the mortality of male (but not female) yearlings occurred between members of large versus small matrilineal groups [17, 18]. Though we investigated the possibility that calf mortality was related to the mother's dominance rank, as it is in some primates (see Dittus 1977, 1979), no such relationship was evident. Nor was there any significant difference in winter mortality between calves

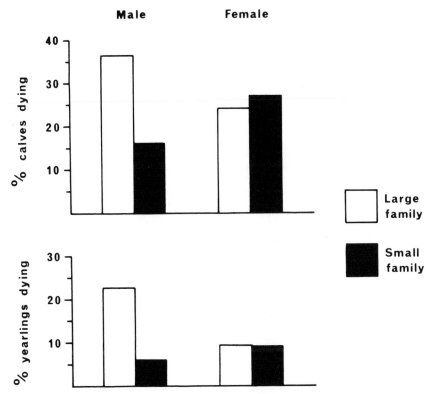

FIG. 5.5. Comparison of the percentage mortality among male and female offspring of hinds belonging to large versus small matrilineal groups. (Large groups were those above the median size for their part of the study area: see chap. 9.) The upper histogram shows the frequency of mortality during the first year of life, the lower the frequency of mortality during the second year.

of previously yeld hinds and those of hinds that had reared a calf in the previous year. We discuss the relationship between winter mortality and the mother's age in the last section.

5.6 THE ULTIMATE CAUSES OF VARIATION IN LIFETIME REPRODUCTIVE SUCCESS

Since calf mortality was strongly influenced by birth date and weight, we predicted that hinds that consistently produced heavy calves and those that conceived early would show improved lifetime reproductive success. Analysis of the relationship between individual differences in calf birth weights and dates and reproductive success in the sample of hinds for which six years of breeding records were available showed that both correlations

were positive, though reproductive success was significantly correlated only with birth date [19].

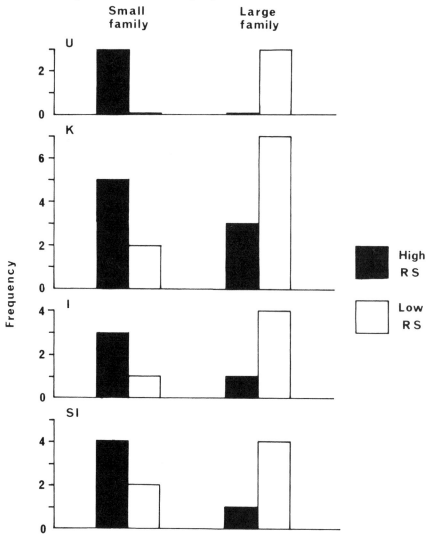

FIG. 5.6. Frequency of animals belonging to small versus large matrilineal groups whose
lifetime reproductive success (RS) was above versus below the median for hinds using that
part of the study area. (U = hinds using the Upper Glen; K = hinds using Kilmory; I = hinds
using the Intermediate Area; SI = hinds using Samhnan Insir).

A mother's weight and body condition probably had an important effect on her reproductive success. Though we could
not measure weight and condition directly, both are known to
be closely related to fecundity and conception date in red deer

(Mitchell and Brown 1974; Mitchell and Lincoln 1973) and also to affect birth weights in other mammals (Berger 1979). In contrast, any effects of the mother's size were probably slight, for the size of hinds is apparently unrelated to their fecundity or conception date (ibid.). However, differences in size may have had some effect, for in the sample of mother/calf pairs shot at Glenfeshie, there was a relationship between the skeletal size (jaw length) of mothers and the size and condition of their daughters, though this did not apply to sons [20].

But what were the causes of variation in maternal condition among hinds? The mother's home range area was clearly important: virtually all aspects of reproductive performance were depressed among hinds using Samhnan Insir compared with those using Kilmory (see chap. 12). In addition, reproductive success and, presumably, maternal condition were apparently influenced by the size of an individual's matrilineal group: in the sample of hinds whose reproductive success could be estimated over six years or more, hinds characterized by high reproductive success (i.e., above the median for the population) were commoner in small matrilineal groups than in large ones [21]. This was not a consequence of differences in reproductive success between animals using different parts of the study area, for the same effect occurred among hinds using all subdivisions (see fig. 5.6). The precise reasons why members of small families were more successful than members of large ones are still obscure. Increased group size could have reduced a hind's reproductive success as a result of increased grazing impact in her home range or because members of large groups more frequently occurred in big parties and increased party size intensified direct competition between individuals (see chap. 10).

5.7 Maternal Age, Parental Investment, and Calf Survival through the Winter

Several evolutionary theories have predicted that parents should invest more in their offspring as their own potential for future reproduction fails (Gadgil and Bossert 1970; Trivers 1972; Pianka 1976; Pianka and Parker 1975; Charlesworth and Léon 1976). In its crudest form, the idea (which we refer to as the theory of terminal investment) suggests that a female that has reached the stage in her life where she is unlikely to raise further offspring

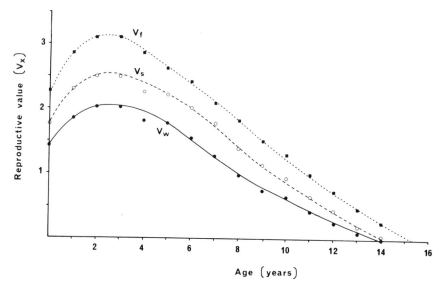

Fig. 5.7. Estimates of reproductive value for hinds of the 1972 cohort at different ages. The top curve (V_f) shows reproductive values calculated taking into account age-specific fecundity. The second curve (V_s) incorporates the probability that hinds of different ages would rear their calves through the summer; and the third (V_w), shows the probability that they would also rear them through the winter. Our methods of calculating reproductive value are described in Appendix 9.

successfully should invest her remaining resources in her existing progeny. In practice, the proportion of her resources a mother should transfer to her offspring in order to maximize her lifetime reproductive success will be affected by age changes in the costs and benefits of energy investment, and no precise prediction concerning the relationship between maternal age and parental investment is yet possible. However, under most conditions, the proportion of her resources a mother should invest is likely to increase toward the end of her life span, when the number of offspring she can expect to rear in the future is low, and some evidence from vertebrate studies supports the existence of a trend of this kind (Pianka and Parker 1975; Tinkle and Hadley 1975).

The usual measure of a female's future reproductive potential is her "reproductive value" (Fisher 1930; Pianka 1978), defined in terms of the number of *female* offspring she can expect to produce in the future (see Appendix 9). Estimates of reproductive value for red deer hinds in our study area (see fig. 5.7, top curve) showed that this followed the usual vertebrate pattern, peaking in young animals and gradually declining as age increased.

Did parental investment by red deer hinds change with reproductive value? Contrary to prediction, calf birth weight and
survival of calves through the summer peaked among the offspring of middle-aged hinds and declined among those of old
mothers (see fig. 5.4). However, postnatal investment may have
increased with maternal age. Unlike summer survival, winter survival was high among calves of hinds of three to six, lower among
the calves of prime-aged hinds of seven to nine, and high again
among the calves of ten- to thirteen-year-olds (fig. 5.8). Moreover,
analysis of kidney fat indexes in calves shot with their mothers
when they were about six months old at Glenfeshie (Invernesshire) showed that calf condition varied with mother's age in a
similar fashion and was significantly correlated with age changes
in winter calf survival on Rhum [22].

High survival among the calves of young mothers was to be
expected on the grounds that young hinds were generally in
superior body condition. However, the low survival of calves of
seven- to nine-year-old mothers and the higher survival of calves
of ten- to thirteen-year-old hinds was surprising, for the former
showed better body condition than the latter (see figs. 2.4, 4.19).

If increased survival among the calves of old mothers was the
result of a tendency for old mothers to invest more heavily in
their offspring, age-related changes in suckling behavior might
be expected. Analysis of the duration of suckling bouts revealed
that mean bout duration showed much the same distribution in
relation to maternal age as calf condition and winter survival (see
fig. 5.9) and was correlated with both variables [23, 24]. One
further line of evidence supported the suggestion that parental
investment varied with maternal age. The ratio of a calf's kidney
fat index (KFI) to the mother's kidney fat index presumably reflects the proportion of a mother's reserves that she has transferred to her calf. Among hinds shot in Glenfeshie, this ratio
rose with increasing maternal age [25] (see table 5.2) and
was negatively correlated with estimates of reproductive value
across hinds of different ages calculated from our Rhum
sample [26].

However, at least four other explanations of the tendency for
old mothers to produce good calves that show high winter survival
rates need to be considered. After the age of twelve, the proportion of hinds that had failed to breed in the previous year rose,

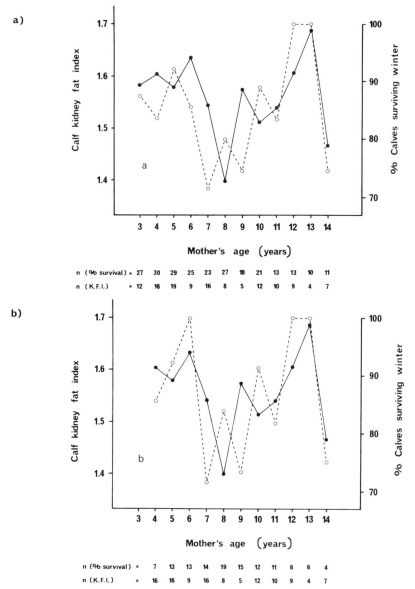

FIG. 5.8. The proportion of calves born to mothers of different ages in our Rhum study area that survived their first winter *(broken line)*. The plots show data collected 1971–79 and exclude all calves dying before their first October. *(a)* Calves born to all hinds; *(b)* calves born to mothers that had reared a calf in the previous year. Mean kidney fat indexes of calves born to mothers of different ages shot between November and February in Glenfeshie are also shown *(solid line)*. Sample sizes are shown below each plot.

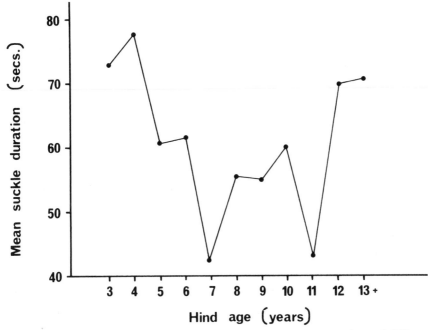

FIG. 5.9. Mean duration of suckling bouts involving calves born to mothers of different ages on Rhum, corrected for differences in age of calf and year of observation.

and this could have accounted for the improvement in condition and survival among offspring born to older mothers. Yet when we examined the distribution of winter survival among hinds that had reared a calf through the previous winter, the same pattern emerged (fig. 5.8*b*).

Another possibility was that more old mothers failed to conceive in the *succeeding* year and thus continued to suckle their calves during the winter months, improving calf condition and survival. However, there was no difference in winter survival between the offspring of pregnant and barren mothers

A third possibility was that there was a positive correlation between the success of hinds in rearing their calves through the winter and their own life span, and that our sample of old hinds was biased by an overrepresentation of good breeders. This was unlikely, for there was no relationship between the frequency with which hinds lost calves during the winter and the age at which they died [27].

TABLE 5.2   Ratio of Calf Kidney Fat Index (KFI) to Mother's Kidney Fat Index in a

						Mother's
	3	4	5	6	7	8
Mean KFI ratio (calf KFI/hind KFI)	0.90	0.88	0.85	0.72	0.85	1.00
N	12	16	19	9	16	8

Source: B. Mitchell, unpublished data.

Finally, it was possible that light calves born to prime-aged mothers were more likely to survive through the summer than light calves born to old mothers, and that they were consequently more likely to die in winter. If this were so, an interaction between summer mortality, maternal age, and birth weight would be expected—but no such interaction was evident [28].

These results raised the question why calf birth weight and summer survival did not vary with maternal age in the same way as calf condition and winter survival. There is no obvious answer to this question. It is conceivable that fetal growth is limited by maternal weight and that these constraints cannot be altered by selection, whereas postnatal growth depends to a greater extent on the mother's fat reserves and the proportion the mother can expend is easier to change. Alternatively, it is possible that selection does not favor increased investment by old mothers in their calves before birth because there is an increased chance that the calf will not survive the neonatal period.

One consequence of the tendency for the offspring of old hinds to show improved winter survival was that the usual measure of reproductive value underestimated the value of old hinds compared with young and middle-aged animals (see fig. 5.5, top curve). Calculations of reproductive value incorporating changes in summer and winter mortality of calves with increasing maternal age (fig. 5.7, lower two curves) showed that a hind's true reproductive value declined more slowly as she aged.

5.8 SUMMARY

1. Lifetime reproductive success in our study area varied from 0 to 13 calves reared per hind, with a median value of 4.5. Differences in the frequency of calf mortality between mothers ac-

Sample of 127 Mother/Calf Pairs Shot in Winter in Glenfeshie

age						
9	10	11	12	13	14+	Overall
1.10	1.01	0.94	1.15	1.09	0.99	0.91
5	12	10	9	4	7	127

counted for a larger proportion of variance in success than differences in fecundity.

2. Individual differences in fecundity were probably a consequence of variation in body weight and condition in autumn and, ultimately, of access to resources.

3. The probability that a calf would survive was related to its birth weight and birth date. Birth weight apparently influenced calf survival through the summer months, and birth date affected survival through the winter. Like fecundity, calf birth weight and date were probably related to maternal condition and the factors that affected this—including the mother's home-range area and the size of her matrilineal group.

4. The most important factors affecting the lifetime reproductive success of hinds were apparently the quality of their home ranges and the size of their matrilineal groups.

5. Some evidence suggests that old hinds may invest more heavily in their offspring than young or middle-aged animals.

Statistical Tests

1. Correlation between lifetime success in seventeen hinds and their age at death.
    Spearman rank correlation coefficient: $r_s = .662$, $t = 3.434$, d.f. $= 15$, $p < .01$.

2. Correlation between lifetime success in seventeen hinds and the proportion of their calves that survived their first year.
    Spearman rank correlation coefficient: $r_s = .523$, $t = 2.314$, d.f. $= 15$, $p < .05$.

3. Correlation between the mean number of calves per season reared to one year old by different hinds and the proportion of each animal's calves that survived their first year.

Pearson product-moment correlation coefficient: $r = .867$, $t = 9.827$, d.f. $= 32$, $p < .001$.

4. Correlation between the mean number of calves per season reared to one year by hinds and differences in fecundity between them after variation in calf mortality (test 3) had been accounted for.

Pearson product-moment correlation coefficient: $r = .445$, $t = 2.979$, d.f. $= 32$, $p < .01$.

5. Correlation between individual differences in fecundity and the proportion of a hind's calves that survived.

Pearson product-moment correlation coefficient: $r = -.286$, $t = 1.688$, d.f. $= 32$, n.s.

6. Comparison between members of large versus small matrilineal groups of the number of calves born per year during their lifetime. (Large groups were those of above the median size for that part of the study area, so the analysis controlled for variation in fecundity between areas).

Mann-Whitney U test: $U = 71.5$, n_1, $n_2 = 16$, $p = .05$.

7. Comparison of the frequency of individuals that bred as three-year-olds among members of large versus small matrilineal groups.

Fisher's exact probability test: $p = .063$ (one-tailed). $N = 31$.

8. Correlation between the mean number of calves reared per season to one year old by hinds in the above sample and differences in (a) calf survival through the summer, (b) calf survival through the winter between mothers.

(a) Pearson product-moment correlation coefficient:
$r = -.657$, $t = 5.006$, d.f. $= 31$, $p < .001$; % variance explained $= 43.2$.

(b) Pearson product-moment correlation coefficient:
$r = -.560$, $t = 3.763$, d.f. $= 31$, $p < .001$; % variance explained $= 31.3$.

9. Comparison of the weights of calves (adjusted for sex of calf, year of birth, and mother's age) that died before October ($N = 33$) versus those that survived to the beginning of the winter ($N = 242$).

Analysis of covariance: $F_{1,273} = 6.376$, $p < .02$.

10. Comparison of the frequency of summer mortality between calves born in different quartiles of the distribution of birth date.

G test: $G = 12.27$, d.f. $= 3$, $p < .01$. $N = 375$.

11. Correlation between the percentage of calves born to mothers of different ages that survived until October and the mean birth weights of each category.

Spearman rank correlation coefficient: $r_s = .706$, $t = 3.155$, d.f. $= 10$, $p < .02$.

12. Comparison of the frequency of winter mortality among calves born before the median birth date for their year versus those born later.

G test: $G = 4.987$, d.f. $= 1$, $p < .05$. $N = 247$.

13. Correlation between the body weight of the mother and the size (jaw length), condition (KFI), and weight of her calf in samples of fifty-eight mothers of male calves and sixty-three mothers of female calves shot, with their calves, in Glenfeshie in winter (B. Mitchell, unpublished data).

Male Calves	$\dagger r$	$*t_{56}$	p value
Mother's weight : calf size	.483	4.124	.001
Mother's weight : calf condition	.425	3.513	.01
Mother's weight : calf weight	.532	4.698	.001
Female Calves	r	t_{61}	p value
Mother's weight : calf size	.319	2.625	.02
Mother's weight : calf condition	.548	5.117	.001
Mother's weight : calf weight	.650	6.680	.001

† Pearson product-moment correlation coefficient.
* Student's t.

14. Comparison between members of large versus small matrilineal groups of the proportion of their calves that died during the first year of life.

Mann-Whitney U test: $U = 87$, n_1, $n_2 = 16$, $.1 > p > .05$. (one-tailed).

15. Comparison of the frequency of male calves born to members of large versus small matrilineal groups that died during the first year of life. (Analysis based on the entire data sample, 1970–80).

G test: $G = 12.60$, d.f. $= 1$, $p < .001$. $N = 235$.

16. G test for interaction between mortality during the first year of life, sex of calf, and size of matrilineal group.

$G = 7.24$, d.f. $= 1$, $p < .01$. $N = 417$.

17. Comparison of the frequency of calves born to members of large versus small matrilineal groups that died during the second year of life. (Analysis based on entire data sample, 1970–80).

  G test: $G = 10.19$, d.f. $= 1$, $p < .01$. $N = 173$.

18. G test for interaction between mortality during second year of life, sex of calf, and size of matrilineal group.

  $G = 4.304$, d.f. $= 1$, $p < .05$. $N = 314$.

19. Correlation between the total number of calves reared to one year old by hinds and (a) the mean birth weight of their calves; (b) the median birth date of their calves, both expressed as deviations from the average in their year of birth.

  (a) Pearson product-moment correlation coefficient:

  $r = .275$, $t = 1.961$, d.f. $= 47$, $.1 > p > .05$.

  (b) Pearson product-moment correlation coefficient:

  $r = -.465$, $t = 3.601$, d.f. $= 47$, $p < .001$.

20. Correlation between the skeletal size (jaw length) of mothers and the skeletal size, condition (KFI), and weight of their calves in samples of fifty-eight mothers of male calves and sixty-three mothers of female calves shot with their calves in winter at Glenfeshie (B. Mitchell, unpublished data).

Male Calves	†r	*t_{56}	p value
Mother's size : calf size	.134	1.015	n.s.
Mother's size : calf condition	.169	1.286	n.s.
Mother's size : calf weight	.011	.079	n.s.
Female Calves	r	t_{61}	p value
Mother's size : calf size	.278	2.261	$< .05$
Mother's size : calf condition	.426	3.652	$< .001$
Mother's size : calf weight	.223	1.768	n.s.

  † Pearson product-moment correlation coefficient.
  * Student's t.

21. Comparison between members of large versus small matrilineal groups of the numbers of individuals whose reproductive success was above versus below the median value for the population as a whole.

  G test: $G = 5.45$, d.f. $= 1$, $p < .02$. $N = 43$.

22. Correlation between the percentage of calves, born to mothers of different ages, that survived their first winter and the mean KFI of each age category.

All calves, Spearman rank correlation coefficient: $r_s = .586$, $t = 2.285$, d.f. $= 10$, $p < .05$.

Calves born to milk hinds only, Spearman rank correlation coefficient: $r_s = .673$, $t = 2.73$, d.f. $= 9$, $p < .05$.

23. Correlation between the mean duration of suckling bouts (corrected for variation in age of calf and year of observation) and calf KFI (across mothers of different ages).

Spearman rank correlation coefficient: $r_s = .636$, $t = 2.48$, d.f. $= 9$, $p < .05$.

24. Correlation between the mean duration of suckling bouts (corrected for variation in age of calf and year of observation) and survival of calves through the winter (across mothers of different ages).

Spearman rank correlation coefficient: $r_s = .582$, $t = 2.146$, d.f. $= 9$, $.1 > p > .05$.

25. Regression of the ratio of calf KFI to maternal KFI on maternal age in hind/calf pairs shot at Glenfeshie (B. Mitchell, unpublished data).

Analysis of variance: $F_{1.125} = 8.45$, $p < .005$.

26. Correlation between the ratio of calf KFI to maternal KFI in Glenfeshie hinds and the reproductive value of mothers of the same ages on Rhum.

Spearman rank correlation coefficient: $r_s = -.593$, $t = 2.327$, d.f. $= 10$, $p < .05$.

27. Correlation between the percentage of years in which hinds reared a calf through the summer but lost it during the winter and their own age at death.

Spearman rank correlation coefficient: $r_s = -.144$, $t = .563$, d.f. $= 15$, n.s.

28. Interaction component between summer mortality of calves, maternal age, and calf birth weight (data 1971–79).

Analysis of variance: $F_{11.234} = .706$, n.s.

6 The Rutting Behavior of Stags

As for the famous "struggle for existence," so far it seems to me to be asserted rather than proved. It occurs, but as an exception; the total appearance of life is not the extremity, not starvation, but rather riches, profusion, even absurd squandering—and where there is a struggle, it is a struggle for *power*. One should not mistake Malthus for nature.

Nietzsche (1889)

6.1 INTRODUCTION

Unlike hinds, stags do not invest heavily in individual offspring, and their reproductive success is limited principally by their ability to gain breeding access to members of the opposite sex. Since hinds aggregate in groups, it is possible for a stag to monopolize access to a considerable number of hinds, and there is intense competition for harems between stags during the autumn rut. A stag's breeding success is closely related to his ability to control the behavior of other animals—of hinds, of young stags that attempt to abduct hinds from the harems of older males, and of other mature males. Sections 6.2 to 6.4 describe the activities of stags during the rut and their interactions with hinds and young stags. Direct competition for hinds is common between mature stags, and fights are frequent and dangerous. In such circumstances stags would be expected to assess their opponents carefully, fighting only in cases where they were likely to win (Parker 1974). The last two sections describe contests between stags and review our previous research on assessment and roaring behavior (Clutton-Brock and Albon 1979).

6.2 RUTTING ACTIVITIES
The Early Rut

In late August, mature stags over five years old become progressively intolerant of each other, and after the middle of September stag groups begin to fragment. One by one, the stags leave their usual ranges, some moving directly to their traditional rutting grounds (Lincoln and Guinness 1973), others spending several days on their own in peripheral areas before moving to their rutting areas. Mature stags seldom attempt to form harems immediately, and, during the first few days after their arrival in their rutting area, their tolerance of other stags varies. During this time some stags systematically pick fights with rivals of similar status. Such fights are seldom caused by a dispute over any specific resource and are generally shorter than subsequent contests.

Research on hormonal changes at the onset of breeding activity in stags shows that testosterone levels rise steeply (see fig. 6.1), triggering first antler cleaning and subsequently seasonal increases in the size of the testes and the development of the mane and neck muscles (Lincoln 1971a; Lincoln, Youngson, and Short 1970; Goss 1963, 1968; Goss and Rosen

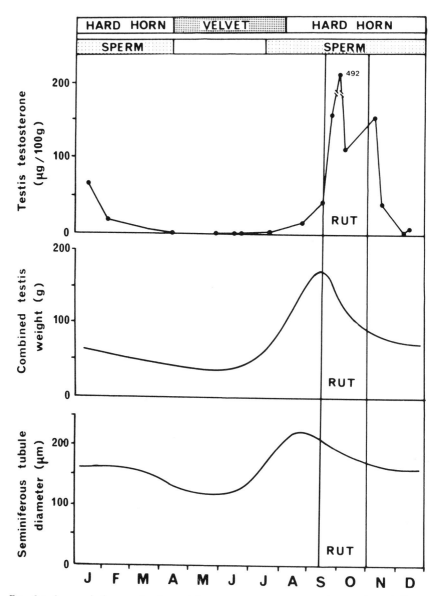

FIG. 6.1. Seasonal changes in the weight, testosterone concentration, and seminiferous tubule diameter of the testes of stags (redrawn from Lincoln 1971a).

1973). However, other hormones are also involved (Fletcher and Short 1974), and the timing of rutting behavior is affected by environmental factors including photoperiod (Jaczewski 1954; Goss 1969a,b), odor from the hinds, and past experience of rutting (Lincoln, Guinness, and Short 1972).

Fig. 6.2. A six-year-old stag with his harem at Kilmory in 1974.

Displays and Interactions

As mature stags spend more of their time associating with hind groups, they move less and less, remaining with their harems in one particular area of the rutting grounds (Lincoln and Guinness 1973) and moving only as their harems move to new feeding areas or to sheltered positions (see fig. 6.2). They begin to display regularly and frequently interact with their hinds. The displays and interaction patterns of stags have already been described in detail (Struhsaker 1967; Bützler 1974; Geist 1982). In our study we used a simplified typology of displays and interactions to make quantification easier. The displays and interactions we examined were as follows.

Roaring: A deep guttural roar (see Clutton-Brock and Albon 1979), Roars are aggregated into bouts of one to ten that are often given on the same exhalation. Roaring rate was the mean number of roars given per minute (see fig. 6.3).

Barking: A series of short barks typically directed at young stags after they have been chased away.

Thrashing: The stag rakes the ground and vegetation with his antlers, often spraying the area he is thrashing with urine and subsequently rolling on it (see fig. 6.4).

Wiping: Harem-holding stags commonly rub their chins, antler pedicles, or preorbital glands on outstanding "landmarks," including posts and rocks (fig. 6.5).

FIG. 6.3. A nine-year-old stag roars at an intruder.

FIG. 6.4. A seven-year-old stag thrashes in *Juncus* marsh.

Fig. 6.5. A seven-year-old stag marks on a log at Kilmory.

Fig. 6.6. A five-year-old stag wallows during mid-September.

FIG. 6.7. A four-year-old stag on the edge of a harem, showing flehmen.

Wallowing: The stag wallows in a pool or peat bog, often also urinating into it and wiping his antorbital glands and antlers on nearby vegetation (fig. 6.6).

Flehmen: After sniffing a hind or the place where a hind has urinated, stags sometimes show flehmen, raising their heads and curling back their upper lips (fig. 6.7).

Sniffing and licking: Stags frequently approach lying hinds and lick the back of their heads and necks, gradually working over the head toward the preorbital region (fig. 6.8). This may be followed by an attempt to sniff and lick around the base of the tail. Bouts of sniffing and licking, which may last several minutes, are usually terminated by the hind's getting to her feet and moving away. After this, the stag commonly sniffs and licks the grass where the hind has rested. Hinds often urinate while lying, and this may permit the stag to scent pheromones released in the urine.

Herding: Stags often head off hinds that are attempting to leave their harems by walking outside them, head held high, with a stiff, prancing gait (see fig. 6.9). The stag's head is typically at an angle to the direction of movement with eyes half closed. If the hind persists in attempts to leave the harem, the stag may

FIG. 6.8. A stag licks the face and orbital area of a lying hind.

FIG. 6.9. Driving in an outlying hind. The stag's nose is lifted, and he looks sideways at the hind.

Fig. 6.10. A nine-year-old stag chivying a mature hind.

threaten her with lowered antlers or by giving a brief scissors
kick with his front legs and may bark at her. Hinds that are
herded are almost always on the periphery of the harem. In
over 90% of observed cases, hinds moved back toward the center
of the harem when herded.

Chivying: Stags frequently chase hinds over short distances within
their harems (fig. 6.10). During these chases the stag trots after
the hind with neck outstretched, sometimes extending his
tongue. Chivies end when the stag stops, apparently losing
interest, and they are often followed by a bout of roaring.

Mounting: Only hinds in estrus allow the stag to mount. Mating
sequences usually involve several mountings over a period of up
to an hour or more (see Morrison 1960; McCullough 1969) (fig.
6.11). Ejaculation can be easily identified by a sudden thrust

FIG. 6.11. Preparing to mount. An estrous hind stands with her back legs splayed as the stag licks her rump.

that jerks the stag's body upright, often throwing the hind forward several paces (fig. 6.12).

Displacing: Stags displace young stags or rivals that approach their harems by walking steadily toward them. We defined displacements as cases where the harem-holding stag approached another stag and the latter retreated.

Approaching: Where one stag moves to within 100 m of another.

Parallel walk: After one stag has approached another, the pair may move into a tense walk in which they move parallel to each other, typically 5–20 m apart (fig. 6.13). At this stage their hair is often raised, and their gait is slow, regular, and stiff. The duration of parallel walks was the total duration of an interaction involving parallel walking, and it usually included several circuits.

Initiating: One stag lowers his antlers, inviting contact. Antler joining often follows so quickly that identifying the initiator of a fight is impossible (fig. 6.14).

Fighting: The two stags lock their antlers and push to and fro, occasionally disengaging, until one individual is driven rapidly backwards (figs. 6.15, 6.16). Fight duration was the period from first antler contact to last contact, including short intervals when stags disengaged before locking antlers again.

FIG. 6.12. A seven-year-old stag serves an eleven-year-old hind. The photograph is taken at the moment of ejaculation when the stag leaps clear of the ground.

FIG. 6.13. Two mature stags parallel walk during the latter part of the rut.

FIG. 6.14. After parallel walking, two stags simultaneously lower their antlers, initiating contact.

FIG. 6.15. Two mature stags fight during the late rut.

FIG. 6.16. A stag leaps up and forward in an attempt to dislodge his opponent.

Fig. 6.17. A harem-holder chases a yearling male from the harem.

Chasing: Stags chase yearlings or young stags away from their harems by running directly at them, often pursuing them till they are more than 100 m from the harem (fig. 6.17). Chases terminate when the harem-holder stops and turns back to his harem. Stags often end a chase with a scissors kick of the fore-legs followed by a bout of roaring. On some occasions stags also chase calves and even yearling females out of the harem, and particular stags do so regularly. Removal of calves from the harem must be of dubious advantage to the stag, since their mothers often follow them when they leave.

Rutting Activities and Age

Only among mature stags five years old or more did a substantial proportion of individuals hold harems: figure 6.19 shows the mean number of stags of different ages that were seen holding hinds in the study area between 25 September and 25 October. The age at which stags started to hold harems slightly preceded the point at which they reached full adult weight (see fig. 2.4), though their breeding success did not peak till they were eight (see below). Studies of other populations indicate that breeding is largely confined to stags of similar ages (Gossow 1971; Mitchell, Staines, and Welch 1977).

FIG. 6.18. At the height of the rut, a group of two- to five-year-old stags collects close to the harem of SAGY on the Kilmory greens.

Stags of two to four years old rarely held harems and spent much of their time on the periphery of the larger harems, frequently attempting to infiltrate or disperse them (see fig. 6.18). Though we never saw such stags mount a hind successfully during the peak of the rut, their depredations could lead to rapid reductions in harem size. Their success in penetrating the harem depended on the position of the harem-holding stag: they were consistently more successful in entering harems if the holder was on the far side of the harem, and they typically approached from the "blind" side (Gibson 1978). A common technique used by young stags was to run through the harem giving alarm barks and thus causing the hinds to scatter. However, in less than 10% of attempts made when harem-holding stags were present did young stags manage to extract hinds successfully, though when the stag was absent they generally did so.

Most stags over eleven failed to hold harems and reverted to

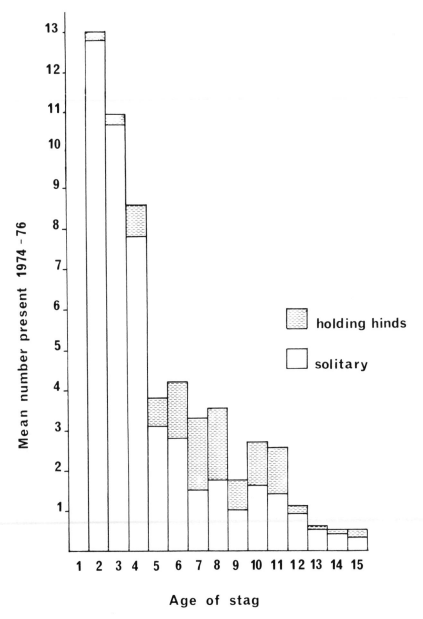

FIG. 6.19. Mean numbers of stags of different ages (in years) in the study area between 1974 and 1976 during the rut.

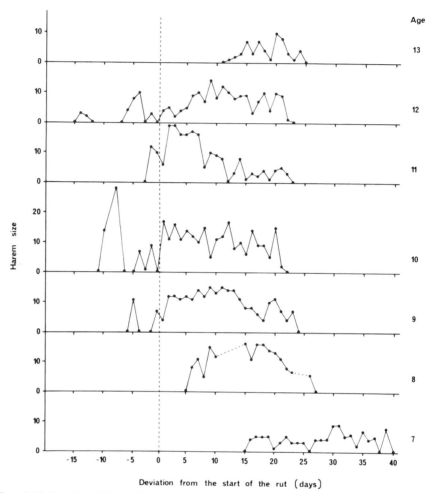

FIG. 6.20. Number of hinds held per day in each of seven years by a successful stag (SAGY). The distribution of harem-holding in each year was standardized around the onset of the peak period of rutting in the population, defined as the first turning point in harem stability.

wandering around the harems of prime stags, abducting an occasional hind. Other studies have shown that both condition and weight typically start to decline at about this age (Mitchell, Staines, and Welch 1977), and most old stags in our study area were in obviously poor condition. Studies of other populations of red deer and elk have shown similar variations in rutting activity between males of different ages (McCullough 1969; Gottschlich 1968).

The timing of harem-holding also varied with the stag's age and prowess. Young and old stags held harems later than those in their prime. For example, figure 6.20 shows the number of hinds

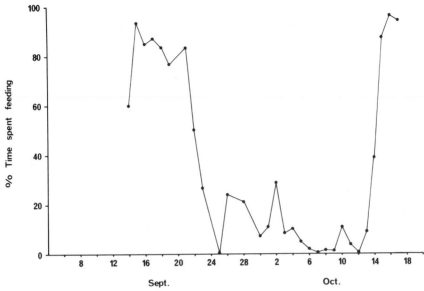

FIG. 6.21. Proportion of time spent feeding in watches of a nine-year-old stag (SAGY) during the 1974 rut.

held per day in each of seven years by a very successful stag (SAGY). As SAGY matured, he held harems progressively earlier relative to the onset of the peak rutting period, and as he aged his rutting period again became later. These changes in the timing of harem-holding were probably partly a direct consequence of changes in condition and partly a result of the fact that less successful fighters were more likely to retain harems when the intensity of competition was reduced.

Activity Budgets of Harem-Holders

When they started to hold harems, the activity budgets of harem-holders showed dramatic changes (see fig. 6.21). The proportion of daytime spent grazing fell from 44% during the summer to less than 5% in the rut, while the proportion of time spent moving and standing inactive increased from 3% and 4% to 15% and 33% respectively (see table 6.1). Nighttime budgets showed similar changes. Young stags and stags without harems did not show as marked a decline in grazing time, though they grazed less than at other times of year. The decline in grazing associated with harem-holding was quickly reversed when the stag ceased to hold a harem, and it increased from a low level to at least as high as at other times of year within 24 h (see fig. 6.21).

TABLE 6.1 Proportion of Time Spent in Different Activities during the Rut (25 September–25 October)

			Lying		Standing		
	N	Grazing	Ruminating	Inactive	Inactive	Moving	Rutting[a]
Day							
Harem-holders ≥ 6 years old	7	4.0	<1.0	36.8	33.1	15.5	9.8
2–5-year-olds	7	35.5	18.0	26.2	9.5	6.8	—
Estrous hinds	4	41.8	24.1	16.7	10.7	4.6	—
Anestrous hinds	10	58.8	21.5	14.8	1.9	1.6	—
Night							
Harem-holders > 6 years old	5	5.6		40.5	41.8	10.7	1.4
Anestrous hinds	Scan data	37.5		56.0	4.4	1.3	—

[a] Times when the stag was engaged in any of the rutting activities listed on p. 00 but was not moving.

Similar changes in the activity patterns of rutting stags have previously been described in elk (Struhsaker 1967;McCullough 1969). Physiological studies indicate that they are probably controlled by hormonal mechanisms, for penned animals maintained on a natural cycle of lighting but kept separate from hinds show a decline in appetite and body weight at the time of the rut (Simpson 1976; Kay 1978; Magruder, French, McEwan, and Swift 1957).

As a result of the high energy costs of rutting and reduced food intake, the body weight and condition of rutting stags shows a rapid decline in September and October (see fig. 6.22), and some individuals lose as much as 20% of their body weight over this period. It is presumably declining condition and associated changes in hormone levels that cause stags to terminate their rutting activities. This typically occurred suddenly, and stags could switch from a high level of activity one day to an almost total cessation of rutting activities combined with intense feeding on the subsequent day (see fig. 6.21). There was apparently a relationship between the body condition of the stag at the beginning of the rut and the number of days he rutted—stags that entered the rut in poor condition sometimes ceased rutting as early as 7 October, several days before the main peak of conceptions.

Quantitative Aspects of Displays and Interactions

Though the rut gave the impression of frantic activity, it was less disorganized than it appeared, and the frequency of most displays and interactions was closely related to variation in the benefits of performing them. Harem-holders displayed and interacted more frequently than animals of the same age that were not defending harems (table 6.2), and their behavior changed rapidly if they lost their harems. A natural experiment occurred when a successfully rutting six-year-old stag ruptured a ligament in his leg in the course of a fight and consequently lost his harem. By the following day he had almost ceased to rut and was feeding intensely (table 6.3).

Among harem-holders, the frequency of most displays and interactions peaked during the first two weeks of October, when the majority of hinds conceived, and varied in relation to the "value" of the harem. For example, harem-holders observed in 1974 roared more frequently on days when their harems were

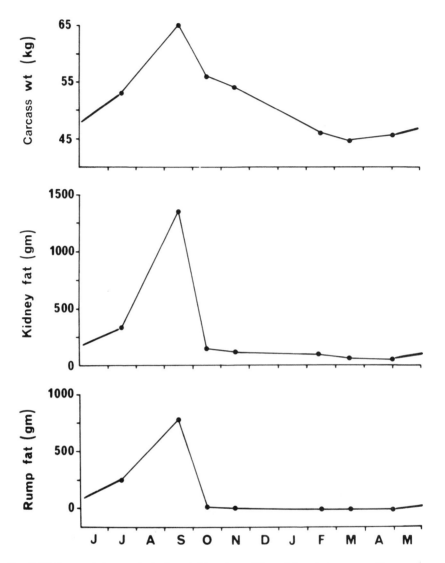

FIG. 6.22. Seasonal changes in body weight and condition of five- to ten-year-old stags shot on Rhum (from Mitchell, McCowan, and Nicholson 1976).

larger than usual than on days when they were smaller [1]. They also tended to roar more, spent less time feeding, and stayed closer to their harems on days when there was an estrous hind in their harem.

The frequency of displays was also related to the probability that the harem-holder would be approached or attacked by other stags. Harem-holders roared more [2] and tended to spend less

TABLE 6.2 Mean Frequencies of Some Common Rutting Activities in Mature Stags (over Five Years)

	Time Period			
	1	2	3	4
Sample size (individuals)	12	14	5	6
Percentage of time spent grazing	84.5	12.0	52.5	82.7
Roars/min	0.1	2.7	0.8	<0.1
Thrashes/h	1.1	1.39	1.0	<0.1
Chivies/h	0.2	7.0	0.2	<0.1
Displacement and chases/h	1.3	2.7	0.7	<1
Nearest-neighbor changes/h	5.3	18.2	7.2	4.5

Note: (1) before the rut (10 September–20 September) (2) during the rut (20 September–20 October); (3) during the rut when solitary; (4) after the rut (after 20 October).

Results are expressed as mean frequencies per minute when the stag was *standing*. Data from the 1974 rut.

TABLE 6.3 Effects of Leg Injury on the Rutting Behavior of a Six-Year-Old Stag

	Before[a]	After[b]
Hinds in harem	7.4	0
Percentage of standing time spent grazing	12.0	65.0
Roars/min	2.8	0.6
Modal distance to nearest male over a year old (m)	150	5
Chivies/h	2.7	0
Herds/h	4.7	0

Note: The injury was probably a ruptured ligament.
[a] Mean frequencies of different activities across the six days before the stag's injury.
[b] Mean frequencies of different activities across the two days after the stag's injury.

TABLE 6.4 Frequency with Which a Successful Nine-Year-Old Stag Chased Stags That Approached His Harem in 1974

	Period of Rut			
Age of Approaching Stag	14 Sept.–21 Sept.	22 Sept.–30 Sept.	1 Oct.–7 Oct.	8 Oct.–16 Oct.
1	0.25[a]	2.6	0.4	0.8
2–3	0.40	3.7	5.4	1.6
4–5	0.80	1.7	1.0	0.1

[a] Chases per hour.

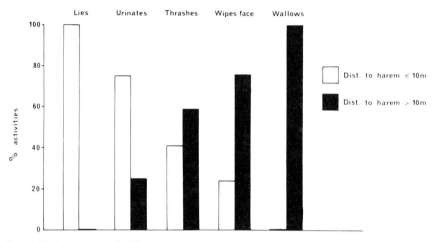

FIG. 6.23. Percentage of different activities by one nine-year-old stag that were performed within 10 m of the edge of the harem versus beyond this distance (1974 data).

time feeding [3] on days when the nearest rutting stag was closer than average. They also roared more frequently and spent more time standing at times of day when their hinds were feeding [4], when the harem was consequently more dispersed and more liable to infiltration by young stags than when their hinds were lying. This was probably related to the fact that other stags were more likely to approach the harem when a high proportion of hinds were feeding [5] (Clutton-Brock and Albon 1979).

The frequency of most rutting activities also had a characteristic spatial distribution. Stags roared infrequently when they were distant from their harems [6], perhaps avoiding advertising that their harems were unprotected. Both lying and urinating tended to occur close to the harem, whereas thrashing, rubbing, and wallowing usually occurred beyond the periphery of the harem (fig. 6.23).

6.3 INTERACTIONS BETWEEN HAREM-HOLDERS AND HINDS

From the first week of October onward, harem-holding stags interacted regularly with the hinds—sniffing, licking, chivying, and herding them (see Appendix 17). On the whole, stags distributed their attentions evenly among the mature members of their harem, though they interacted with calves, yearlings, and two-year-olds less frequently than with adult hinds. Stags also paid particular attention to hinds that had recently joined their harems, and the frequency with which they licked and chivied them was higher than average. Stags holding larger harems did not compensate by chivying hinds more often, and the frequency with which individual hinds were chivied or licked was negatively related to harem size [7]. However, on days when particular stags held large harems, they tended to stay closer to them and spent less time feeding.

When a hind was in estrus, the harem-holder directed most of his attentions toward her, interacting more frequently with her than with other hinds. Stags typically mounted hinds in estrus several times before ejaculating and remained close to them afterward, subsequently showing progressively more interest in other hinds. Several sequences of mounting sometimes occurred with the same hind, particularly late in the season when the estrous behavior of hinds was more pronounced.

As we described in chapter 4, the stability of harems varied throughout the rut and was highest during the period of peak conception when the stags were rutting most intensively.

6.4   Interactions between Harem-Holders and Young Stags

During the early stages of the rut, harem-holding stags chased the male yearlings out of their harems. The latter hung around, attempting to get back to their mothers, and were regularly threatened by the harem-holders until they learned to keep their distance. The extent to which harem-holders tolerated yearling males within the vicinity of a harem appeared to depend on the development of the yearling's secondary sexual characteristics: in cases where two or more yearling males were within 50 m of a harem and the stag moved out to chase or displace them, the harem-holder first chased the one with the longest knobs or spikes on over 90% of occasions. These interactions were common features of the rut, and during the early rut they occurred as often as five times per hour on average (see table 6.4).

The reaction of a harem-holder to an approacher depended on the latter's age. Whereas yearlings were often allowed within 25 m of the harems, harem-holders usually reacted to an approach by a four- or five-year-old when he was over 100 m from their harem (fig. 6.24). In addition, harem-holders displaced or chased older approachers farther from their harems than younger stags, and, when a harem-holder was approached by several young males at

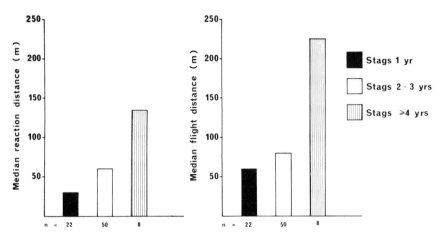

FIG. 6.24. Median distance of approaching stag to harem when the holder began the chase or displacement (reaction distance) and when he ended the interaction (flight distance). The number of interactions in the sample is shown below each histogram.

the same time, he almost invariably chased the oldest or best-developed one first.

The age of the approacher was also related to the type of interaction that followed. When approached by stags of two to three years (which were appreciably smaller than older animals), the defending stag usually ran directly at them and chased them away. In contrast, when approached by stags of over four years, harem-holders usually walked toward them and rarely ran at them immediately (fig. 6.25). The advantage of this behavior was presumably that it gave stags the opportunity to impress the rival with their size and strength, giving him time to assess them and avoiding the chance that he would precipitate an unnecessary fight.

6.5   INTERACTIONS BETWEEN MATURE STAGS
The Form of Contests

Relationships between harem-holding stags were almost invariably hostile, and fights were common. Continuous watches of individual stags suggested that most mature stags fought at least once every five days during the rut and were usually involved in about five fights during each breeding season (Clutton-Brock et al. 1979). During the peak rutting period, most fights involved mature stags (see fig. 6.26), of which at least one was in possession of a harem, though young stags sometimes fought each other and

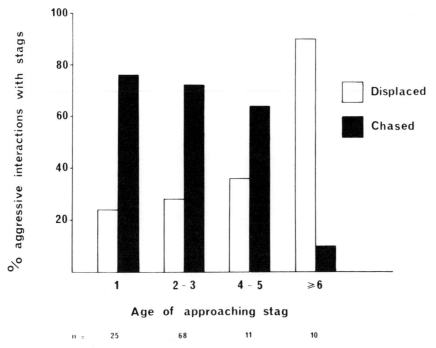

FIG. 6.25. Proportion of aggressive interactions involving harem-holding stags and approaching stags of different ages in which the approacher was chased versus displaced (see text) by the harem-holder.

occasionally challenged harem-holders, especially in the later stages of the rut.

Between 1971 and 1976, we observed 107 fights involving 72 different stags (Clutton-Brock et al. 1979). The typical course of events was that a challenging stag approached to within approximately 200 m to 300 m of a harem-holder and the two roared at each other for several minutes, after which the approacher usually withdrew. If the approaching stag came within 100 m of his opponent, the contest was more likely to escalate first to a further exchange of roars and then, in the majority of cases, to a parallel walk, with the two contestants moving tensely up and down, typically at right angles to the direction from which the approacher came. At any moment during a parallel walk, either stag might invite contact by turning to face his opponent and lowering his antlers. Opponents almost always accepted this invitation, turned quickly, and locked antlers. Both animals would then push vigorously, attempting to twist the opponent to gain the advantage of

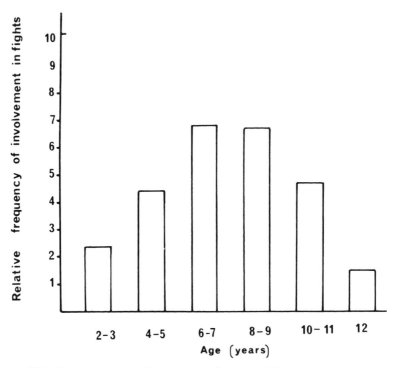

Fɪɢ. 6.26. Relative frequency of involvement of stags of different ages in fights during the rut. Relative frequency was calculated by dividing the number of cases where an animal in each age grade was involved by the mean number of animals of that age grade present in the study area (1974–76).

any slope. In the course of longer fights, contestants frequently separated for a few seconds at a time, rejoining after one of the pair invited contact again by lowering his antlers. Fights lasted until one of the pair was pushed rapidly backward, broke contact, and ran off. Winning stags seldom pursued losers for more than 10 to 20 m, though if a stag slipped in the course of a fight his rival would immediately attempt to horn him in the flank, rump, or neck, and there was no evidence of dangerous attacks being inhibited in such situations.

For a sample of fifty cases where an approaching stag came to within 100 m of his opponent, figure 6.27 shows the number that led to roaring contests, parallel walks, and fights. Fights that were not preceded by roaring contests or parallel walks mostly occurred either where there was an obvious inequality in fighting ability between the contestants or where an intruding stag had

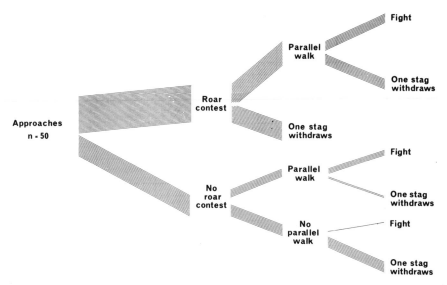

Fig. 6.27. Number of approaches to within 100 m involving two stags six years old or older that led to roaring contests, parallel walks, and fights, in a sample of fifty observed in 1977. Each line represents one case.

usurped a harem in the temporary absence of its holder and was discovered on the latter's return. In such cases the previous holder almost invariably challenged the intruder without preliminary displays.

Fighting Success

Analysis showed a close relationship between fighting success and age (see fig. 6.28): fighting success peaked between the ages of seven and ten years and declined rapidly in stags of over eleven. There was thus a close relationship between age changes in fighting success and those in body weight (see chap. 2), which peaked over the same period.

The association between age, weight, and fighting success was supported by examination of the outcome of fights between individuals of different ages. Before the age of five, substantial increments in body weight occurred each year, whereas after this growth rates were slower. Consequently, when a stag of five years or less fought an older individual, there was usually a substantial discrepancy in weight between them. This was not necessarily the case when stags more than five years old fought. As would be expected if body weight played an important part in fighting

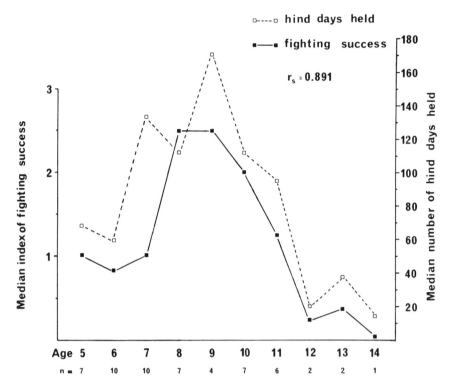

FIG. 6.28. Median fighting success (see Appendix 10) and holding success in relation to age in stags over five years old. Figures below each point show the number of individuals sampled (from Clutton-Brock et al. 1979).

success, stags five years old or less that fought older individuals mostly lost, while success was not closely related to age among stags six years old and older (Clutton-Brock et al. 1979).

The Costs of Fighting

Fighting was both dangerous and costly. Between 1974 and 1976 we observed the following injuries to mature stags rutting in the study area: permanent blindness of one eye (two cases); temporary blindness of one eye (one case); damaged front leg causing permanent lameness (one case); temporary lameness in one leg (two cases); major antler breakage (twenty-six cases). In previous years, injury frequency was not recorded systematically, but one broken foreleg, five cases of temporary lameness, and five eye wounds were noted. In our total sample of 107 observed fights, two cases of permanent injury occurred. Taking into account the number of stags in the study area, this indicated that about 23%

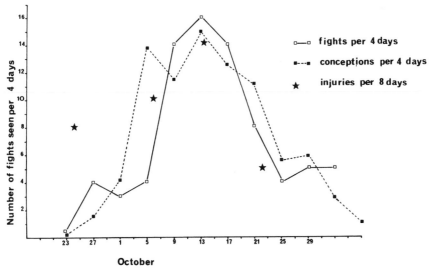

FIG. 6.29. Distribution of fights during the rut. The figure compares the number of fights per four-day period with the number of calves conceived in each period, calculated by backdating from calf birth dates, assuming a gestation length of 235 days (from Clutton-Brock et al. 1979).

of stags over the age of five years showed some sign of injury during the rut each year and that up to 6% were permanently injured (Clutton-Brock et al. 1979). Since most stags rutted for three to five years during their lifetime and natural mortality among mature stags was low (fig. 2.5), this suggested that virtually all stags may be slightly injured at some stage and as many as 20% may sustain permanent injury during their lifetime. However, in later years the frequency of major injuries fell (see chap. 13), and the percentage of all mature stags that died in the study area before 1980 and carried a permanent injury was approximately 10%. Although injuries occurred throughout the rut, the majority (58%) happened in the ten days between 8 October and 18 October, coinciding closely with the frequency of fights observed (see fig. 6.29).

The frequency of rutting injuries is known to be high in other deer populations. In two samples of Russian red deer, mortality from rutting injuries accounted for 13% and 29% of all adult male deaths (Heptner, Nasimovitsch, and Bannikov 1961), while in German populations about 5% of stags die from fighting each year (Müller-Using and Schloeth 1967). In a sample of mature mule deer observed over a single breeding season, 19% showed some sign of injury (Geist 1974a), while in reindeer and moose rutting mortality is also a common cause of death in adult males (Bergerud 1973, 1974a,b; Pielowski 1969). Since even slightly injured individuals probably run a considerably higher risk of winter mortality (Craighead et al. 1973) and predation by large carnivores (see Estes and Goddard 1967; Kruuk 1972; Schaller 1972), and since most studies have been carried out in areas where the densities of large carnivores were artificially low, the true costs of fighting in terms of mortality may be even higher than these figures suggest.

In some cases, though not in all, injuries had an important effect on reproductive success. The seven-year-old stag that ruptured a ligament during a fight in the study area (see table 6.3) not only lost his harem immediately but his rutting success was severely affected in subsequent years. We have also observed several cases where stags that broke off a substantial part of their antlers immediately lost a large part of their harems, and artificial removal of large parts of the antlers has similar effects (Lincoln 1972). The effects of smaller injuries were less obvious. Lameness often appeared to reduce rutting success temporarily, while loss of antler points (the majority of cases of antler breakage) apparently had little effect.

Fighting was also costly because young stags sometimes dispersed the entire harems of both contestants. Though hinds gradually returned to the rutting areas after they had been dispersed and though young stags seldom, if ever, mated with hinds they had abducted, a prolonged fight could lead to a reduction in the size of the harem of the winner that lasted several days—as well as to the loss of individual estrous hinds that were chased into the harems of other mature stags. The benefits of fighting to harem-holders were consequently reduced, both because their harems were likely to be infiltrated while they were fighting and because fights involving two harem-holders tended to be long and the probability that they would lose a sizable proportion of their hinds was consequently high (Clutton-Brock et al. 1979). In the sample of fights for which we were able to collect detailed information, harem-holders that challenged their neighbors and won gained a smaller number of hinds than solitary animals that did so [8]. Harem-holders that beat approaching challengers were more likely to lose hinds on occasions when the challenger was another harem-holder than when he was a solitary stag [9], probably because fights between harem-holders were usually prolonged.

Finally, fighting was costly because of the energy expenditure involved. We were unable to measure these costs in the Rhum study, but stags which had recently fought usually appeared exhausted, and on several occasions they were quickly challenged by other rivals.

Adaptive Aspects of Fighting

On account of the costs of fighting, there were likely to be strong selection pressures favoring individuals that minimized the frequency with which they were involved in fights, so far as this was compatible with maximizing their reproductive success (see Maynard Smith and Price 1973; Maynard Smith 1976). As would be expected, stags generally appeared reluctant to fight, especially when they were unlikely to gain by doing so: for example, stags that were challenged had little to gain and much to lose by fighting with other stags and seldom initiated contact when approached.

Our detailed study of fighting (Clutton-Brock et al. 1979) showed that both the frequency and the duration of fights varied with the probable benefits of fighting. Fights were most frequent during the first two weeks of October, coinciding with the peak period of conceptions (fig. 6.29), and both the duration of fights (fig. 6.30) and the frequency of injuries peaked at the same time. In addition, stags fighting over hinds fought for longer than did solitary stags, and fights between two harem-holders tended to be longer than those between harem-holders and solitary stags.

Challenging stags would also be expected to assess their opponents carefully before fighting them, avoiding contests with opponents they were unlikely to beat (Parker 1974). There was firm evidence that stags could identify gross differences in fighting ability: stags five years old or less (which seldom won fights against older animals) seldom fought individuals more than a year older than they [10], whereas among mature stags there was no tendency for stags to avoid fighting with older animals [11]. However, this did not indicate that mature stags fought at random, and fights between individuals whose positions in the rutting "hierarchy" were more than two ranks apart were less common than would have been expected by chance [12].

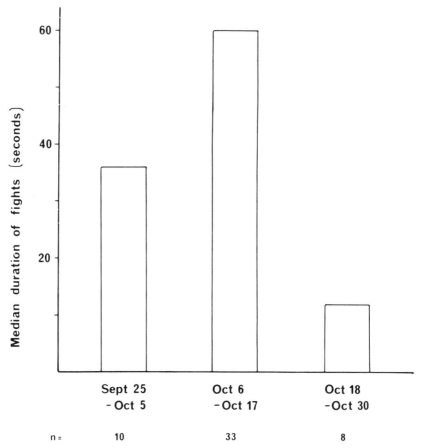

FIG. 6.30. Median duration of fights involving hinds during different periods of the rut (excluding fights ≤ 1 sec in length) (from Clutton-Brock et al. 1979).

6.6 ASSESSMENT

Evidence that stags usually fought with individuals of approximately the same fighting ability as their own indicated that rivals assessed each other before fighting. How did they do this?

It has been widely suggested that mature stags may assess each other by their antler sizes (Beninde 1937; Bubenik 1968; Geist 1971*b*, 1982; Henshaw 1968, 1971; Topinski 1974), an argument that rests on the claim that antler size or complexity is correlated with fighting success or dominance. However, among stags over five years old in our study area, neither fighting success nor reproductive success was related to antler length. Both measures were correlated with antler weight (see chap. 7), but the most likely explanation was that fighting success, point number, and antler weight were all related to individual differences in body size and weight (Clutton-Brock 1982). A second line of evidence often cited in support of the idea that stags assess each other on their antler size—that experimental removal of antlers reduces fighting success and is quickly recognized by rivals (e.g., Espmark 1964; Lincoln 1972)—has no bearing on whether or not natural variation in antler size is related to fighting success or is used in assessment (ibid.).

The weakness of the evidence does not indicate that relative antler size is *not* used in assessment. However, there is at least one reason it would be surprising if competing stags relied principally on antler size. As in other seasonal breeders (e.g., Jarman 1979), the fighting ability of individual stags changes during the course of the rut as their body condition declines, and individuals vary in the timing of their decline. Consequently, a stag that assessed its opponents on criteria that did not vary with changes in body condition during the rut would make many incorrect decisions (Clutton-Brock and Albon 1979).

Are criteria available that provide a more sensitive indication of a stag's fighting ability? Our investigation of the displays that precede fights suggested that the form and frequency of many displays reflects variation in condition and fighting ability. Before fights, stags typically engaged in periods of intense roaring. Playback experiments (see fig. 6.31) showed that their rate of roaring during these exchanges was related to the number of roars they heard and indicated that the two contestants attempted to outroar each other (Clutton-Brock and Albon 1979). Moreover, the mean and maximum roaring rates of individuals in these exchanges were correlated with their fighting success [13], and we were usually able to predict which of two stags would win a fight from their roaring behavior (see Clutton-Brock and Albon 1979). Unlike anatomical criteria such as antler size, roaring rates varied throughout the course of the rut as stags lost condition and became exhausted, and consequently they provided a sensitive criterion for assessing fighting ability. Studies of vocal displays in other vertebrates reveal relationships between the form or frequency of displays and the fighting ability or condition of the displayer (Morton 1977), and at least one study has produced experimental evidence that calls are used to assess rivals (Davies and Halliday 1978).

Assessment probably continued during parallel walks (Clutton-Brock and Albon 1979), which could be as short as 5 sec or as long as 30 min. Only a proportion of the parallel walks that we observed were followed by fights, and in many cases they terminated with the withdrawal of the approaching animals. Like roaring contests, they were commonest when both contestants held harems (and therefore had much to lose by fighting). They were also associated with situations where the outcome of the fight was uncertain: they occurred more frequently between well-matched opponents than between those that obviously differed in fighting ability [14], and longer parallel walks were less likely to be followed by fights than shorter ones [15]—but if fights did follow, these were more likely to be protracted [16].

Many other criteria may also have been used in assessment. In red deer stags, as in many other mammals (Pond 1978), reserves of fat and muscle tend to be in prominent positions, and it is conceivable that individuals may be able to assess the condition of their opponents by direct observation. Moreover, the concentra-

Fig. 6.31. A harem-holder replies to prerecorded roars. In most cases the roaring apparatus was set up farther from the stag and in a better-concealed position.

tion of pheromones in urine and other secretions is probably related to body condition and testosterone levels, and it is not inconceivable that a stag's fighting ability can be predicted from his smell. Like display prowess, these characters probably change with the stag's body condition and fighting ability during the course of the rut.

Fights usually ended when the loser was driven rapidly backward and ran away. The winner seldom followed and in most cases quickly returned to his harem and drove out the young stags that had infiltrated it. In the 5 min immediately following a fight, winners roared frequently while losers roared little [17] and typically moved away almost silently. Such marked differences in roaring rate did not occur after indecisive contests [18], when both stags retreated, roaring as they went. Similar "triumphs" have been observed in Bewick's swans: pairs that have just won encounters with other pairs perform elaborate wing-flapping ceremonies (Scott 1978) that may attract the attention of neighboring pairs, both revealing the identity of the winners and displaying the fact that even after an exhausting encounter, they have sufficient energy reserves left to display energetically. The function of such displays may be to reduce the probability that the winner will be challenged again.

6.7 SUMMARY

1. Stags spent most of the year in bachelor groups, which broke up in mid-September when individuals moved separately to their traditional rutting areas.
2. Only stags that had reached full adult weight were able to hold large harems during the period of peak conceptions, though young and old stags held harems before and after this date.
3. Rutting stags fed little and displayed regularly. Both the frequency and the intensity of many rutting displays were related to the probability that other stags would attack them or try to infiltrate their harems.
4. Fights were frequent and dangerous. Harem-holders appeared to minimize the number of fights they were involved in, and both the frequency and duration of fights were related to the value of the resources at stake.
5. Before committing themselves to a fight, stags assessed their rivals in roaring contests and parallel walks.

Statistical Tests

1. Comparison of roaring rate (roars/min) in stags on days when they held six or fewer versus more than six hinds.
    Wilcoxon matched-pairs test: $T = 7, N = 8, p < .02$.
2. Comparison of roaring rate (roars/min) in harem-holding stags on days when the harem nearest to their own was less than the mean for their period of rut versus days when the nearest harem was greater than the mean.
    Wilcoxon matched-pairs test: $T = 0, N = 8, p < .01$.
3. Comparison of percentage of daytime spent feeding by harem-holding stags on days when the harem nearest to their own was less than the mean for their period of rut versus days when the distance to the nearest harem was greater than the mean.
    Wilcoxon matched-pairs test: $T = 5, N = 8, .1 > p > .05$.
4. Correlation between the mean rate of roaring by five harem-holding stags across different hours of the day and the mean proportion of hinds seen feeding per hour in quarter-hourly scans of their harems.
    Spearman rank correlation coefficient: $r_s = .677, t = 4.215$, d.f. $= 21, p < .001$.
5. Correlation between the mean proportion of hinds seen feeding in the harems of five stags across different hours of the day and the total number of times that stags over one year old were seen approaching these harems to within 100 m.
    Spearman rank correlation coefficient: $r_s = .542, t = 2.954$, d.f. $= 21, p < .01$.
6. Correlation between the roaring rate (roars/min) for one rutting stag and his distance from the edge of his harem.
    Spearman rank correlation coefficient: $r_s = -.96, N = 6$, $p < .02$.
7. Correlation between the mean number of times per hour that hinds were chivied by their holding stag and the size of the harem.
    Spearman rank correlation coefficient: $r_s = -.64, t = 2.65$, d.f. $= 10, p < .05$.
8. Comparison of number of hinds gained by solitary stags that approached and beat a harem-holding stag versus harem-holding stags that approached and beat another harem-holding stag.

Mann-Whitney U test: $U = 29, n_1 = 13, n_2 = 9, .1 > p > .05$.

9. Comparison of number of occasions on which holding stags that beat solitary challengers lost hinds versus the number of occasions on which holding stags which beat holding challengers did so.

Fisher exact probability test: $n_1 = 10, n_2 = 7, p < .05$.

10. Comparison of observed versus expected frequencies with which stags of four to five years fought individuals within one year of their own ages. Expected frequencies were calculated from the numbers of stags of different age classes in the study area, taking into account differences in the frequency that members of each age class were involved in fights.

$\chi^2 = 17.5$, d.f. $= 1, p < .001. N = 82$.

11. Comparison of observed versus expected frequencies with which stags of more than five years fought individuals within one year of their own ages.

$\chi^2 = 1.40$, d.f. $= 1$, n.s. $N = 119$.

12. Comparison of observed versus expected frequencies of fights between stags whose ranks on fighting success were more than versus less than two ranks apart.

$\chi^2 = 18.2$, d.f. $= 1, p < .001. N = 26$.

13. Correlation between (a) the mean and (b) the maximum rates at which stags roared during contests and their fighting success. Spearman rank correlation coefficient:

(a) $r_s = .800, t = 3.77$, d.f. $= 8, p < .01$.
(b) $r_s = .761, t = 3.32$, d.f. $= 8, p < .02$.

14. Comparison of frequency of parallel walks in cases where two stags of seven years or older approached each other versus cases where a five- to six-year-old approached a harem-holder of seven or more years.

$\chi^2 = 17.94$, d.f. $= 1, p < .001. N = 50, 36$.

15. Comparison of the duration of parallel walks that were followed by fights with the duration of those not followed by fights.

Mann-Whitney U test: $z = 2.29, n_1 = 25, n_2 = 29, p = .022$.

16. Correlation between the duration of parallel walks and the duration of the fights that followed them in cases where stags both parallel-walked and subsequently fought.

Spearman rank correlation coefficient: $r_s = .498, t = 2.694$, d.f. $= 22, p < .02$.

17. Comparison of roaring rate (roars/min) in winning versus losing stags during the five minutes after the end of decisive fights.

Wilcoxon matched-pairs test: $T = 2.5$, $N = 17$, $p < .01$.

18. Comparison of differences in roaring rate between winners and losers after decisive versus indecisive fights.

Mann-Whitney U test: $U = 8$, $n_1 = 17$, $n_2 = 4$, $p < .02$.

7 Reproductive Success in Stags

Just as man can improve the breed of his game-cocks by the selection of those birds which are victorious in the cockpit, so it appears that the strongest and most vigorous males, or those provided with the best weapons, have prevailed under nature.

Charles Darwin (1871)

7.1 INTRODUCTION

As we described in the previous chapter, stags can monopolize breeding access to considerable numbers of hinds, and competition for harems is intense. As a result, the reproductive success of stags varies widely. In sections 7.2 to 7.4 we describe differences in harem size, reproductive success in particular years, and lifetime reproductive success between stags. Sections 7.5 and 7.6 examine the proximate and ultimate causes of variation in reproductive success, and section 7.7 compares variance in reproductive success and its causes in stags and hinds. The final section discusses relationships between these differences and sexual dimorphism in body size and early growth.

7.2 HAREM SIZE

A count of all harems on Rhum in October 1974 showed that they ranged from one hind to more than twenty (see fig. 7.1), and in recent years successful stags have gathered even larger harems in our study area as a consequence of the increase in population density (see chap. 12). As a crude measure of the success of stags in monopolizing hinds within particular seasons, we totaled the numbers of hinds one year old or older that each stag was seen to hold each day in rut censuses between 20 September and 30 October—a measure we refer to as the number of hind/days held.

Both harem size and the number of hind/days held by stags were closely related to age, and both measures peaked in stags of seven to eleven years and fell in stags of less than six or more than eleven years old (see fig. 6.28). Though age differences accounted for a large proportion of the variation in harem size, individual differences between stags were also important. For example, figure 7.2 shows the number of hinds held per day by three different nine-year-olds. As this indicates, successful stags not only held larger harems but tended to hold harems for longer than unsuccessful ones. Stags that rutted successfully in one season usually did so in successive years [1] until their performance declined with age.

7.3 VARIATION IN REPRODUCTIVE SUCCESS WITHIN SEASONS

Did the number of hinds a stag was able to collect in a particular season reflect his reproductive success? This need not necessarily have been the case, for a considerable proportion of hinds failed to breed each year, either because they were below breeding age

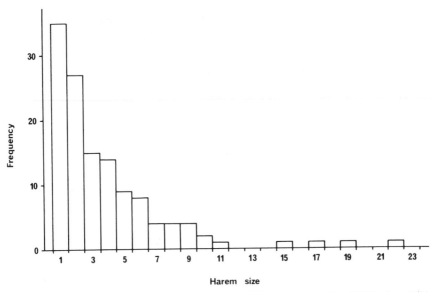

FIG. 7.1. Distribution of harem sizes on the whole of Rhum between 3 and 9 October 1974 (from Lincoln and Guinness 1977).

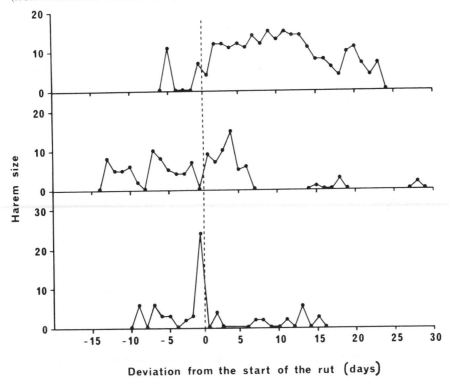

FIG. 7.2. Number of hinds held per day by three different nine-year-old stags, showing the range of variation in success at the same age.

or because they were in poor body condition (see chap. 5). Moreover, the duration of harem-holding of particular stags was generally shorter than the duration of the conception peak (see figs. 4.3, 7.2), so that a proportion of the hinds in a stag's harem conceived either before or after his period of holding.

To estimate the number of fertilizations achieved by each stag in each year, we multiplied the number of hinds he was observed to hold on each day of the rut by the probability that they would conceive (see Gibson 1978; Gibson and Guinness 1980a). In this calculation, the probability of conception on each day was the proportion of hinds of different ages that conceived each year multiplied by the fraction of them that conceived on each day (calculated by backdating from calf birth dates, using a standard gestation length). Since conception time was affected by the hind's reproductive status (see chap. 5), the proportion of animals conceiving was calculated separately for three classes of hinds: first-breeders (usually two or three years old at the time of the rut), yeld hinds (those without a calf at foot at the time of the rut), and milk hinds. By this method we estimated the reproductive success in each of three years (1971, 1974, and 1975) of all stags five years old or older that used the study area during the rut. Estimates of the number of calves fathered by each stag based on this method were closely correlated with the number of estrous hinds seen with each stag [2] as well as with estimates of the number of calves sired by each stag, derived by backdating from the birth date of each individual calf (see below) [3]. They were also correlated with cruder measures of reproductive success such as the number of hind/days held by each stag [4].

These estimates of reproductive success indicated that, in any year, nearly 50% of stags over four years old failed to breed, and that only the most successful individuals (approximately the top 5% of breeders) sired more than four calves (fig. 7.3). Variation in the estimated number of fertilizations achieved by different stags was thus considerably less than variation in the number of hind/days held, principally because stags had the opportunity to breed only with some of the hinds in their harem (see above).

7.4   LIFETIME REPRODUCTIVE SUCCESS

Though some stags were more successful than others at particular stages of their life span, this did not necessarily show that their

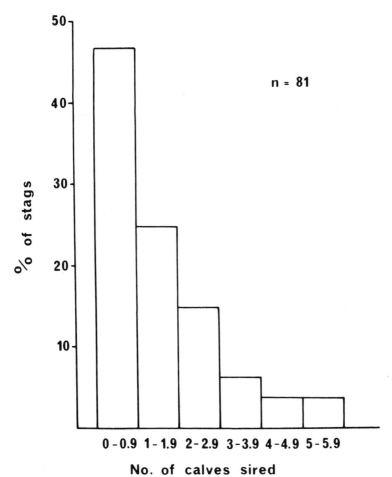

FIG. 7.3. Estimates of the number of calves sired per season by stags over five years old that rutted in the study area in 1971, 1974, or 1975. Data from different seasons were combined to produce this frequency distribution (analysis from Gibson and Guinness 1980*a*).

lifetime reproductive success was greater: studies of mountain sheep suggest that animals that are highly successful in breeding competition may have shorter life spans than less successful individuals (Geist 1971*b*) and it was possible that a reduction in life span could have compensated for increased success in the peak breeding years.

Between 1971 and 1979, we were able to measure the reproductive success of thirteen stags from the beginning of their effective reproductive life spans at the age of five to their deaths. In analysis of lifetime reproductive success we preferred to use a method of estimating the number of fertilizations achieved by each stag that was based on the conception dates of individual calves, since this allowed us to take into account variation in calf survival and to produce results comparable with our estimates of variation in lifetime reproductive success among hinds. We estimated the conception date of each calf born into the study population by backdating from its date of birth, using a mean gestation length of 236 days for male calves and 234 for females (see chap. 8). From the rut censuses, we then identified the stag whose harem the mother was in on the estimated conception date and on each of five days on either side of this date (= one standard deviation of gestation length on either side of the mean). To allow for variation in gestation length, each of these stags was "credited" with one fertilization multiplied by the proportion of these eleven days that the mother was in his harem—unless the calf died before it was one year old, in which case it was ignored. For example a male calf, BIG8, was born on 6 June 1978 and survived throughout his first year. Backdating by 236 days (the mean gestation length for male calves), we estimated that he was conceived on 13 October 1977. On this day his mother was recorded in the harem of RGRS, and she was in the harem of the same stag on a further five days during the eleven-day period between 8 and 18 October inclusive. Consequently, we estimated the probability that RGRS had fatherd BIG8 as 6/11 or .545. Of the remaining five days BIG8's mother was with BOSS for three days and SX95 for two days—as a result we credited BOSS and SX95 with .275 and .182 respectively. In this way we estimated the number of surviving calves fathered by each stag in each year. This method had the advantage that it allowed for any consistent difference in mortality between calves fathered by different stags, but in practice the results it gave were virtually identical to those produced by the coarser method used to estimate reproductive success within particular seasons.

Among the sample of thirteen stags, our estimates of lifetime reproductive success varied from 0 to 24 calves surviving to one year old, with a mean of 6.38. There was no evidence that calf mortality varied systematically between the offspring of different stags. Differences in lifetime reproductive success were correlated both with the average number of hind/days held per year by each stag and with life span, though the former accounted for a considerably larger proportion of the variation than the latter [5]. In contrast to mountain sheep (see above), there was no evidence that a stag's lifespan was negatively correlated with his reproductive success: correlations between the average and maximum number of hind/days held per year and the stag's age at death were not significant, but both coefficients were positive [6].

7.5   THE PROXIMATE CAUSES OF VARIATION IN LIFETIME REPRODUCTIVE SUCCESS

Our analysis of lifetime reproductive success indicated that the total number of surviving calves a stag fathered during his lifetime was closely related to the number of hinds he was able to collect and defend during his peak years of breeding activity. What, then, were the factors influencing the numbers of hinds a stag could monopolize?

The proximate causes of differences in reproductive success within particular breeding seasons were individual differences in mean harem size, in the duration of harem-holding, and in the timing of the stag's rut. Harem size and holding duration were positively correlated [7], and the total number of hind/days held was independently related to both measures [8]. The effects of average harem size were slightly stronger than those of holding duration [9], and harem size and holding duration jointly accounted for 78.6% of variation in annual reproductive success (Gibson and Guinness 1980b).

The time of harem-holding was also related to reproductive success: stags that rutted before and after the peak of conceptions showed lower success than those that rutted during the peak. As we have described, these were usually young stags or old animals past their prime. Dominant stags showed considerable synchrony in the timing of their rut.

Reproductive success was also affected by the area where the stag rutted. At Kilmory and Samhnan Insir, where large numbers of hinds collected on the greens, the reproductive success and harem size of stags was higher than in the Intermediate Area or the Upper Glen, where there were fewer hinds. Similar relationships between the quality of grazing and the size of breeding groups occur in many other species (Gosling 1974; Jarman 1974; Kitchen 1974).

The principal determinant of differences in harem size, holding duration, and the time of harem-holding was fighting success. Stags both gained and lost considerable numbers of hinds as a result of fights (see Clutton-Brock et al. 1979), and this had an immediate influence on the average size of a stag's harem and on the duration of his rut. As would be predicted, fighting success among prime stags was correlated with measures of reproductive success, both across age categories [10] (see fig. 6.28) and within them [11]. However, this was not the case among young stags of five or six or among stags over eleven years old [12], whose rutting behavior was less consistent than that of prime stags (Clutton-Brock et al. 1979).

Fighting success also affected the site of a stag's rutting area. The best fighters in the study area almost invariably rutted on the greens, either at Kilmory or at Samhnan Insir, while less successful stags rutted in the Intermediate Area and the Upper Glen, or moved between rutting sites. For example, SAGY, the most successful stag that rutted in the study area during the past eight years, began his career by rutting on the peripheral greens at Kilmory, held a central position on the main greens behind the beach during his prime (see fig. 6.20), and again rutted on the periphery of the greens during his last two years.

7.6 FACTORS INFLUENCING FIGHTING SUCCESS AND REPRODUCTIVE SUCCESS

The study indicated that a stag's fighting success played a major part in deciding his reproductive success, and studies of other polygynous mammals have produced similar results (Buechner 1961; LeBoeuf 1974). But what determined fighting ability?

Body condition was evidently important. Stags in poor condition seldom rutted successfully and often failed to secure harems, and studies of other polygynous ungulates have produced similar results: for example, territorial impala rams have thicker necks than nonterritorial males (Jarman 1979). This indicated that factors that affected body condition during the winter probably influenced reproductive success. An individual's condition was presumably influenced by the quality and quantity of food available to him, but there was no significant relationship between fighting success in the rut and the abundance of greens in the individual's home range or the number of other stags using it, though reproductive success was significantly correlated with social rank during the previous winter (see chap. 10).

Body size and weight evidently affected reproductive success.

As we have described, stags that were considerably lighter than their opponents usually lost. Though not all large stags were good fighters, individuals that were consistently successful in fights and held large numbers of hinds in several seasons were almost always of more than average size. Evidence for a relationship between body size and reproductive success was provided by the fact that lifetime reproductive success among stags was well correlated with mean antler weight [13], which is known to be related to body size and weight in red deer (Huxley 1926; 1931; Hyvarinen, Kay and Hamilton 1977; Clutton-Brock et al. 1979). In addition, Suttie (1979) has demonstrated a direct correlation between body weight and social dominance in young stags. Studies of a wide variety of other vertebrate species in which fights are decided by contests involving pushing have also found that fighting success or dominance is related to body size (dairy cattle: Schein and Fohrman 1955; Bouissou 1972; reindeer and caribou: Espmark 1964; Bergerud 1974a; toads: Davies and Halliday 1977), though this is not invariably the case (Lott 1979).

If body size affected the fighting ability and reproductive success of adult stags, early development was likely to have played an important part in deciding an individual's success, for studies of many mammals show that individuals that grow slowly during the first months of life can seldom compensate fully by subsequent increases in growth rate and thus show reduced adult size and weight (Gunn 1964a,b, 1965; Schinkel and Short 1961; Williams, Tanner, and Collins 1974; Fraser and Morley Jones 1975; Russell 1976; though see also Allden 1970), and this is probably also the case in red deer (Suttie 1980).

Investigation of relations between early growth and subsequent reproductive success among thirteen male calves born in 1972 supported this conclusion. The length of their cleaned antlers at sixteen months old (which could be judged by eye without difficulty and is related to body size and weight at the same age [14]) was correlated with the number of antler points they carried as six-year-olds [15] and their reproductive success at the same age [16]. The consistency of breeding success across years within older cohorts (see above) suggests that these differences will be maintained and that variation in the early development of these individuals will be related to their lifetime reproductive success.

Consequently, factors that affected growth during a stag's first year of life probably had an important influence on his eventual breeding success. What were these likely to have been? The antler length of yearling males again provided a useful measure of early development. In a sample of one hundred yearlings born from 1970 to 1976, antler length at sixteen months was significantly related to birth date [17] and weight [18]: male calves born after the median birth date or below the median birth weight for their part of the study area had smaller antlers as yearlings than those born before the median date or above the median weight.

This suggested that factors influencing the birth weight and date of calves probably had an important influence on their subsequent reproductive success. As we described in chapter 5, birth date, birth weight, and postnatal growth are apparently influenced by the mother's weight and body condition and vary with her age, reproductive status, and home-range area. In contrast to some other vertebrates (see Garnett 1981; Berger 1979), there is little evidence that the mother's body size is important, for in samples of mother/calf pairs shot in winter there were no significant relationships between the body size of hinds and the size, condition, or weight of their sons (see chap. 5, test 20).

7.7   REPRODUCTIVE SUCCESS IN STAGS AND HINDS
Variation in Lifetime Reproductive Success

Thus the extent of variation in reproductive success, the causes of this variation, and the distribution of reproduction throughout the life span all differed between stags and hinds. To produce an estimate of variance in lifetime reproductive success among stags that was comparable to that for hinds (see chap. 5), it was necessary to allow for the number of individuals that failed to reach effective breeding age (five years). In our study area, 36% of all stags born died before the age of five. To take these animals into account, we added five "zero" scores to the thirteen animals we had followed throughout their life spans and smoothed the distribution by reiteration (see Appendix 11).

In this sample of stags, lifetime reproductive success varied from 0 to 24 calves surviving to one year, with a median value of 4.2 (see fig. 7.4). This compared with an estimated range of 0 to 13 calves surviving to one year for hinds in our study area. The shape of the distribution for stags again prevented us from calculating the statistical variance for the entire sample. However,

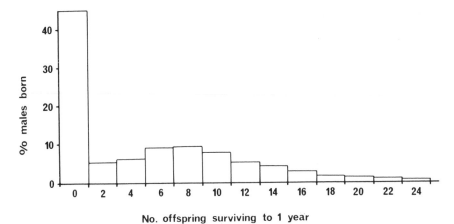

Fıg. 7.4. Distribution of estimated lifetime reproductive success (calculated in terms of calves surviving at one year old) of stags that rutted in the study area between 1972 and 1979. The analysis assumed that stags did not breed successfully (1) before they were five years old, (2) outside the study area, and (3) on days when they were not holding a harem. Though occasional exceptions may have occurred to all three assumptions, these were probably too rare to have any substantial effect on the results.

when the sample was restricted to stags that had fathered at least one calf that survived its first year, the mean was 7.64 and the variance 38.73 compared with a mean of 7.38 with a variance of 6.05 for hinds. Reproductive success was thus considerably more variable among stags than among hinds, and our results confirmed previous studies comparing variance in reproductive success between the sexes within particular breeding seasons (Howard 1979; Sherman 1976; Payne 1979).

Although lifetime reproductive success varied more widely among stags than among hinds, the difference between the sexes was less pronounced than might have been expected on the basis of variance within seasons. This was partly because reproductive success in stags was closely related to age: most individuals held harems successfully for no more than four years. In addition, sex differences were reduced because certain hinds were consistently successful breeders, and this increased the extent to which lifetime success differed between individuals (see chap. 5). Since pronounced age variation in reproductive success among males and consistent individual differences in success among females are commonplace among polygynous mammals, this indicates that breeding sex ratios within years systematically overestimate the strength of sexual selection.

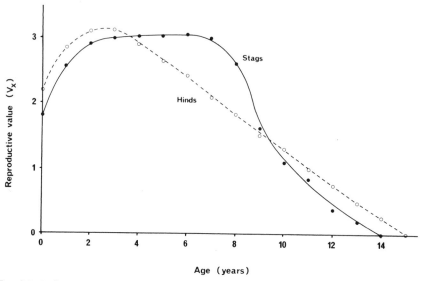

Fig. 7.5. Estimates of reproductive value (Pianka 1978) for stags of different ages (*solid line*) compared with reproductive value in hinds (*broken line*). For both hinds and stags, reproductive value is calculated in terms of the number of female offspring *surviving to one year old* that parents of different ages can expect to produce in the future. Measures of reproductive value for stags are approximate, since they did not take into account changes in population density in the study area. The method by which they were calculated is described in Appendix 9.

Reproductive Value

The sexes also differed in the distribution of reproductive success thoughout their life spans. Calculation of reproductive value for stags of different ages (see fig. 7.5) showed that this started to fall later than that of hinds but declined more sharply after the age of seven. Unlike hinds, there was no evidence that stags invested more heavily in breeding activities in their declining years—old stags were involved in fewer fights and held smaller harems than stags of seven to ten years (see figs. 6.26, 6.28). However, it is possible that their investment in breeding activity per season was greater but that their declining capabilities reduced their success: in bull reindeer there is some evidence that the difference between prerut and postrut weight tends to increase toward the end of the life span (Leader-Williams and Ricketts 1982). Unfortunately, no similar data are available for Scottish red deer.

Factors Affecting Reproductive Success

Both the proximate and the ultimate factors affecting reproductive success thus differed between stags and hinds (see table 7.1).

TABLE 7.1   Summary of Sex Differences in Reproductive Behavior and in the
Factors Affecting Reproductive Success in Stags and Hinds

Parental investment: Stags invest less in individual offspring than hinds do.

Reproductive behavior: Stags compete intensely during the rut, fight frequently, and engage
in elaborate displays that may allow them to assess their rivals. Hinds show little
evidence of competition during the rut and seldom engage in escalated contests
outside it.

Reproductive success: Varies more widely among stags than among hinds both within seasons
and across the life span.

Proximate factors affecting reproductive success: Success of stags limited by access to opposite
sex and related to harem size, reproductive life span, and fighting ability. Success in
hinds not limited by access to stags but related to differences in life span, calf
mortality, and fecundity.

Ultimate factors affecting reproductive success: Success of stags related to body size, early
growth, and maternal investment during the first year of life. Success of hinds not
closely related to size or (by implication) to investment before weaning, but strongly
affected by resource access during adulthood and the size of the matrilineal group
to which they belong.

Breeding life span and reproductive value: Effective breeding life span in stags typically restricted to three to five years between the ages of six and eleven, usually peaking
between the ages of seven and ten years. Effective breeding life span in hinds can be
as long as twelve years and extends from the age of three to fifteen or older.
Consequently, reproductive value of hinds starts to fall first, while that of stags
declines more rapidly after the age of six.

Morphological differences: Stags develop larger body size, antlers, and other "secondary"
sexual characteristics associated with fighting or display, including manes, seasonal
increases in size of neck muscles, and thicker skin on the forequarters.

Early growth: Young stags grow faster, probably have higher metabolic rates, lay down less
body fat, and grow faster during their first winter.

In stags, reproductive success was limited by access to the opposite
sex and was related to harem size, holding duration, fighting ability, and, less closely, life span. Among hinds, variation in reproductive success was principally related to differences in life
span and in calf mortality, while variation in fecundity was less
important (see chap. 5). Social rank outside the rut (assessed from
threats between individuals) was not significantly correlated with
reproductive success in hinds, but a correlation existed among
stags.

Among the ultimate factors influencing the reproductive success of stags, body size was evidently important. Since adult size
was correlated with early growth rates, this suggested that maternal investment during the period of gestation and lactation probably influenced reproductive success, and our results supported
this conclusion. Body condition also had a major effect on a stag's
reproductive success in particular years, but the reproductive success of stags was not related to their home-range area or quality.

TABLE 7.2   Mean Kidney Fat Indexes (KFI) and Carcass Weights of Young
Red Deer Stags and Hinds Shot at Glenfeshie, Invernesshire

	Age (years)			
	0.5	1	2	3
Stags				
KFI	1.5	1.5	1.9	3.6
Carcass weight (kg)	26	38	49	67
Hinds (nonlactating)				
KFI	1.6	2.0	3.5	3.8
Carcass weight (kg)	24	36	47	51

Source: Mitchell, Staines, and Welch (1977) and B. Mitchell, unpublished data.

In contrast, the reproductive success of hinds was influenced by
their home range area and the size of their matrilineal group but
was not closely related to their body size, suggesting that differences in maternal investment were less important than among
stags.

7.8   SEXUAL DIMORPHISM AND THE FACTORS AFFECTING
REPRODUCTIVE SUCCESS

Differences in body size and weaponry between stags and hinds
presumably represent adaptations to the contrasting factors affecting reproductive success in the two sexes: as we describe in
chapter 13, both differences are more strongly developed in ungulates that show a high degree of polygyny. In addition, the same
selection pressures may explain some of the contrasting patterns
of growth in males and females. If large body size has important
advantages for stags and adult size is inevitably related to early
growth, we might expect to find adaptations favoring faster early
growth rates in young males compared with young females. In at
least two respects, young stags appear to sacrifice safety for rapid
growth to a greater extent than young hinds. As in many other
mammals (see Glucksman 1974), the fat reserves of growing stags
are smaller than those of growing hinds (table 7.2). Fat acts as an
insurance against winter food shortage, and there is a close association between sex differences in fat deposition and differences in
mortality (see sec. 12.4), but it is energetically costly to accumulate
and inefficient to convert (Blaxter 1961; Pond 1978). By reducing
the proportion of their nutritional intake used for fat deposition,

young males presumably increase their growth rates. Second, although both male and female calves reduce their growth rates at the onset of winter (see p. 74; presumably an adaption to reducing the danger of mortality), there is some evidence that the reduction is more pronounced in females than males [19], and the degree of size dimorphism between male and female calves tends to increase during the course of the winter.

7.9   SUMMARY

1. Harem size varied between stags from one to more than twenty hinds. However, because the duration of harem-holding was generally shorter than the conception peak and only a proportion of hinds conceived, reproductive success did not vary so widely, and few stags fathered more than four calves in a season.
2. Some stags failed to breed during their lifetime, while our estimates indicated that others fathered as many as twenty-five offspring surviving to one year of age.
3. Variation in reproductive success among stags was caused by differences in harem size, duration of harem-holding, rutting area, and life span. Of these, differences in harem size and the duration of harem-holding were most important and were closely related to fighting success.
4. Fighting ability and reproductive success were related to body size and condition. Size was affected by early development, and stags that showed advanced development at one year old were the most successful breeders at the age of six.
5. As the theory of sexual selection predicts, lifetime reproductive success was more variable among stags than among hinds, and both the immediate and the ultimate factors influencing reproductive success differed between the sexes. Sex differences in body size, weaponry, and growth rates are probably adaptations to the contrasting factors affecting reproductive success in males and females.

Statistical Tests

1. Concordance of number of hind/days held during the peak rut by nine individual stags born between 1965 and 1970 at ages six through nine inclusive.

   Kendall coefficient of concordance: $W = .700$, $\chi^2 = 22.40$, d.f. $= 8$, $p < .01$.

2. Correlation between the estimated number of hinds fertilized by different stags and the number of hinds seen in estrus in each stag's harem in 1971, 1974, and 1975.

   Spearman rank correlation coefficient: r_s = .797, N = 59, $p < .001$ (analysis from Gibson and Guinness 1980a,b).

3. Correlation between the number of hinds fertilized by different stags estimated by the method described on p. 146 and the number estimated by backdating from the birth date of individual calves in each year assuming a standard gestation length of 235 days.

   Spearman rank correlation coefficient: r_s = .945, N = 48, $p < .001$ (analysis from Gibson and Guinness 1980a,b).

4. Correlation between the estimated number of hinds fertilized by different stags and the mean number of hind/days held by stags rutting in the study area in 1971, 1974, and 1975.

   Spearman rank correlation coefficient: r_s = .921, N = 81, $p < .001$ (analysis from Gibson and Guinness 1980a).

5. Percentage of variance in estimated lifetime reproductive success attributable to (a) mean hind/days per year, (b) life span of the stag.

   (a) Spearman rank correlation coefficient: r_s = .938, t = 8.99, d.f. = 11, $p < .001$; % of variance accounted for, 88.0.

   (b) Spearman rank correlation coefficient: r_s = .577, t = 2.34, d.f. = 11, $p < .05$; % variance accounted for, 33.3.

6. Correlations between (a) mean hind/days held per year, (b) maximum hind days held per year and age at death.

   (a) Spearman rank correlation coefficient: r_s = .499, t = 1.670, d.f. = 11, n.s.

   (b) Spearman rank correlation coefficient: r_s = .438, t = 1.618, d.f. = 11, n.s.

7. Correlation between mean harem size and the period for which harems were held by stags rutting in the study area, 1971, 1974, and 1975.

   Spearman rank correlation coefficient: r_s = .82, N = 81, $p < .001$ (analysis from Gibson and Guinness 1980b).

8. Partial regression between the estimated number of hinds fertilized by different stags and (a) their mean harem size, (b) their duration of harem holding.

   (a) $F_{1.78}$ = 235.02, $p < .001$.

   (b) $F_{1.78}$ = 33.45, $p < .001$.

   (Analysis from Gibson and Guinness 1980b).

9. Standardized partial regression coefficients for the relationship between the estimated number of hinds fertilized and mean harem size (b_1') and the relationship between the estimated number of hinds fertilized and duration of harem-holding (b_2').

   $b_1' = .52$, $b_2' = .44$ (analysis from Gibson and Guinness 1980b).

10. Correlation between mean fighting success of stags five or older belonging to different age grades and the mean number of hind/days held by stags in each category.

    Spearman rank correlation coefficient: $r_s = .891$, $t = 5.55$, d.f. $= 8$, $p < .001$.

11. Correlation between fighting success and number of hind/days held by (a) seven- to eight-year-old stags, (b) nine- to ten-year-old stags, 1971–75.

    (a) Spearman rank correlation coefficient: $r_s = .786$, $N = 7$, p (one-tailed) $< .05$.

    (b) Spearman rank correlation coefficient: $r_s = .708$, $N = 8$, p (one-tailed) $< .05$.

12. Correlation between fighting success and number of hind/days held by (a) five- to six-year-old stags, (b) eleven- to fourteen-year-old stags, 1971–75.

    (a) Spearman rank correlation coefficient: $r_s = .17$, $t = .52$, d.f. $= 9$, n.s.

    (b) Spearman rank correlation coefficient: $r_s = .537$, $N = 7$, n.s.

13. Correlation between mean antler weight and lifetime reproductive success in stags.

    Spearman rank correlation coefficient: $r_s = .818$, $t = 4.025$, d.f. $= 8$, $p < .01$.

14. Correlation between antler weight and body weight in yearling stags shot on Scarba (calculated from B. Mitchell, unpublished data).

    Pearson product-moment correlation coefficient: $r = .584$, $t = 3.45$, d.f. $= 23$, $p < .01$.

    $y = -100.6 + 4.27x$.

15. Correlation between the length of the antlers of yearlings born in 1972 and the number of antler points they carried in 1978 at six years of age.

    Spearman rank correlation coefficient: $r_s = .624$, $t = 2.646$, d.f. $= 11$, $p < .02$.

16. Correlation between the length of antlers of yearlings born in

1972 and the number of hind/days held during the 1978 rut at six years of age.

   Spearman rank correlation coefficient: $r_s = .730$, $t = 3.702$, d.f. $= 12$, $p < .01$.

17. Comparison of numbers of yearling males with antlers longer than average (the median for those individuals born in the same part of the study area in the same year) born before versus after the median birth date for their cohort.

   G test: $G = 3.876$, d.f. $= 1$, $p < .05$. $N = 75$.

18. Comparison of numbers of yearling males with antlers longer than average (see test 17) born above versus below the median birth weights for their cohorts.

   G test: $G = 6.135$, d.f. $= 1$, $p < .02$. $N = 52$.

19. Regression of jaw length of *(a)* male and *(b)* female calves on date of death among animals shot between November and February in Glenfeshie (B. Mitchell, unpublished data).

   (a) $y = 162.4 + .063x$, $t = 2.013$, d.f. $= 74$, $p < .05$.
   (b) $y = 161.9 + .017x$, $t = 3.627$, d.f. $= 72$, $p < .001$.

8 Parental Investment in Male and Female Offspring

At every moment in the game of life, the masculine sex is playing for higher stakes.

G. C. Williams (1975)

8.1 INTRODUCTION

So far, we have ignored the fact that parents produce two types of offspring—males and females. How should they be expected to allocate the resources at their disposal to their male and female offspring so as to maximize their own reproductive success? Current theory suggests that, where variance in reproductive success is greater in one sex and is influenced by the amount of parental investment the individual receives, parents should invest more heavily in individual offspring belonging to the sex that shows the higher variance (Trivers and Willard 1973; Maynard Smith 1980). As we have shown, variance in lifetime reproductive success is greater among stags and is probably more closely related to early growth and parental investment than in hinds (see chap. 7). Consequently, red deer would be expected to invest more heavily in individual sons than daughters.

Since parental investment by stags ends at fertilization, only hinds have the opportunity to invest more heavily in individual sons. Theoretically, they might accomplish this in one of two ways: either they might vary the sex ratio of their progeny so as to produce sons at times when they are in superior body condition and can afford the costs of heavy energy expenditure on their offspring, perhaps selectively aborting or resorbing male fetuses when in poor condition (Trivers and Willard 1973); or they might transfer a larger proportion of their resources to individual sons than daughters (see Reiter, Stinson, and Le Boeuf 1978). In sections 8.2 and 8.3, we test each of these possibilities.

Mothers should also "decide" what proportion of sons and what proportion of daughters to produce. In section 8.4 we describe Fisher's theory of the evolution of the sex ratio (1930) and examine the extent to which red deer conform to its predictions.

8.2 ADAPTIVE VARIATION IN THE SEX RATIO?

What evidence was there that the birth sex ratio of calves varied with maternal condition in our study population? Apparently, very little. Table 8.1 shows the sex ratios of calves born to different categories of mothers: none of these comparisons revealed statistically significant variation in the sex ratio. However, two trends should be noticed. First, the highest sex ratios were found among the offspring of four-year-old first-breeders, which, as a category, were probably in better body condition than any other

TABLE 8.1   Birth Sex Ratios of Calves Born to Different Categories of Hinds

	Age of Mother (years)						
	3	4	5–6	7–8	9–10	11–12	13+
Percentage of males	52.6	59.7	56.0	58.3	53.8	58.8	69.0
N	38	67	109	84	65	57	29

	Reproductive Status of Mother				
	First Breeders		Yelds		Milks
	3-year-olds	4-year-olds	Summer	True	
Percentage of males	52.6	64.3	60.3	59.3	56.4
N	38	42	68	59	218

	Birth Order						
	1	2	3	4	5	6	7+
Percentage of males	56.3	53.5	65.1	51.4	55.6	68.2	50
N	87	58	43	37	27	22	26

	Previous Breeding Sequence				
	One Female	One Male	Two Females	One Female and One Male	Two Males
Percentage of Males	51.6	56.8	42.9	57.1	62.5
N	37	31	8	14	7

	Time of Conception	
	Early (Before Median Date for Year)	Late (After Median Date for Year)
Percentage of Males	62.9	53.5
N	210	202

	Year of Birth								
	'71	'72	'73	'74	'75	'76	'77	'78	'79
Percentage of males	60.0	69.2	50.0	53.1	54.2	53.4	55.9	59.3	59.3
N	25	39	32	49	48	58	59	54	59

Continued

TABLE 8.1   *(continued)*

	Mother's Home Range			
	Upper Glen	Kilmory	Intermediate Area	Samhnan Insir
Percentage of Males	59.2	56.0	59.8	57.1
N	49	168	127	98

age group (see Mitchell, McCowan, and Nicholson 1976; Mitchell, Staines, and Welch 1977). Second, there was a tendency, which approached statistical significance, for more males to be born early in the season [1]. Similar differences in birth timing between male and female progeny have been found in several seal species (Coulson and Hickling 1961; Stirling 1971). There was no suggestion in our data that birth sex ratio varied between mothers with home ranges of varying quality, between members of large versus small matrilineal groups, or between milk and yeld hinds. Although some previous studies have claimed that yeld hinds produce more male calves (Davies 1931; Darling 1937), none have demonstrated a significant difference, and other studies have found trends in the opposite direction (Miller 1932; Flook 1970).

One possible explanation of the conflicting evidence is that the attempt to relate maternal condition directly to the sex ratio is naive. If selection favors increased investment in male offspring and males are consequently more costly to rear than females, we might expect the sex of sequential offspring to alternate. However, there was no evidence in our data that birth sex ratio differed according to the sex of the calves reared previously by the same mother (see table 8.1).

8.3   COMPARATIVE INVESTMENT IN INDIVIDUAL SONS AND DAUGHTERS

The alternative method by which parents might invest more in male than female offspring would be to distinguish between offspring (or fetuses) of different sexes, allocating a larger share of their available resources to individual sons than daughters. Comparison of maternal investment in male versus female calves suggests that this was the case in our population (Guinness, Albon, and Clutton-Brock 1978; Clutton-Brock, Albon, and Guinness 1981). Male calves were born heavier than females [2] (means 6.72

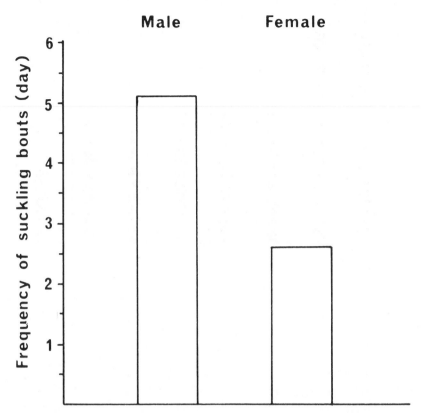

FIG. 8.1. Mean frequency of suckling in seventeen male calves and ten female calves observed for one day each in July and August.

and 6.22 kg respectively); their gestation lengths were longer [3] (males 236.1 days ± 4.75; females 234.2 ± 5.06 days); and, by analogy with studies of calves reared on deer farms (Blaxter et al. 1974), males probably grew faster. Similar differences between the sexes have been observed in a wide variety of studies of other sexually dimorphic mammals (Defries, Touchberry, and Hays 1959; Short 1970; Glucksman 1974; Nordan, Cowan, and Wood 1970; Krebs and Cowan 1962; Dittus 1979; Robbins and Moen 1975; McEwan and Whitehead 1970; Benedict 1938; McEwan and Wood 1966; Mitchell, Staines, and Welch 1977), though it is not yet known whether they occur in monomorphic or monogamous species.

Differences in the size and growth of male and female calves suggested that males probably obtained more milk from their mothers than females, and comparisons of sucking behavior sup-

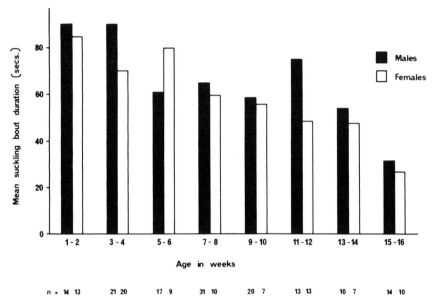

FIG. 8.2. Mean duration of suckling bouts to male and female calves observed in July to October. The number of bouts in each sample is shown below the histograms.

ported this conclusion. Male calves sucked more frequently than females [4] (see fig. 8.1) and also tended to suck longer (see fig. 8.2): though there was no significant difference in suckling-bout duration in the relatively small sample of data collected in continuous watches, when suckling bouts timed during the course of censuses were allocated to different fortnights from birth to sixteen weeks, the mean length of bouts involving male calves was longer than the mean length of those involving females in seven out of eight periods [5].

Unfortunately our data were inadequate to determine whether the timing of weaning differed between the sexes, as in some other dimorphic animals (Reiter, Stinson, and LeBoeuf 1978). The more frequent sucking of male calves was not associated with any tendency for them to be rejected less frequently by their mothers. Indeed, when the effects of the calf's age were taken into account, there was a trend for males to be rejected *more* frequently than females that was significant among the offspring of barren hinds [6].

Their increased size and more frequent sucking indicated that

TABLE 8.2 Fecundity of Hinds That Had Reared Male versus Female Calves into the Previous Winter

Year	Males	Females	N
1971	81.8	90.9	33
1972	92.8	100.0	24
1973	88.0	92.3	40
1974	92.9	88.2	31
1975	58.8	83.3	39
1976	82.6	88.9	41
1977	66.7	75.0	44
1978	66.7	27.8	51
1979	29.2	60.0	44
1980	28.6	47.4	47
1981	16.7	35.7	52

the energy costs to the mother of producing and rearing male calves were greater than those of producing and rearing females. To test this prediction, we compared the subsequent reproductive performance of hinds that had reared male versus female calves in the previous year. Hinds that had reared a male were more likely to be barren the following year than those that had reared a female calf (table 8.2) [7]. A further indication that the body condition of hinds was usually poorer in years following breeding seasons when they had reared a male was provided by comparison of their calving dates: hinds calved, on average, eleven days later in years following breeding seasons when they had supported a male calf than in years following seasons when they had supported a female calf [8]. Since late-born calves were more likely to die than those born early in the breeding season (see chap. 5), this provided additional evidence of increased investment in individual sons.

Other studies of lactation also indicate that males may be more costly to produce and rear in other dimorphic mammals. Male caribou calves are born heavier than females, grow faster, and have a higher milk intake (McEwan 1968; McEwan and Whitehead 1971). And in elephant seals, male pups are born heavier and grow faster than females, are weaned later, and tend to be more persistent in their attempts to obtain milk after weaning (Reiter, Stinson, and LeBoeuf 1978).

The immediate cause of the increased energy cost of rearing male calves was probably that their more frequent demands stimulated their mothers to produce more milk: a relationship between the demands of offspring and the milk yield of nursing mothers has been demonstrated in several domestic animals (see Alexander and Davis 1959; Arnold and Dudzinski 1978). This does not mean that a functional explanation was not necessary, for the physiological responses of hinds to the demands of their calves has presumably been subject to natural selection (see Trivers 1974).

8.4   TOTAL INVESTMENT IN MALE AND FEMALE PROGENY

As we have already argued, mammalian mothers would be expected to produce sons and daughters in the ratio that maximizes their own reproductive success. Ecologists have sometimes been puzzled that polygynous mammals, including red deer, produce approximately equal numbers of male and female offspring when a single male is able to fertilize many females. In such species, would it not be advantageous for breeding females to conceive more daughters than sons? The argument is appealing because it is immediately obvious that this would represent the most economical use of resources by the population. However, it is fallacious, since natural selection does not maximize the reproductive performance of populations (see chap. 1), and the question must be restated in terms of individual advantage—what proportion of sons and daughters should an individual produce to maximize his or her own reproductive success?

   The answer was first provided by Fisher (1930), who argued that, where the cost of producing sons and daughters was the same, the average parent should produce equal numbers of both. The reasoning behind Fisher's conclusion was based on the fact that the mean reproductive success of members of each sex must be inversely related to their frequency in the population. For example, suppose a genotype that caused its carriers to produce an excess of females had spread through a population as a result of some chance event. Because the population would include a high proportion of females, the mean reproductive success of individual males would exceed that of females. Consequently, selection would favor individuals that produced a higher proportion of male offspring. As male-producing genotypes spread, their advantage would decline, and when the population contained a preponderance of males there would again be selection for genotypes producing female-biased sex ratios. A stable situation would be reached when the average parent produced equal numbers of sons and daughters.

Where the costs of producing sons and daughters differ, Fisher's argument predicts that parents should divide their *investment* equally between the two sexes, producing fewer of the more costly sex. The reason for this can also be understood intuitively. Suppose a species existed in which sons cost twice as much as daughters, but the sex ratio of offspring produced was equal. In these circumstances the benefit gained by a parent who produced a son would be less, per unit investment, than the benefit gained by one who produced a daughter. Consequently, a genotype that produced daughters would be favored by selection and would spread until a situation was reached where the average parent invested equally in offspring of the two sexes.

It is important to notice that Fisher's theory predicts that investment in sons and daughters should be equal *at the end of the period of parental investment.* For example, if the costs of rearing sons and daughters were the same but sons were more likely to die before the end of investment, the average cost to the parents per son conceived would be less than that per daughter (because investment in sons would be more frequently terminated prematurely), but the cost per son *reared* would be greater (on account of the "wasted" investment in sons that died). In these circumstances, parents that invested equally in sons and daughters should produce more males at birth, but the sex ratio should be female-biased at the end of the period of investment. Conversely, increased mortality among members of one sex *after* the termination of parental investment should have no effect on the optimal sex ratio for parents, since reduced chances of survival in one sex will be compensated for by a proportional increase in the reproductive success of the survivors (Leigh 1970).

Since hinds invested more heavily in male than female calves before weaning, we initially predicted that the sex ratio at weaning would be female-biased (Clutton-Brock, Albon, and Guinness 1981). However, contrary to this prediction, sex ratios of twelve-month-old calves were not female-biased—in contrast, they showed a (nonsignificant) tendency to be biased toward males (see table 8.3). A similar result has been found in elephant seals (LeBoeuf and Briggs 1977), where there is also some evidence

TABLE 8.3   Sex Ratios (% Males) of Calves Born in Different Years at Birth and at the End of Their First and Second Winters

Year of Birth	Sample Size	Sex Ratio at Birth	Sex Ratio at End of First Winter (May)	Sex Ratio at End of Second Winter
1971	25	60	57	59
1972	39	69	64	64
1973	32	50	43	43
1974	49	53	41	40
1975	48	54	52	52
1976	58	53	56	54
1977	59	56	61	59
1978	54	59	51	52
1979	59	59	55	56
Mean of Years		57.0	53.3	53.2

that male pups may cost more to rear than females (see above).

One explanation of the absence of female-biased sex ratios at weaning in red deer is that the conception sex ratio is immutably fixed at unity and subsequent manipulations of the sex ratio of progeny by selective abortion, resorption, or starvation reduces a parent's reproductive success (Maynard Smith 1980). If selection favors increased investment in individual sons, such a situation could lead to increased total investment in males (ibid.). However, two recent studies of rodents indicate that adaptive manipulation of sex ratios can occur during gestation and lactation (Gosling and Petrie 1982; McClure 1981), while some evidence suggests that fetal and birth sex ratios can vary adaptively between species (Trivers, Seger, and Hare unpublished; Clutton-Brock and Albon 1982).

An alternative explanation is that hinds continue to invest in their daughters (but not in their sons) after the period of weaning. As we describe in chapter 9, daughters usually occupy ranges and core areas overlapping those of their mothers, while sons disperse from the mother's range between the ages of two and four years. The evidence that hinds belonging to large matrilineal groups show reduced reproductive success (see chap. 5) suggests that there may be important costs involved in permitting daughters to remain in the home range—which could offset increased investment in sons before weaning. Thus the pattern of maternal investment could follow the curves shown in figure 8.3, with in-

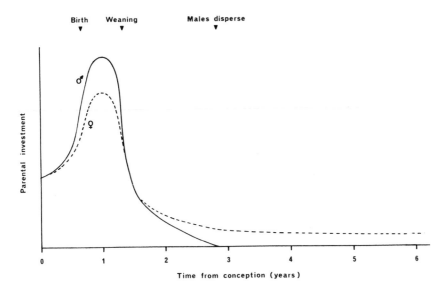

Fig. 8.3. Hypothetical temporal distribution of maternal investment in sons versus daughters.

creased preweaning investment in sons being counterbalanced by increased postweaning investment in daughters. A similar argument, which has been used to explain the adaptive significance of male-biased postweaning sex ratios in bush-babies *(Galaginae)* (Clark 1978), is that, where siblings of one sex compete more intensely with *each other* than siblings of the other sex, parents should produce more of the latter (see Hamilton 1967). There are good reasons for thinking that selection might favor hinds who invested more heavily in sons before weaning and more heavily in daughters afterward, for early growth probably has a greater effect on the reproductive success of stags than that of hinds, while access to optimal habitat during adulthood probably affects the reproductive success of hinds more than that of stags because lactation has such high energetic costs.

However, although the reduced reproductive success of members of large matrilineal groups *suggests* that the presence of relatives in her home range affects a hind's reproductive performance, there is an obvious loophole in the argument. If large matrilineal groups occurred in parts of the study area where population density was high, the negative association between matrilineal group size and reproductive success could have been an effect of population density rather than of the number of resident relatives per se. Obviously, the latter contributes to population density, but it was still possible that it was the impact of unrelated animals rather than of relatives that depressed a hind's reproductive success. In particular, if it were the case that a hind whose female relatives left her home range did not benefit from increased access to resources because these were eaten by unrelated intruders, the presence of female relatives would not represent a continuation of parental investment (Clutton-Brock, Albon, and Guinness 1981).

Our analysis of the relationship between matrilineal group size and reproductive success removed the effects of variation in population density between the four parts of the study area (see sec. 5.6), but it was still conceivable that there was an association between the size of a hind's matrilineal group and the frequency with which members of matrilineal groups other than her own used her range *within* parts of the study area. To examine this possibility, we extracted the frequency with which matrilineal relatives of all kinds and members of other matrilines were seen in the core area of each hind in our sample. On average, the frequency with which nonrelatives were seen in a hind's core area exceeded the frequency with which relatives were seen there. Using these data, we regressed our estimates of each individual's reproductive success on the frequency with which (*a*) members of her own matriline and (*b*) members of other matrilines were seen in her core area (see p. 48) over a four-year period (1975–78). If the tendency for hinds belonging to large matrilineal groups to show low reproductive success was a consequence of the frequency with which members of other matrilines used their ranges, and this happened also to be correlated with frequency of use by resident relatives, negative correlations between reproductive success and the frequency with which both relatives and nonrelatives were seen in the core area should have been apparent. On the other hand, if the negative association between matrilineal group size and reproductive success was a consequence of the presence of relatives, only the first relationship should have been negative. As the second hypothesis predicted, the reproductive

TABLE 8.4   Summary of Influence of Calf's Sex on Maternal Behavior and
Early Development

Birth Weight: Higher for male calves.
Gestation length: Longer for male calves.
Growth rates: Probably faster for males, especially where food is abundant.
Suckling frequency: Higher for males.
Suckling duration: Slight tendency for suckling bouts involving male calves to be longer than those involving females.
Rejections: Suckling attempts by male calves tended to be rejected more frequently than those by females.
Maternal investment: Mothers that had raised a male calf were more likely to be barren the following year than those that had raised a female calf. In addition, hinds calved later (relative to their average calving date) in years following seasons when they had reared a male calf than after they had reared a female.

success of hinds was negatively correlated only with the frequency with which relatives were seen in the hind's core areas [9], implying a cost that is specific to resident *relatives.*

It is not yet clear why the presence of relatives but not of members of other matrilines depresses a hind's reproductive success. In particular, there were no significant differences in the proportion of relatives versus nonrelatives that were seen feeding in core areas. The most likely explanation is that relatives had access to certain key resources from which members of other matrilines were excluded. Our current work is focusing on this problem.

8.5   SUMMARY

1. Evolutionary theories predict that, where reproductive success varies more widely among males than among females and is influenced by parental investment, parents should invest more heavily in individual sons than daughters but should produce more daughters than sons. Increased investment in sons could be achieved either by varying the sex ratio of their progeny in relation to their body condition or by allocating a greater proportion of their resources to sons.

2. There was no evidence that the sex ratio of red deer calves in our study area varied with factors likely to affect the mother's body condition, though there was a tendency for male calves to be born earlier in the season than females.

3. However, mothers suckled male calves more frequently and for longer periods than females. After rearing a male calf, they were more likely to be barren in the next breeding season and,

if they did conceive, to do so later than after years when they had reared a female calf.

4. Evolutionary theory predicts that, where offspring of one sex cost more to produce, the sex ratio should be biased against them by the end of the period of parental investment. However, there was no evidence that the sex ratio at weaning was female-biased. The most likely explanation is that parental investment in daughters (but not sons) continues after weaning because daughters continue to use their mothers' ranges.

Statistical Tests

1. Comparison of sex ratio of calves born before the median calving date for their year versus the sex ratio of those born after this.

   G test: $G = 3.739$, d.f. $= 1$, $.1 > p > .05$. $N = 391$.

2. Comparison of birth weights (corrected for changes across years) of male versus female calves.

   Analysis of variance: $F_{1.303} = 9.166$, $p < .005$.

3. Comparison of gestation lengths of hind versus stag calves, in six years (excluding all over 251 days: see Guinness, Gibson, and Clutton-Brock 1978).

   Binomial test: $x = 0$, $N = 6$, $p = .032$ (one-tailed).

4. Comparison of frequency of suckling bouts in male versus female calves observed in July and August (1975 and 1977 data).

   Mann-Whitney U test: $U = 34$, $n_1 = 17$, $n_2 = 10$, $p < .02$.

5. Comparison of number of two-week intervals, starting from birth, within which the duration of suckling bouts involving male calves exceeded those involving female calves.

   Sign test: $x = 1$, $N = 8$, $p < .035$ (one-tailed).

6. Comparison, across months, of the frequency with which suckling attempts by male versus female calves were rejected by their mothers. Sample restricted to barren mothers.

   Wilcoxon matched-pairs test: $T = 2$, $N = 11$, $p < .01$.

7. Comparison across years of the proportion of mothers that had previously reared male versus female calves that failed to calve in the subsequent year.

   Wilcoxon matched-pairs test: $T = 13$, $N = 11$, $p < .05$ (one-tailed).

8. Comparison of the relative date of conception of individual hinds (measured as a deviation from the median for the population in each year) following winters when they had supported a male calf versus years following winters when they had supported a female calf.

Wilcoxon matched-pairs test: $z = 2.66$, $N = 32$, $p < .01$.

9. Correlation and regression between measures of lifetime reproductive success in hinds and the frequency with which (a) members of their own matrilineal group and (b) members of other matrilineal groups were seen in their core area over a four year period.

(a) Pearson product-moment correlation coefficient: $r = -.343$, $t = 2.249$, d.f. $= 38$, $p < .05$; $y = 8.130 - 0.0033x$.

(b) Pearson product-moment correlation coefficient: $r = .091$, $t = 0.563$, d.f. $= 38$, n.s.

9 The Structure of Social Groups in Hinds and Stags

The outstanding feature of the hind group is its cohesion which, doubtless, is derived from the stability of the family.... The stag company is a number of egocentric males and is a very loose organization.

F. Darling (1937)

9.1   INTRODUCTION

Most previous studies of the social behavior of red deer and elk
have concentrated on the rut, when the behavior of the two sexes
shows obvious contrasts. During the rest of the year, the social
behavior of stags and hinds is superficially similar—both form
unstable parties that vary in size and membership from hour to
hour. However, it has seldom been possible to compare the
structure of parties or to determine whether the regularity of
association differs between the sexes. In fact, opinion is divided
on whether any form of long-term association between individuals
occurs at all. Some workers have argued that stable clans exist
among hinds, whose members aggregate in kin-based parties that
are unstable from day to day (Darling 1937; Graf 1956; Altmann
1952, 1956; Mutch, Lockie, and Cooper 1976; McCullough 1969;
Gossow 1974; Franklin, Mossman, and Dole 1975; Franklin and
Lieb 1979), while others have stressed the irregular nature of all
social ties except those between mothers and their dependent offspring (Harper 1964; Schloeth 1961; Knight 1970; Craighead et
al. 1973; Craighead and Shoesmith 1966). At least three studies
have produced quantitative data demonstrating that the frequency of association between mature hinds is not high (Schloeth
1961; Craighead and Shoesmith 1966; Knight 1970), but such
evidence does not preclude the presence of consistent groups: for
example, if a population were divided into matrilineal clans whose
members formed small parties of unstable membership but
showed no strong preference for associating with particular individuals within the clan, average coefficients of association (Cole
1949) could still be low despite the presence of social divisions
within the population. In addition, the majority of studies have
relied on data collected from small numbers of marked animals
within much larger populations. Since one animal is usually immobilized at a time, marked animals are likely to belong to different groups, and measures of association between them may
consequently underestimate the frequency of association between
individuals in the population as a whole.

In this chapter we examine the development of social relationships and the structure and regularity of association between individuals of both sexes. Section 9.2 describes the sizes of
stag and hind parties and the environmental factors affecting
party size. Section 9.3 describes variation in association between

mothers and their calves, while section 9.4 outlines the development of association patterns between mothers and their daughters, and section 9.5 examines whether long-term groups exist among hinds. Section 9.6 describes association patterns between stags.

9.2 PARTY SIZE AND COMPOSITION
Segregation between the Sexes

As in most red deer populations, hinds and stags were segregated for most of the year. Most hinds spent 80–90% of their time in parties that did not include any stags over three years old. Among stags, the frequency with which individuals were seen in association with hinds declined from about 90% in yearlings to 60% in two-year-olds and 20% in three-year-olds, stabilizing at between 10 and 20% among mature stags (see fig. 9.1).

Comparing our figures with those of other studies of red deer shows that the degree of segregation between the sexes varies between populations. Values range from a study of a Crimean red deer population in which only 18–29% of stags and 50–56% of hinds seen were in segregated parties (Yanushko 1957) to Jackes's study of a Scottish population where over 90% of animals seen were in segregated parties (Jackes 1973; Mitchell, Staines, and Welch 1977). Segregation occurs in both forest- and moorland-living populations (Mitchell, Staines, and Welch 1977) but is reduced where animals of both sexes are attracted to artificial feeding sites (Altmann 1956; Gossow 1974).

Variation in Party Size

The size of parties seen in our study area ranged from two animals to more than eighty (see fig. 9.2). Since, in a finite population, small parties are inevitably commoner than large ones, the party size in which the average animal finds itself most frequently provides a more meaningful estimate of grouping tendencies than mean party size (Jarman 1982): in the comparisons described below, we describe party size of the median individual in each category (i.e., party size for all individuals seen were ranked, and we took the median figure). For convenience, we refer to this as "median party size."

In both sexes, median party size was smaller in winter than in summer (see table 9.1), though stags were found in slightly smaller

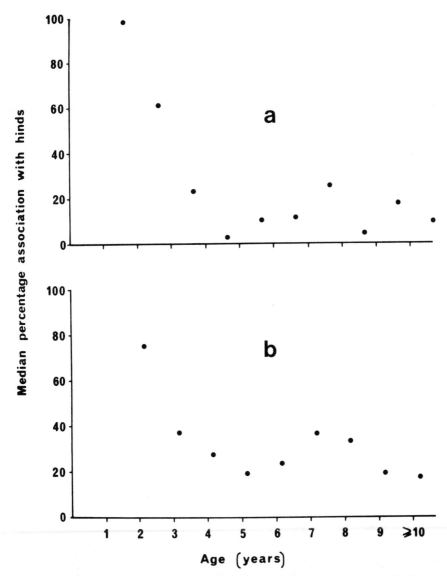

FIG. 9.1. Age changes in the frequency with which stags associated with hinds. The figure shows the median percentage of observations of different individuals in (a) winter (January–March) and b) summer (May–July) when they were in the same party as hinds. Data from 1975.

parties than hinds during the winter months [1]. Seasonal changes in party size were evidently not a consequence of differences in the extent to which particular plant communities were used: party sizes in summer were larger on all plant communities (fig. 9.3).

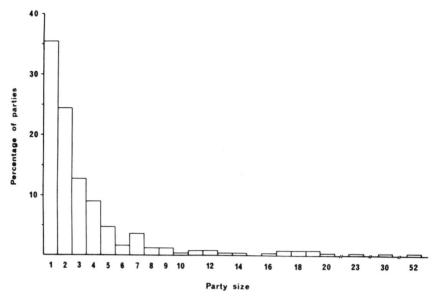

FIG. 9.2. Distribution of party size during census in July 1975.

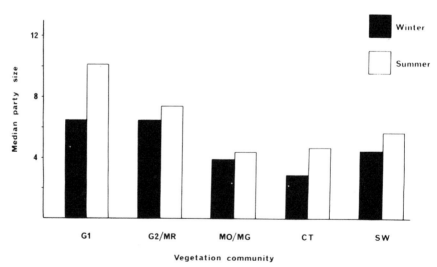

FIG. 9.3. Median party size for hinds and stags four years old or older seen standing on different plant communities in summer (May–July) and winter (January–March).

In both sexes and both seasons, median party size was largest on the short greens and lowest on heather moorland. These differences may have reflected the smaller area of the greens, though they may also have arisen because food was comparatively evenly distributed and indirect competition was reduced (see Jarman 1974; Clutton-Brock and Harvey 1978). Increased party sizes during the summer may also have been a response to reduced competition. Similar seasonal changes in grouping patterns have been demonstrated in other deer populations (Hirth 1977; Dzieciolowski 1979) as well as among other mammals living in unstable groups (Clutton-Brock 1977).

TABLE 9.1   Median Party Size for Hinds and Stags in Summer and Winter

	Summer[a]	Winter[b]
Hinds ≥ 4 years	6.88	5.24
Stags ≥ 4 years	6.92	4.0

Note: See text for explanation of party size.

[a] May–July.

[b] January–March.

Median party size increased during the course of the day, and the largest parties were usually seen in the late afternoon and early evening. These changes coincided with diurnal variation in feeding behavior and apparently occurred because the animals fed on the short greens during the afternoon and evening (see chap. 11). Party size also varied with weather conditions: groups were largest in calm weather (wind speed < force 4, Beaufort scale) and smallest in high winds, when the animals sought out small pockets of sheltered ground [2].

Party size on Rhum was considerably smaller than in many mainland populations, where parties of more than one hundred animals are not uncommon (Lowe 1966; Staines 1974). As Lowe has suggested, it seems likely that the smaller party size on Rhum is related to the small scale of the topography and to the fact that patches of food and shelter are small and widely dispersed (see Wiens 1976; Clutton-Brock and Harvey 1978). Most forest-dwelling red deer populations are found in even smaller parties than on Rhum (Picton 1960; Knight 1970; Mitchell, Staines, and Welch 1977), a trend that follows the general tendency for deer species living in open country to be found in larger groups than those living in closed habitats (see chap. 13).

9.3 ASSOCIATION BETWEEN MOTHERS AND CALVES
Changes Related to Calf Age

After calves were born (see chap. 4) the frequency with which they were seen in the same party as their mothers increased from about 40% in the first week after birth (when the calf was left alone much of the day) to nearly 90% during the calf's first autumn (see figure 9.4) and subsequently began to decline. Within the party, calves were usually seen within 10 m of their mothers, though this was not invariably the case and it was unsafe to assume that the hind closest to a calf was necessarily its mother.

When the mother calved again, there was usually a marked decline in the frequency of association with her previous calf. Although the frequency with which yearlings associated with their mothers increased again later in the summer, they did not regain their former levels. A similar reduction in the frequency with which yearlings associated with their mothers occurred again at the end of their second year, and after this both daughters and sons were seen outside their mothers' parties with increased frequency (F. E. Guinness, unpublished data).

Variation with Mother's Subsequent Reproductive Status

The mother's subsequent reproductive status had a pronounced effect on the frequency with which she associated with her calf. Calves of hinds that failed to conceive during the rut following their birth (barren mothers) continued to associate closely with their mothers throughout their first two years of life, whereas calves whose mothers became pregnant showed a gradual decline in association frequency from the age of about six months and a pronounced drop following the birth of the next calf. If a mother calved again but the calf died shortly after birth, this decline was not so pronounced, especially among male calves (Guinness, Hall, and Cockerill 1979). Barren mothers also tended to be closer to their calves within the party. Before the birth of the next calf, this difference was evident only among mothers of male calves [3], but after the next calf's birth it occurred both among mothers of males and among mothers of females [4]. The reason for these differences was probably that barren mothers continued to suckle

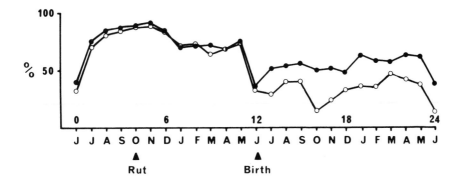

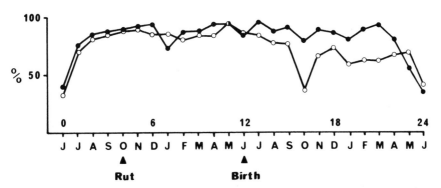

FIG. 9.4. Frequency of association between male *(open circles)* and female *(solid circles)* calves and their mothers during the first two years of life. The upper plot shows association patterns for calves whose mothers produced another calf in the year following their birth (pregnant mothers), the lower plot, the figures for calves whose mothers failed to produce a calf the following year (barren mothers) (from Guinness, Hall, and Cockerill 1979).

their existing calves (see chap. 4), thus increasing the advantages to the calf of remaining close to them.

Sex Differences

From the first month of life, male calves associated less with their mothers than did female calves (Guinness, Hall, and Cockerill 1979) (see fig. 9.4). This difference was pronounced after the end of their first year, among both sons whose mothers gave birth again the following year and sons of mothers that did not. In addition, the mean distance within parties between mothers and their sons was greater than between mothers and their daughters [5]. Similar differences have been found in other ungulate species (Pratt and Anderson 1979).

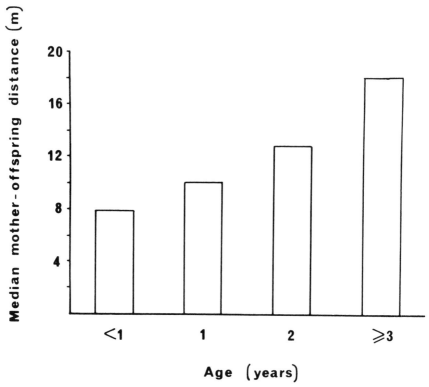

Fig. 9.5. Median distance between mothers and their daughters of different ages when both were members of the same party. $N = 51$ dyads.

The marked decline in association between male yearlings and their mothers during their second rut was a consequence of the behavior of rutting stags. However, this was evidently not the cause of the final decline in association between mothers and their sons, for after the rut sons associated with their mothers nearly as frequently as before it. Association did not fall to low levels until after the second calving season following their birth (fig. 9.4).

9.4   Association between Hinds
Daughters and Mothers

As a daughter's age increased, the frequency with which she associated with her mother gradually declined [6], and there was an increase in the average distance between mothers and daughters at times when they were in the same party [7] (see fig. 9.5). The immediate cause of the decline in association was that, after the

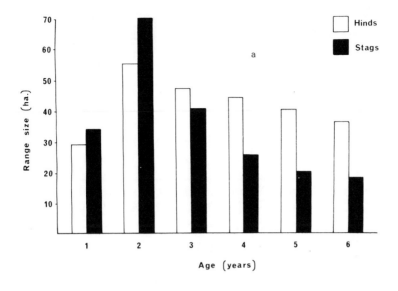

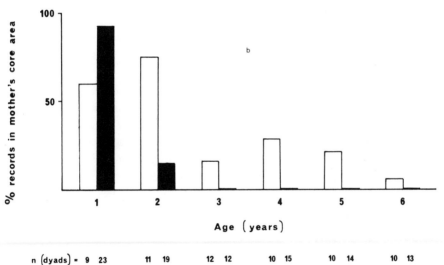

n (dyads) =   9   23   11   19   12   12   10   15   10   14   10   13

Fig. 9.6. Changes in the ranging behavior of daughters and sons with increasing age: (a) total range size in subsequent years of animals born in 1972; (b) proportion of quadrats within their core areas that were also within their mothers' core areas.

end of their second year, hinds began to range more widely (fig. 9.6). During their fourth and fifth years, daughters gradually adopted core areas that typically overlapped the core areas of their mothers, and their range size declined. On average, hinds older than three years shared about 30% of their core areas with their mothers (fig. 9.6), though in very few cases did they move out of their mothers' ranges. Since 1971 only three hinds have left the study area permanently.

Individual Differences

Association between mothers and their daughters varied widely, and these differences were stable across years [8]. One of the factors affecting association frequency was the mother's age: daughters of older mothers associated with them less frequently than daughters of younger ones [9]. A likely explanation was that older mothers normally had several daughters already resident in their home range, and that younger daughters consequently had alternative companions to their mothers. This suggestion was supported by the fact that the frequency of association between daughters and their mothers was *negatively* related to the number of female offspring the mothers had reared previously, when the effects of the daughters' and mothers' ages were taken into account [10].

The frequency of association between daughters and their mothers was related to the dominance rank of the mothers: both in the Intermediate Area and at Samhnan Insir, daughters of dominant mothers associated with them more frequently than did those of subordinate mothers, though this was not the case at Kilmory [11]. This effect could not have been a product of an association between maternal dominance and age, since association frequency was negatively, not positively, related to the mothers' age. It may have occurred because the offspring of dominant hinds gained more from proximity to their mothers than did those of subordinate ones.

Association with Other Relatives

Hinds also associated frequently with relatives other than their mothers. In particular, sisters were commonly seen in the same group, as were aunts and nieces (see fig. 9.7). Unlike association frequency between mothers and their offspring, association between sisters was not significantly related to the age of the hinds, the age (or existence) of their mothers, or dominance rank. Studies of social behavior in sheep show that close relationships between animals of similar ages that are formed early in life tend to persist (Hunter and Davies 1963), and it seems likely that dif-

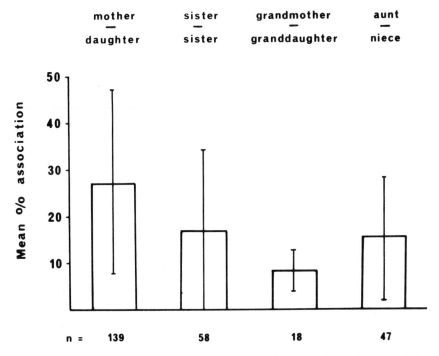

FIG. 9.7. Mean frequency of association between hinds three years old or older and different relatives (data for 1975 and 1976). The sample of dyads is shown below each histogram; extending lines show standard deviations.

ferences in the strength of social bonds between sisters may be a consequence of events during adolescence.

9.5   MATRILINEAL GROUPS AMONG HINDS

As we have already described, hinds tend to associate with their mothers and sisters more frequently than with animals that do not belong to the same matriline. However, this does not provide a formal answer to whether long-term groups exist. Cluster analysis provides one technique for identifying groups in such circumstances (Jardine and Sibson 1968; Slater 1974; Morgan et al. 1974). In this analysis the degree to which all pairs of variables in a set (in this case, the number of individual deer in the study area) are associated is determined. Pairs that are least disassociated are then linked, and the analysis proceeds to identify which individuals, or pairs, are least disassociated with each cluster, linking these in their turn and building a dendrogram in which the level at which two individuals are tied corresponds to the extent to which they are associated with each other. The presence of groups is revealed by clusters of variables that are linked to each other at low level.

To define clusters, it is necessary to select a level of linkage as a cutoff point. If this is too high, a large number of individuals are classified on their own; if it is too low, all animals are classified as belonging to a small number of clusters, and important distinctions are obscured. In our analysis, the level of association selected to define clusters was, for each year, that which gave the maximum number of clusters of two or more individuals. The actual levels used in each year are shown in Appendix 18 and varied from 22% to 25.5% association between years.

With these techniques, we examined the frequency of association between all hinds one year old or older using the study area in each year between 1973 and 1976. This showed that consistent clusters were evident. The majority of these were small: over 80% included fewer than four animals, though clusters of up to seven animals occurred. In almost all cases these consisted of matrilineal relatives; for example, the mean coefficient of relatedness (Bertram 1976) estimated for the thirty-three clusters defined in 1976 was .35. Since our methods assumed that older hinds of unknown affinity were unrelated to the individuals they were associated with and ignored patrilineal relatedness, the real figure must have been even higher. Only a quarter of the clusters contained members known to belong to different matrilines, and in the majority of cases where members belonged to two or more matrilines it was possible that the matriarchs were related. Clusters were stable from year to year

Thus our analysis of ranging behavior and of association patterns revealed the presence of stable matrilineal groups whose members shared a common range. The core areas of these groups (i.e., the smallest areas that accounted for 65% of sightings of all members of the matriline) overlapped extensively but rarely coincided perfectly with each other (see fig. 9.8).

Is the presence of consistent matrilineal groups in some populations of red deer and elk but not in others a genuine difference, or is it a consequence of differences in sample size and analytical techniques? There seems a good chance that it may be a result of differences in methodology. The most extensive data available on individual association are from Knight's study of a Montana elk

FIG. 9.8. Core areas of matrilineal hind groups using the study area. The plot shows up the four parts of the population: Upper Glen: groups 19, 4, 24; Kilmory: groups 9, 16, 23, 18, 25, 10, 15, 11, 8, 20, 17; Intermediate Area: groups 6, 3, 5, 1; Samhnan Insir: groups 7, 12, 13, 14, 2, 21, 22.

population of about thirteen hundred animals, 15% of which were individually recognizable (Knight 1970). Knight measured the frequency of association between marked individuals using Cole's (1949) coefficient of association and, finding that the frequency of high association coefficients between individuals was low, argued that this demonstrated the absence of any long-term social groups. However, when we selected 15% of our study population of hinds at random and examined association coefficients between individuals using the same analytical techniques, association coefficients were no higher than in the Montana herd, despite the presence of long-term social groupings.

9.6   ASSOCIATION AMONG STAGS
Dispersal

Association between mothers and their yearling sons declined rapidly during the calving season, usually falling to less than 20%. During the following winter, the frequency with which male yearlings were threatened by their relatives rose (chap. 10), and this was followed by an increase in the size of their ranges in their third year (fig. 9.6), which subsequently declined as they spent progressively more time in one of the areas regularly frequented by stags. In almost all cases, this was outside the mother's core area, and both core area overlap and association frequency between mothers and their sons over two years old was minimal. This process was probably influenced by hormonal changes in the males, for two males castrated as calves associated closely with their mothers throughout their lives, adopting core areas that closely coincided with their mothers'.

About one-third of male calves born in the study area before 1973 left the study population and adopted home ranges elsewhere on the island before their fifth winter. Young stags born and reared elsewhere also immigrated into our study population, and about 25% of all stags between the ages of six and eight resident in the study area in 1979 were immigrants.

Rhum is too small to provide an indication of how far dispersing stags are likely to move from their natal area. However, of male calves tagged by the Red Deer Commission in mainland study sites, 70% were subsequently recovered more than 2 km from their birthplaces, and the greatest distance any individual moved was 22 km (Red Deer Commission Report for 1978). This contrasted with hind calves, most of which were recovered within 2 km of their birthplaces.

On Rhum, there was no obvious way of predicting which male yearlings would disperse from the study area and which would remain within it: the two categories did not differ in their birth dates, ages, home range areas, subsequent reproductive status of their mothers, or their knob or spike size at sixteen months. However, birth weights of dispersers were heavier than those of non-dispersers [12].

Stags sometimes adopted ranges in the same areas as their older male sibs, but our sample size was too small to allow us to investigate whether they did so more often than would have been expected on a random basis. We suspect not—observation of social relations between male sibs provided no indication that they recognized each other or that their social relationships differed in any way from those between other animals (Appleby 1981).

Of stags wintering in the study area, some rutted inside it and others rutted elsewhere (see chap. 6). There was no obvious difference in the antecedents, growth, dominance rank, or harem size of those that dispersed to rut versus those that stayed in the study area. Of the latter, about 15% rutted in the same part of the study area where they were born.

Stag Groups

The majority of stags that remained within the study area adopted core areas in one of three places—the Laundry Greens at Kilmory, the Fank Greens in the Upper Glen, and the flats of the Intermediate Area—and in each of these localities several parties of stags were usually present. However, stags commonly used more than one of these areas, sometimes spending a large proportion of their time in one, then shifting to another. As a result, stag "groups" were looser and less well defined than hind groups. Cluster analysis of stag association data showed fewer close linkages between stags. Similar differences in the regularity of association between males and females have been noted in other populations of red deer (Gossow 1974) as well as in other polygynous ungulates (Underwood 1981).

A stag's closest associates were usually animals of similar age [13], partly because younger animals spent more time on the periphery of groups while older ones spent more time in central positions (see also Mitchell, Staines, and Welch 1977). Aging stags were often seen alone. The membership of stag parties changed form hour to hour as individuals joined or left. Relationships between parties were relaxed, and we did not observe either aggressive interactions between parties or cases where one party displaced another.

9.7   FUNCTIONAL CONSIDERATIONS

As in many other mammals (see Greenwood 1980), there is a fundamental difference in the structure of male and female groups in red deer (see table 9.2). Hinds typically adopt ranges and core areas overlapping those of their mothers and associate with their matrilineal relatives throughout their lives. In contrast, stags disperse from their mothers' home ranges, forming loosely structured herds that consist largely of individuals not closely related to each other. The difference in dispersal between stags and hinds is associated with sex differences in mother/calf relationships evident from the first months of life.

These results raise four fundamental questions. Why do red deer aggregate at all? Why are the sexes segregated for much of the year? Why do hinds associate with relatives? And why do stags disperse from their mothers' ranges?

TABLE 9.2   Sex Differences in Association

Mother/Calf Association: Hind calves associate more frequently and more closely with their mothers than stag calves do.

Dispersal: Sons disperse from mothers' ranges between the ages of two and three. Daughters adopt core areas overlapping those of their mothers.

Association between adults: Hinds associate more regularly with the same individuals than stags do.

Group structure: Hinds associate most frequently with matrilineal relatives, while stags usually associate with unrelated animals of similar ages.

Why Do Red Deer Aggregate?

Since the frequency with which individual deer are interrupted by other individuals while feeding increases with the size of group they are in (p. 207), there are presumably important advantages to group living (Alexander 1974). What are these likely to be?

Investigations of social behavior in a wide range of vertebrate species suggest that individuals benefit from sociality in three main ways: being in a group affects their ability (1) to detect, avoid, or defend themselves against predators (Patterson 1965; Lack 1968; Vine 1971; Treisman 1975a,b), including biting flies (Duncan and Vigne 1979); (2) to find, handle, or exploit food resources or to defend them against conspecies (Krebs, MacRoberts, and Cullen 1972; Wilson 1975; Kruuk 1972; Cody 1971; Thompson, Vertinsky, and Krebs 1974; Wrangham 1980); and

(3) to gain, retain, or monopolize access to mates (Brown 1975; Wrangham 1975). However, there is little consensus concerning the relative importance of these consequences of grouping.

In the case of red deer, it is difficult to believe that social feeding increases the individual's ability to find food, for animals consistently use the same feeding sites. Grouping may allow individuals to displace others from food sources (see Miller and Denniston 1979; Wrangham 1980), and the results described in chapter 8 suggest that residents may have excluded members of other matrilineal groups from access to key resources, though we seldom observed obvious displacements between parties.

Indirect evidence supports the suggestion that grouping in red deer is an adaptation against predation. Deer in our study area were more nervous and alerted more frequently when feeding in small groups than when feeding in large ones [14]; and hinds, which mostly had calves with them, alerted more frequently than stags when in small groups [15]. A similar relationship between alerting frequency and group size has been found in wild sheep (Berger 1978). Second, across the Cervidae, group size is largest in species living in open country that rely on escape or defense to avoid predation and smallest in forest-dwelling species that rely on crypsis (see chap. 13). Finally, though it seems unlikely that increased protection from biting flies is the only reason deer aggregate, individuals attacked by warble flies or botflies often run into groups of other animals (S. Temple, n.d.), and group size tends to increase on days when fly activity is high.

Why Do the Sexes Segregate?

Why do stags and hinds segregate? The answer is not obvious, for there are many seasonally breeding ungulates where the sexes associate throughout the year (Jarman 1974; Estes 1974). While several possible explanations exist (e.g., Geist and Bromley 1978), the most likely one for the Rhum red deer was that nutritional requirements differed between stags and hinds and this encouraged them to use different areas.

Several lines of evidence support this conclusion. On the Scottish mainland, segregation is associated with the use of different habitat types between the sexes (Watson and Staines 1978); and on Rhum the degree of segregation varied between different plant communities [16] (see fig. 9.9). The extent to which stags

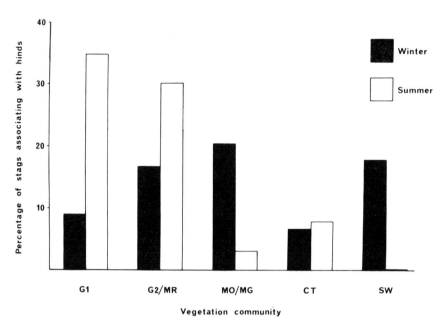

FIG. 9.9. Degree of segregation between stags and hinds when feeding on different plant communities in summer (May–July) and winter (January–March). The figure shows the percentage of stags four years old or older seen on each community that were associating with hinds.

associated with hinds declined between the ages of three and five (fig. 9.1), and this coincided with changes in their feeding behavior (see chap. 11). Finally, segregation apparently increased where food was short: it was greater in winter than in summer and was more pronounced at Samhnan Insir than at Kilmory. We return to this topic in chapter 11.

Why Do Hinds Associate with Relatives?

Why do hinds associate with their mothers and other matrilineal relatives? The existence of social bonds between mothers and offspring far outlives the period of lactation. An important benefit of close association with relatives demonstrated by other studies is the reduction of direct interference while feeding: several studies have shown both that individuals are more likely to tolerate relatives feeding close to them (e.g., Clutton-Brock and Harvey 1976; Kurland 1977) and that younger animals are protected from threats of older ones by the proximity of their parents (Scott 1978). As we describe in chapter 10, similar effects occur in red deer. Variation in the frequency with which offspring associated

TABLE 9.3   Comparison of the Social Relationships and Activity Budget of an Orphaned Two-Year-Old (FEE5) with Those of Two Other Hinds of the Same Age (CLE5, YHI5) Whose Mothers Were Still Alive

	FEE5	CLE5	YHI5
Threats received (per hour)	0.57	0.22	0.11
Percentage of time spent on peripheri of party	74.4	34.7	16.7
Frequency of leaving party (per hour)	0.947	0.373	0.105
Percentage of time spent moving	6.9	4.6	2.6
Percentage of time spent standing	66.2	59.5	53.1

with their mothers supports this explanation of association between relatives: association declined as the daughter gained rank with age; and daughters of dominant mothers (which presumably afforded more effective shelter) spent more time with them than daughters of subordinate mothers. In addition, some (though not all) the orphans in the population were regularly threatened by other hinds: for example, compared with two-year-old hinds whose mothers were living, one two-year-old orphan we observed received more threats, spent more time on the periphery of parties, left parties more frequently, and spent more time standing and moving (see table 9.3).

Being threatened is probably disadvantageous for at least three reasons: the subordinate animal is displaced from a food source; it is forced to spend time and energy moving to another site; and the threat itself may be damaging. Threats by hinds may well be more dangerous than they initially appear, particularly where calves are involved. In one case, a female calf was disturbed by a visitor and joined a mature hind. The latter kicked her across the back, forcing her to the ground and rupturing her right shoulder ligament. The calf's body condition quickly deteriorated after this, and she died five days later. Adult hinds not infrequently show the marks of kicks or bites on their flanks and ears, and a significant proportion of hinds shot in the annual cull had broken ribs.

However, this does not answer the more fundamental question why hinds should tolerate their relatives, allowing them to share access to their core area and range when their presence evidently has heavy costs (see chaps. 5 and 8). The most likely explanation is that individual hinds that leave the range of their matrilineal group are unable to join parties of unrelated animals and that their chances of survival in any environment where predators are common is consequently low. In addition, they may be excluded from key resources and may have little chance of finding access to adequate food supplies to allow them to reproduce successfully (see Clark 1978). Hinds that tolerate their relatives may not only increase their inclusive fitness (see Hamilton 1964a,b) but also benefit directly if the presence of other group members helps them to exclude members of other matrilineal groups from particular resources within their range (see Wrangham 1982).

Why Do Stags Disperse from Their Mothers' Home Ranges?

Evidence from studies of other species that dispersal is associated with a decline in the efficiency of food exploitation as well as in an individual's ability to escape predation (Lack 1954; Brown 1975; Pollock 1977) indicates that it must yield important compensating benefits. It seems likely that these benefits are often connected with the avoidance of incestuous breeding, since breeding with close relatives in normal, outbred populations depresses reproductive success (Bulmer 1973; Willis and Wilson 1974; Greenwood, Harvey, and Perrins 1978; Ralls, Brugger, and Ballou 1979).

There are probably at least two advantages to males in dispersing. First, dispersal presumably reduces the chances of incestuous breeding, which is known to depress reproductive success in outbred populations (Bulmer 1973; Willis and Wilson 1974; Greenwood, Harvey, and Perrins 1978; Ralls Brugger and Ballou 1979). Studies of a wide variety of mammals living under natural or seminatural conditions have shown that most members of one sex (and in a few cases members of both sexes) disperse before breeding (Bischof 1975; Packer 1979a,b; Greenwood 1978) and that incestuous matings are rare in wild populations. It has recently been suggested that this may not be the case among fallow deer, on the grounds that daughters frequently accompany their mothers to the same rutting areas where they were themselves conceived and that breeding bucks return to the same areas in successive years (Smith 1978, 1979). However, the probability of father/daughter mating depends on the length of the breeding life spans of bucks and the duration of their rut, and the fact that a proportion of females return to the area where they were con-

ceived could have a negligible effect on the probability of inbreeding (see chap. 4).

Second, male yearlings and two-year-olds are less frequently
tolerated by matrilineal relatives than females of the same age
while the heavily used greens, which form the focus of the ranging behavior of most hind groups, may be less attractive to stags,
which may be able to minimize their energy expenditure in times
of food shortage by feeding on plant communities where food is
more abundant, though of lower nutritional quality (see chap. 11).

The costs of dispersing are also likely to be lower among stags
than hinds. Not only is variation in reproductive success among
members of a stag population less closely related to differences in
home range quality than among hinds (see chap. 7), with the
effect that a stag has less to lose by quitting his natal area, but
there is no evidence that members of stag groups attempt to dis-
courage new recruits from joining them.

9.8 SUMMARY

1. Except during the rut, stags and hinds seldom associated with
   each other. Individuals of both sexes were most commonly
   found in parties of four to seven animals (excluding calves).
2. The frequency with which calves associated with their mothers
   peaked during their first autumn and subsequently declined.
   Male calves associated with their mothers less than did female
   calves, and calves whose mothers became pregnant in the rut
   following their birth associated with them less than did calves
   whose mothers failed to conceive again.
3. Daughters usually adopted ranges that overlapped those of
   their mothers. Mothers, daughters, and sisters commonly associated with each other, and it was possible to identify and
   define long-lasting matrilineal groups within the hind population.
4. Stags dispersed from their mothers' ranges between the ages of
   two and three and mostly joined stag groups in the vicinity.
   Members of stag groups shared overlapping ranges, but individuals associated with each other less regularly than did hinds.
5. The functional significance of sex differences in grouping is
   discussed. It seems likely that the principal function of grouping in red deer is to detect or avoid predators. Segregation
   between the sexes may be a consequence of their different food

requirements, which may also help explain why stags, but not hinds, disperse from their natal area. Related hinds probably associate with each other because this reduces the frequency of interference while feeding.

Statistical Tests

1. Comparison of median group size between stags and hinds in winter.
    G test: $G = 15.05$, d.f. $= 1, p < .001$. $N = 67, 86$ individuals.
2. Comparison of group sizes between days when wind speed was $<$ force 4, Beaufort scale with days when it was $\geq$ force 4.
    G test: $G = 9.56$, d.f. $= 1, p < .01$. $N = 15$ days.
3. Comparison of mean mother-calf distance between mothers of male calves that produced a calf in the following year and those that did not. Data for the month of May (i.e., before the sib's birth).
    Mann-Whitney U test: $U = 1, n_1 = 6, n_2 = 6, p < .002$.
4. Comparison of mean mother-calf distance between mothers that produced a calf in the subsequent year and those that did not. Data for June–August (i.e., after the sib's birth).
    Mothers of female calves, Mann-Whitney U test: $U = 2, n_1 = 3, n_2 = 11, p < .05$.
    Mothers of male calves, Mann-Whitney U test: $U = 0, n_1 = 6, n_2 = 6, p < .001$.
5. Comparison of mean distance between mothers and male versus female calves during their first summer.
    G test: $G = 4.46$, d.f. $= 1, p < .05$. $N = 41$ dyads.
6. Regression of percentage association between mothers and daughters three years old or older on daughter's age (data for 1975 and 1976).
    $y = 0.3757 - 0.0326x$.
    Analysis of variance: $F_{1,137} = 6.808, p < .001$.
7. Comparison of mean distance between daughters of different ages (calves, yearlings, two-year-olds, and three-year-olds or older) and their mothers when both were in the same group.
    Friedman two-way analysis of variance: $\chi_r^2 = 29.925$, d.f. $= 3, p < .01$. $N = 51$ dyads.
8. Concordance of percentage association between daughters (1971 cohort) and their mothers across years, 1973 to 1976.

Kendall coefficient of concordance: $s = 220.0, N = 7, k = 4,$
$p < .05.$

9. Multiple regression of percentage association between daughters three years or older and their mothers on daughter's age (x_1) and mother's age (x_2). Data from 1975 and 1976.
$y = 0.506 - (0.0218)x_1 - (0.0166)x_2.$
Analysis of variance: $F_{2.136} = 13.496, p < .001.$

10. One-way analysis of variance in deviations in association frequency from the values predicted by daughter's age and mother's age. The analysis compared deviations between mothers that had previously reared different numbers of female offspring from zero to four or older.
Analysis of variance: $F_{4.130} = 2.979, p < .05.$

11. Correlation between mother's dominance rank and the frequency of association with daughters three years old or older. The association measure used was the deviation from the values predicted from the mother's age and the daughter's age using the multiple regression equation shown in test 9. Results were calculated separately for hinds using each part of the study area in 1976. In the Upper Glen, sample size was too small, and at Kilmory there was no clear dominance hierarchy.
Intermediate Area, Spearman rank correlation coefficient:
$r_s = .785, t = 4.915,$ d.f. $= 15, p < .001.$
Samhnan Insir, Spearman rank correlation coefficient:
$r_s = .347, t = 1.526,$ d.f. $= 17, .1 > p > .05.$

12. Comparison of number of dispersers versus nondispersers born above and below the median birth weight for their year.
G test: $G = 4.63,$ d.f. $= 1, p < .05. N = 44.$

13. Correlation between the ages of 99 stags allocated to ten year classes and the mean ages of the closest associates of stags in each age category.
Spearman rank correlation coefficient: $r_s = .889, t = 5.491,$
d.f. $= 8, p < .001.$

14. Comparison of the rate of alerting in hinds and stags when feeding in groups of six or fewer animals compared with groups of more than six.
Hinds, Wilcoxon matched-pairs test: $T = 1, N = 9, p < .02.$
Stags, Wilcoxon matched-pairs test: $T = 22, N = 19, p < .02.$

15. Comparison of rate of alerting in hinds versus stags when in groups of six or fewer animals.

Mann-Whitney U test: $U = 27$, $n_1 = 9$, $n_2 = 18$, $p < .02$.

16. Comparison of frequency of association with hinds by stags of five years or more on different plant communities in summer (May, June, July) and winter (January, February, March).

Summer, G test: $G = 70.26$, d.f. $= 4$, $p < .001$. $N = 578$ groups.

Winter, G test: $G = 12.36$, d.f. $= 4$, $p < .02$. $N = 400$ groups.

10 Social Interactions among Hinds and Stags

Where a species is of social habit, I would emphasize the necessity of taking sociality fully into account in observing and interpreting behaviour. The life-history of the red deer would be an empty and meaningless thing divorced from the sociality which is the very foundation of their existence.

F. F. Darling (1937)

10.1 INTRODUCTION

Our analysis of grouping patterns in the Rhum deer population showed a fundamental difference in the structure of hind and stag groups: hinds regularly associated with the same individuals, who were usually matrilineal relatives, whereas the frequency of association between individual stags was lower, and members of the same party were seldom closely related. Studies of other mammals have revealed pronounced differences in social relationships between matrilineal relatives compared with those between unrelated animals (see Kurland 1977). Do similar contrasts in social behavior exist between the members of stag and hind groups? Few previous studies have attempted to quantify interaction patterns between deer, for the rates of interaction are usually low, and it is impracticable to investigate many of the questions that can be answered in more frequently interacting animals, such as primates (see Chalmers 1979). The exceptions have mostly described the social behavior of animals at feeding sites (e.g., Lincoln, Youngson, and Short 1970; Gossow 1971; Bützler 1974; Wiersema 1974) and are open to the objection that the high rates of interaction that occur when food supplies are artificially clumped create regular hierarchies even when these do not exist under more natural conditions (Gartlan 1968; Rowell 1974).

In this chapter we describe and compare social interactions among hinds and stags. Section 10.2 describes the frequency of agonistic interactions in each sex, and section 10.3 examines the distribution of threats between individuals. In section 10.4 we briefly compare nonagonistic interactions, and the final section discusses the adaptive significance of variation in interaction patterns between the sexes.

10.2 THREAT TYPES

Among hinds, the commonest forms of agonistic interaction were nose and ear threats—in the former, one hind poked her head toward another, usually at the neck, shoulder, or rump, and often exhaled loudly, while in the latter a hind laid back her ears while moving toward another animal. Displacements, where one hind walked steadily toward another, were also common. More intense threats included kicking, usually with a single foreleg, but sometimes with both (see fig. 10.1), and boxing, when two hinds reared

Fig. 10.1. A mature hind giving a scissors kick while chasing her yearling son.

on their hind legs and slapped at each other's heads or shoulders with their hooves (see Geist 1982). Bites were also common and were usually directed at the head, neck, or shoulders: they often occurred as the finale of a gradually escalating series of threats and were frequently followed by a chase or an attempt to butt the opponent. Of a sample of 414 interactions between members of different matrilines, 33.5% consisted of nose or ear threats, 29.5% of displacements, and 37.0% of kicks or bites (excluding threats to calves). For quantitative analysis, we divided threats and displacements between hinds into two categories: nose threats, ear threats, and displacements; and bites, kicks, and butts. We refer to the first group as "mild" threats and the second as "severe" threats on the grounds that they often involved physical contact and were usually associated with rapid avoidance by the recipient.

Extensive descriptions of social interactions between stags can be found in several previous studies (Bützler 1974; Burckhardt 1958; Geist 1982). In our population, displacements (where one stag walked directly toward another individual) were commoner than all other categories of "threats." In displacements the dominant animal sometimes used a stiff, pronounced gait, often accompanied by a concentrated sideways glance. In more intense threats, stags raised their heads and pointed their chins at their opponents, sometimes curling back the upper lip and hissing or grinding their teeth at the same time (see fig. 10.2). When a dominant stag approached a subordinate and the latter failed to move away, the dominant usually responded with a nod of his head, directing his antlers toward the subordinate, often flattening his ears or rolling his eyes at the same time. Stags sometimes kicked at other individuals, either with one or with both forelegs (see fig. 10.3), and occasionally chased them, attempting to horn them in the flank, but such events were rare. Unlike hinds, stags rarely tried to bite each other but, when in velvet, would rear on their hind legs and box with their forefeet (see fig. 10.4). Our analysis of stag threats does not differentiate between types of interaction, including displacements, and for convenience we refer to all of them as "threats."

It is important to distinguish between threats and sparring interactions. Particularly in late summer, young stags frequently locked antlers (see fig. 10.5) and turned and twisted their heads (see Bützler 1974; Geist 1982). These sparring bouts seldom involved vigorous pushing and rarely developed into fights. After the age of five, the frequency with which stags were involved in sparring bouts declined with age. In the Kilmory stag group, sparring bouts were usually initiated by the dominant member of the pair, though they were generally terminated by the subordinate.

10.3 Threat Frequency
Environmental Factors Affecting Threat Frequency

Why do deer threaten each other? Threats seldom drove the recipient away from the locality. However, they often removed the recipient from a particular feeding site, and a close connection between threat rates and competition for access to preferred feeding sites is indicated by comparisons of the frequency of threats in different contexts. Among both hinds and stags the frequency of threats was highest during times of food shortage. Threat rates were higher in winter than in summer: for example, the mean rate of threats given by the ten hinds observed at Kilmory in winter was 0.37 per h compared with 0.15 per h for the five hinds observed in the same area in summer, while for stags at Kilmory threat rates were 1.8 threats/h in February and March

FIG. 10.2. A mature stag giving a high chin threat to an opponent. The stag's upper lip is curled back and he is hissing at his opponent.

FIG. 10.3. A seven-year-old stag scissors kicks at another animal, displacing him from his feeding site.

compared with about 1.0 threats/h in June and July (Appleby 1980). In addition, among stags, the proportion of cases in which the threatener subsequently fed at the same site as the animal he had just displaced increased during the winter months (see fig. 10.6). Similar increases in threat rate in winter have been reported in other deer species (e.g., Ozoga 1972).

Fig. 10.4. Two mature stags boxing shortly after antler casting in April.

FIG. 10.5. Two three-year-old stags sparring in the dunes.

The rate at which deer threatened other individuals increased with party size. In hinds, the number of threats given during summer rose from 0.35 threats/h in parties of three to five animals to 0.51 threats/h in parties of more than ten. In stags, threat rates rose from 1.06 threats/h in parties of three to five to more than 1.98 threats/h in parties of more than ten animals. One of the reasons for the increase in threat frequency with party size was that nearest neighbor distances declined as party size increased [1]. Threat rates were also particularly high when hinds or stags were wallowing and there was room in the wallow for only one animal (see also Gossow and Schürholz 1974).

Sex Differences

As the previous results indicate, there was a marked difference in threat frequency between hinds and stags. Most comparisons showed that stags threatened other individuals about three times as frequently as hinds did: for example, in the continuous watches of nine hinds and nineteen stags carried out in the summer of 1975, hinds gave, on average, 0.46 threats/h whereas stags gave 1.56 [2]. Similar differences existed in winter and have been observed in other populations (Franklin and Lieb 1979).

Sex differences in threat frequency were not a result of differences in the plant communities used by stags and hinds, for they occurred between animals using the same communities (see table

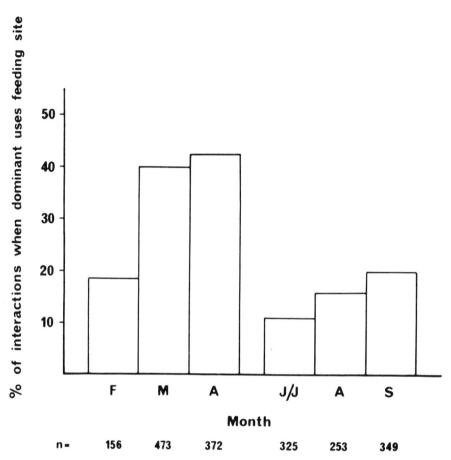

FIG. 10.6. Proportion of all threats or displacements in which the dominant animal immediately fed at the same site used by the subordinate after displacing him (from Appleby 1980).

10.1). Nor were they a consequence of any tendency for nearest neighbor distances to be smaller among stags—in fact, these were consistently smaller among hinds than among stags on all plant communities [3], perhaps because of the lower rate of threats (see fig. 10.7). And it seems unlikely that they occurred because competition for food was more intense between stags than between hinds, for they were as pronounced in summer as in winter. To understand the reasons for these differences, it was necessary to consider the distribution of threats between different categories of individuals.

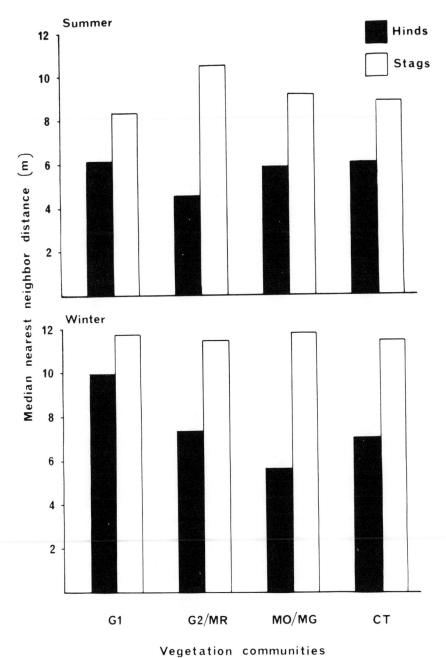

FIG. 10.7. Nearest neighbor distances in hinds and stags when feeding on different vegetation communities. (Calves were excluded as nearest neighbors.) Based on data collected in continuous watches during July and August 1975 and February and March 1976 (sampling interval 5 min; watches for each sex in each season are combined).

TABLE 10.1   Mean Threat Rate among Hinds and Stags Feeding on Different Plant Communities in Summer (July/August)

	Short Greens	Long Greens and *Juncus* Marsh	*Molinia* Grasslands	*Calluna* Moorland
Hinds	0.58[a]	0.08	0.67	0.41
Stags	1.76	0.86	2.08	1.90

[a] Threats given per hour.

10.4   THE DISTRIBUTION OF THREATS BETWEEN HINDS

The most striking aspect of the distribution of threats among hinds was that threats between matrilineal relatives were uncommon. Table 10.2 shows the number of threats directed at matrilineal relatives versus other animals by hinds over two years old. Since most threats were directed at the threatener's nearest neighbor and matrilineal relatives generally associated with each other, these figures must have greatly underestimated the frequency with which other animals were threatened compared with the time they were accessible. In addition, threats given by hinds to matrilineal relatives were less often intense than those directed at other animals.

Young hinds gave fewer threats and received more than older animals (table 10.3), apparently because there was an age related dominance hierarchy among hinds: individuals of up to two years

TABLE 10.2   Number of Threats Received by Hinds from Matrilineal Relatives Compared with Other Animals

Age of Recipient	Matrilineal Relatives	Other Animals	Total Number of Animals in Age Category in Population
Calves (both sexes)	6	54	45
Female yearlings	5	29	16
Male yearlings	20	57	24
2-year-old hinds	10	81	15
3–6-year-old hinds	7	129	54
7–10-year-old hinds	4	32	26
≥ Eleven-year-old hinds	1	16	16

Note: Includes threats of all types seen during the course of routine observations during one year (1977–78) that were received from daughters, sons, mothers, sisters, brothers, aunts, nieces, or granddaughters.

TABLE 10.3 Comparative Numbers of Threats Given and Received by Different Categories of Individuals

	Calves	Yearlings		Hinds			
		Females	Males	2-Year-Olds	3–6-Year-Olds	7–10-Year-Olds	≥11-Year-Olds
Threats given per head	.98[a]	1.44	1.46	3.4	3.98	4.54	4.25
	(44)[b]	(23)	(35)	(51)	(183)	(118)	(68)
Threats received per head	2.98	3.69	3.67	6.13	2.74	1.35	1.06
	(134)	(59)	(88)	(92)	(148)	(35)	(17)

Note: Threats between matrilineal relatives were excluded from this sample. The analysis was based on a total of 441 threats observed throughout 1977. Comparative threat rates were calculated by dividing the total number of threats by the numbers of each age/sex class in the study population.

[a] Mean threats per head.

[b] Total number of threats.

old were almost always subordinate to older animals, and among hinds of three years and older dominance rank was also correlated with age [4]. Similar results have been found in other herds (Hall 1978; Franklin, Mossman, and Dole 1975; Franklin and Lieb 1979). The causes of dominance rank among hinds were obscure. A hind's rank was not significantly correlated with the rank of her mother or her sisters and did not vary with her reproductive status.

Was a hind's rank related to her reproductive success? A correlation might be predicted on the grounds that dominance presumably increased access to preferred feeding sites, and studies of primates have demonstrated correlations between the rank of females and their reproductive success (e.g., Dunbar and Dunbar 1977). However, when age effects were taken into account there were no significant correlations between the rank of hinds and any measure of their reproductive success (see chap. 5).

10.5 The Distribution of Threats between Stags

Stags rarely associated with matrilineal relatives, and their higher threat rates compared with hinds probably reflected this difference. Where two brothers did occur in the same group, there was no evidence that they either associated more closely or threatened each other less frequently than other animals.

Dominance relationships among stags grazing under natural conditions were consistent from year to year and were no less regular than among animals attracted to artificial feeding sites (Appleby 1980). Studies of other ungulates grazing on natural food supplies have produced similar results (Clutton-Brock, Greenwood, and Powell 1976).

The hierarchy in winter was more linear among stags than among hinds: in only 2.7% of stag dyads was an animal of higher rank usually dominated by an individual of lower rank, while this occurred in 15% of hind dyads. Within dyads, the direction of interactions was also more consistent among stags: cases where a dominant both threatened another animal and was threatened by it accounted for under 2% of all dyads, while among hinds they accounted for over 9% (Appleby 1981).

The regularity of the hierarchy declined during the period of antler casting and growth, since older and more dominant stags cast earlier than younger animals and in some cases became sub-

ordinate to them. Where reversals of previous dominance relationships occurred, the newly dominant animal often appeared
to initiate interactions with the stag it had beaten. Stags that had
recently cast apparently avoided conflict by keeping away from
the main group for some days, especially when they were among
the first to have cast.

What factors affected the dominance rank of stags? Several
lines of evidence suggested that body weight was important.
Among stags of less than five, rank was closely related with age,
and there were few cases where a younger animal dominated an
older one. In contrast, among stags that had reached adult body
weight (animals of five years and older), rank was not correlated
with age but was related to size (Appleby 1981). Studies of the
relation between body weight and rank in captive stags have also
shown that the two measures are closely correlated (Suttie 1979),
and a relationship between dominance rank in adulthood and
measures of early growth would consequently be expected. As
predicted, there was a significant correlation between a stag's
dominance rank at the age of six and his antler length as a yearling [5].

Several studies have argued that antler size or complexity exerts
an important influence on dominance rank (Geist 1971*b;* Topinski 1974). However, there is little firm evidence that this is the
case, and most attempts to relate rank to antler size have con-
founded differences in antler size with differences in age and
body size (Clutton-Brock 1982).

What effects, if any, did dominance rank in winter groups have
on a stag's subsequent reproductive success? Among the Kilmory
stags, winter dominance was correlated with reproductive success
in the subsequent rut [6]. However, rutting success was also correlated with rank in the *following* winter, and the relationship
between rutting success and winter dominance rank may have
arisen through a common dependence of both measures on body
weight. Nevertheless, high-ranking stags were able to displace
subordinates from feeding sites, and experimental fertilization of
patches of the short greens showed that dominant animals were
able to monopolize these patches when the deer were allowed
access to them (Appleby 1980). Presumably, this had some effect
on their condition in the following year and may have contributed
to their reproductive success.

10.6   AFFILIATIVE BEHAVIOR

If differences in the frequency of agonistic interactions between hinds and stags are mainly a result of the greater tolerance hinds show to their matrilineal relatives, we might also expect the frequency of affiliative interactions to be higher among hinds than among stags.

As figure 10.7 shows, nearest-neighbor distances were consistently less between hinds than between stags, and hinds often tolerated other individuals feeding within a few centimeters of their heads. In addition, hinds sometimes groomed each other, licking and nibbling the hair of their subjects, particularly around the neck, head, and ears (see fig. 10.8). Grooming was invariably unidirectional and was uncommon compared to that in equids and primates (see Clutton-Brock, Greenwood, and Powell 1976; Sparks 1967). Grooming by hinds was confined to other hinds and their own dependent offspring, and, with the exception of those in estrus during the rut, hinds rarely groomed stags over two years old. In contrast, stags virtually never groomed each other— exceptions were rare and involved young stags one to three years old.

Finally, hinds sometimes gave a short, staccato alarm bark when they were disturbed (Darling 1937). This was almost always given by adults (though not necessarily by older animals) and instantly alerted other members of the party. As in many other mammals, males rarely gave alarm barks, though young stags occasionally did so (see Hirth and McCullough 1977; Sherman 1977).

10.8   FUNCTIONAL CONSIDERATIONS

Social relationships between members of hind and stag groups thus differed in a variety of ways (see table 10.4). Hinds threatened each other less frequently than did stags, and their dominance relationships were less predictable. Dominance rank was age-related among hinds, but among stags it was correlated with age only among animals that had not yet reached adult body size—after this, differences in body size were apparently more important than age. Nearest-neighbor distances were shorter among hinds, and they sometimes groomed each other, whereas stags rarely did so. The lower threat rates and closer association among hinds were at least partly a consequence of the fact that

FIG. 10.8. A hind grooms her two-year-old daughter.

Table 10.4   Sex Differences in Social Relationships

Threat rate: Higher among stags than hinds. Intense threats more frequent.
Dominance relationships: More predictable among stags than among hinds.
Factors affecting dominance rank: Dominance rank among hinds related to age. Among stags
    five years old or younger, rank closely correlated with age; among mature stags,
    rank related to differences in body size, weight, and early growth.
Grooming: Hinds groom matrilineal relatives. Stags seldom groom each other.
Nearest-neighbor distances: Greater among stags than among hinds.
Alarm calls: Hinds occasionally give alarm barks when disturbed; stags rarely do so.

they usually associated with matrilineal relatives and that threats
between relatives were scarce. But why did hinds tolerate the close
proximity of their relatives? As many studies of social relationships among other species have argued, it seems likely that
some form of kin selection is involved. Perhaps the most likely
explanation is that close association has mutual benefits in terms
of both predation avoidance and shelter from aggressive interactions (see p. 194), but that the number of animals that can feed
close to each other is limited by feeding interference. In such
circumstances individuals that shared the benefits of close association with their relatives would probably be favored by selection
(Wrangham 1982).

Our results also raise questions concerning the functional
significance of dominance. Among hinds, dominance was not correlated with estimates of reproductive success, whereas among
stags winter dominance and reproductive success in the subsequent rut were significantly correlated. Did this represent a
genuine difference in the importance of dominance rank in the
two sexes? Probably not, for the association between reproductive
success and winter dominance in stags was likely to have resulted
because both variables were related to body size. The absence of
this relationship among hinds may merely have reflected the fact
that, in contrast to stags, social dominance and reproductive success did not have the same causes. This does not say that being
dominant had no advantages. High rank may have reduced the
energy costs of food collection in both sexes, and the absence of
significant correlations between dominance and reproductive
success may only indicate that other factors are more important to
reproductive success.

10.8 SUMMARY

1. Individuals threatened each other when competing for feeding sites. Threat frequency increased with party size and was higher in winter than in summer.
2. Hinds threatened each other less frequently than did stags. This was apparently because hinds seldom threatened their matrilineal relatives and commonly associated with them.
3. An age-related dominance hierarchy existed among hinds. Dominance rank among stags was age-related among animals that had not yet reached adult body size but not beyond this age.
4. Dominance relationships were less regular and less consistent among hinds than among stags. Apart from age, there were no obvious correlates of dominance among hinds. The rank of stags was probably related to their body size and to their early growth rates and was positively correlated with reproductive success.

Statistical Tests

1. Relationship between median nearest-neighbor distance and group size among hinds feeding in winter on short greens.
    Spearman rank correlation coefficient: $r_s = -.683$, $N = 9$, $p < .05$.
    Similar results were observed for different vegetation types as well as in different seasons.
2. Comparison of mean number of threats given per hour by hinds versus stags during continuous watches in summer (1975).
    Mann-Whitney U test: $U = 32$, $n_1 = 9$, $n_2 = 19$, $p < .02$.
3. Comparison of nearest-neighbor distances among stags and hinds in summer and winter on different plant communities.
    Summer, G test: Total, $G = 25.844$, d.f. $= 4$, $p < .001$. $N = 937$
    Winter G test: Total, $G = 63.634$, d.f. $= 4$, $p < .001$. $N = 1,030$.
4. Relationship between dominance rank (calculated for each area using the same index used for calculating the dominance of stags during the rut) and age of hind (data for 1977–78).
    Kilmory, Spearman rank correlation coefficient: $r_s = .430$, $t = 2.720$, d.f. $= 32$, $p < .02$.

Intermediate Area, Spearman rank correlation coefficient: $r_s = .715$, $t = 4.686$, d.f. $= 21$, $p < .001$.

Samhnan Insir, Spearman rank correlation coefficient: $r_s = .393$, $t = 1.760$, d.f. $= 17$, $.1 > p > .05$.

In all three areas, these hierarchies were concordant across years, 1975–78 inclusive.

5. Correlation between the length of the antlers of yearlings born in 1972 and their dominance rank in winter groups in 1979 at six years of age.

Spearman rank correlation coefficient: $r_s = .680$, $t = 2.93$, d.f. $= 10$, $p < .02$.

6. Correlation between rutting success (number of hind/days held) and dominance rank during the preceding winter in stags five years old or older (from Appleby 1980).

Spearman rank correlation coefficient: $r_s = .46$, $t = 2.538$, d.f. $= 24$, $p < .02$.

7. Correlation between rutting success (number of hind/days held) and dominance rank during the following winter in stags of five years or older (from Appleby 1980).

Spearman rank correlation coefficient: $r_s = .52$, $t = 2.982$, d.f. $= 24$, $p < .01$.

11 Feeding Behavior and Habitat Use

The female has to expend much organic matter in the formation of her ova, whereas the male expends much force in fierce contests with his rivals, in wandering about in search of the female, in exerting his voice, pouring out odoriferous secretions, etc.

Charles Darwin (1871)

11.1 INTRODUCTION

Individuals may be selected to propagate their genes, but their success in doing so depends on their ability to collect and process sufficient nutrients to cover their own energetic requirements and to provide a surplus large enough to allow them to reproduce. In most temperate habitats, food is scarce during the winter months and superabundant in spring and early summer (Moen 1973, 1978). The appetites of northern cervids are related to this boom and bust economy: in white-tailed deer (Silver et al. 1969; Holter et al. 1976; Short 1975), red deer (Simpson 1976; Kay 1978), reindeer (McEwan and Whitehead 1970), and moose (Gasaway and Coady 1974) metabolic rate and food intake *decline* during winter when the animals are losing weight (Moen 1973). This reduction occurs in animals maintained on an ad libitum diet and is probably triggered by changes in day length (Simpson 1976; Kay 1978; Moen 1978; Kay and Staines 1981). Presumably, this mechanism helps to reduce the energetic costs of feeding in situations where the quantity or quality of food does not allow individuals to cover their daily requirements, though it is not clear why such an inflexible system is adaptive.

In dimorphic cervids, the maintenance requirements of males are likely to be substantially greater than those of females on account of their larger body size. However, direct comparisons of the nutritional requirements of males and females are complicated by differences in the extent and timing of the costs of reproduction.

Very approximate estimates of energy requirements for Scottish red deer (Anderson 1976) suggest that total energy requirements of milk hinds are substantially greater between May and July than in the winter months (see fig. 11.1). Both energy and protein requirements of females are typically slight in the first two-thirds of the gestation period and increase logarithmically during the last third (Moen 1973; Verme 1965). Costs of lactation are higher than those of pregnancy and peak during the fourth to sixth week after parturition (Moen 1973, 1978). In captivity, lactating red deer hinds will eat more than twice their maintenance requirements (Arman et al. 1974; Kay and Staines 1981), and similar results have been obtained for rodents (Daly 1979).

Estimates of the nutritional requirements of males are more rudimentary. In red deer, the costs of rutting are evidently high,

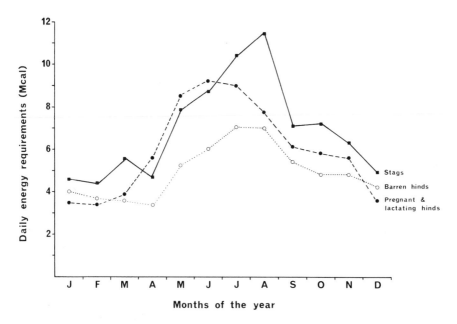

Fɪɢ. 11.1. Estimated daily energy requirements of red deer stags, barren hinds, and pregnant and lactating hinds in different months (from Anderson 1976). Estimates do not take into account energy losses as a result of winter weather, and winter figures are based on fasting metabolic rates established for American deer: they consequently represent only a rough approximation of the energy requirements of the three classes.

and stags must also expend considerable amounts of energy and protein on the annual growth of antlers and neck muscles during the summer months, though no serious attempt has been made to measure either of these costs. Anderson's guesstimates of the energy requirements of Scottish stags (fig. 11.1) suggest that, during the rut, the metabolic requirements of stags are even greater than those of hinds at the peak of lactation while, during the summer months, their total energy requirements are approximately similar to those of lactating hinds.

How do seasonal changes in the nutritional requirements of hinds and stags affect their feeding behavior? In this chapter we compare the feeding behavior of the two sexes, beginning with detailed aspects of grazing behavior (sec. 11.2) and subsequently describing the proportion of time spent in different activities throughout the day (sec. 11.3) and the extent to which they used different altitude levels (sec. 11.4) and plant communities (sec. 11.5). Section 11.6 compares variation in day range length and home range size, and section 11.7 describes the use of shelter—

one of the principal factors affecting the distribution of both sexes from day to day. In the final section, we discuss the significance of differences in feeding behavior between stags and hinds.

11.2 INGESTION AND RUMINATION

Differences in ingestion rate and rumination rate between the sexes were probably small. Bite rate did not differ and varied little between plant communities (table 11.1): it typically lay between 50 and 60 bites/min. Though we could not measure bite size, if this was related to incisor breadth (as seems likely among animals feeding on the shorter plant communities), it was likely to have varied little between stags and hinds: the mean incisor breadth of hinds five years old or older in our study area was 40.5 mm, while that for stags was 43.7 mm, a difference of less than 8%. As studies of other grazing herbivores have shown (Allden and Whittaker 1970), bite size typically increases with the standing crop of the vegetation selected and indirect evidence suggested that this was the case in red deer (see below). Rates of rumination were also similar in stags and hinds (see table 11.2) and were intermediate between those of sheep (approximately 100 chews/min: England 1954) and cattle (50–70 chews/min: Morgan 1951).

11.3 ACTIVITY BUDGETS
Grazing Bouts

In both sexes, grazing was aggregated into bouts (see Appendix 14) whose length varied from 10 to over 200 min. The duration of daytime grazing bouts was shorter in summer than in winter and did not differ significantly between the sexes: in July and August, median grazing bout length was 73 min for hinds and 67 min for stags, whereas in February and March median daytime bout lengths increased to 122 min for hinds [1] and 144 min for stags [2]. Nighttime grazing bouts in hinds in summer tended to be shorter than daytime ones, averaging 45 min, while in stags they were longer, averaging 82 min [3]. Nighttime bouts in winter were considerably shorter than daytime bouts in both sexes, averaging 45 min in hinds and 43 min in stags [4, 5].

In summer both stags and hinds usually fitted four to six grazing bouts into the seventeen h of daylight and fed between one and three times at night. In winter, as grazing bout length increased, the number of daytime bouts dropped to two or three,

TABLE 11.1   Mean Bite Rate in Mature Stags and Hinds Feeding on Different
Plant Communities before and after the Spring Growth Period

	Short Greens	Long Greens	*Juncus* Marsh	*Molinia* Grassland	*Calluna* Moorland
Hinds					
Before[a]	61 (7)	—	57 (5)	54 (6)	—
After[b]	65 (6)	64 (7)	58 (4)	66 (4)	—
Stags					
Before	68 (6)	54 (5)	63 (3)	57 (8)	—
After	63 (5)	57 (9)	60 (5)	50 (3)	57 (3)

Note: Bite rate means bites per minute. Estimates were based on 4-min sampling periods.
The number of individuals sampled is shown in parentheses.
[a] April/May.
[b] June/July.

and the animals fed three or four times at night. Previous studies
of Continental red deer have produced similar estimates (Bubenik
and Bubenikova 1967).

Diurnal Variation in Grazing

Both the duration and the frequency of grazing bouts varied
throughout the day. In summer the longest bouts usually started
either just before dawn or in the middle to late afternoon, running through to the early hours of darkness, and bouts during the
middle of the day tended to be shorter. In both sexes, feeding
bouts in the later afternoon were the most regular as well as being
the longest, and the proportion of time spent grazing increased
throughout the day, peaking between 1700 and 2100 (see fig.
11.2). In winter, relatively less time was spent feeding during the

TABLE 11.2   Ruminating Behavior in Mature Stags and Hinds in Summer

	Hinds (mean)	Stags (mean)
Chews/min	85.3	84.6
	(34)	(28)
Chews/bolus	62.3	64.1
	(35)	(30)
Bolus duration (sec)	47.9	47.8
	(34)	(30)

Note: The number of individuals sampled is shown in parentheses.

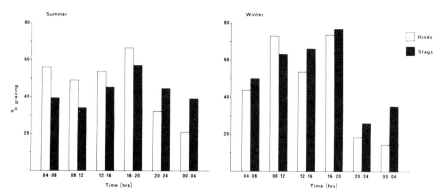

FIG. 11.2. Mean percentage of time spent grazing during different periods of the day by hinds and stags in summer and winter based on continuous watches.

hours of darkness and more time was spent in the first part of the morning, but again grazing peaked between 1700 and 2100.

Similar diurnal patterns in feeding behavior have been recorded in several other red deer populations (Bubenik and Bubenikova 1967; Georgii and Schröder 1978; Jackes 1973) as well as in elk (Craighead et al. 1973), moose (Belovsky and Jordan 1978; Belovsky 1981), reindeer (Thomson 1971), white-tailed deer (Michael 1970), and sheep (Hughes and Reid 1951). Regular feeding plays an important part in stabilizing the environment of microorganisms in the rumen and in maximizing digestive efficiency (van de Veen 1979; Kay and Staines 1981), while the tendency to feed relatively more during the daytime in winter and less at night (which also occurs among sheep: see Tribe 1948) probably reduces heat loss.

Daytime Grazing Budgets

Milk and yeld hinds differed in the proportion of time they spent grazing. In July and August, mature hinds spent on average 56% of daytime grazing, 22% ruminating, 12% lying inactive, 5% standing inactive, and 3% moving. Milk hinds spent more time grazing than yelds [6] (means 57% and 51% respectively), though grazing time did not increase in proportion to the difference in their nutritional requirements (see fig. 11.1). The percentage of daytime spent grazing by hinds increased in late winter (February and March) to 62% [7], while time spent ruminating fell to 16% [8], lying inactive to 4% [9], and standing inactive to 2% [10]. By March, milk and yeld hinds no longer differed in the proportion of daytime they spent grazing (means

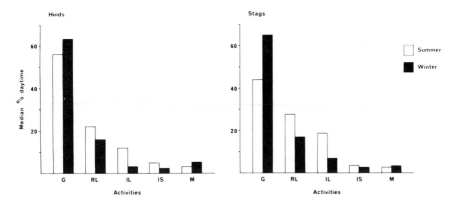

G · grazing, RL · ruminating lying, IL · inactive lying, IS · inactive standing, M · moving

Fig. 11.3. Activity budgets for hinds and stags in summer and winter. Histograms show the median percentage of time spent per day in different activities. Medians were based on nine days' observation of different individual hinds in summer and eighteen of different stags, and eleven days' observation of different hinds in winter and eleven of different stags.

61% and 62% respectively), though barren mothers tended to graze more than pregnant ones [11] (76% versus 58%)—perhaps because they mostly continued to suckle their calves (see chap. 4).

Compared with hinds, stags spent a smaller proportion of daytime grazing in summer [12] (44%) and more time lying inactive [13]. In winter, their grazing time increased to 65% [14], which did not differ significantly from that of hinds (see fig. 11.3).

These comparisons should not be allowed to disguise the fact that the proportion of daytime spent feeding also varied widely from day to day: the yeld hind observed for twenty-four days in summer spent from 47% to 74% of daytime grazing on different days. These differences were correlated with the proportion of her feeding time spent on short greens [15] (see fig. 11.4), presumably because biomass on the greens was low and this reduced her rate of food intake. However, feeding on the greens allowed her to reduce the amount of time she spent ruminating [16], and the total amount of time she spent grazing plus ruminating was less variable.

Nocturnal Grazing Budgets

Seasonal changes in the proportion of daytime spent grazing contrasted with changes in nocturnal grazing. In the eleven watches of hinds carried out in July and August, individuals aver-

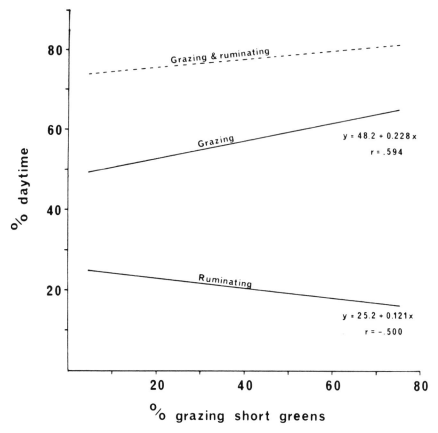

FIG. 11.4. Regression of percentage of daytime spent grazing or ruminating on the percentage of grazing time spent on short greens by a single hind observed for twenty-four days in July and August 1976.

aged 26% of nighttime spent grazing (33% for milk hinds, 21% for yelds), and in winter they averaged 17%. Similar changes occurred among stags, but in both seasons stags grazed more at night than hinds: the five stags observed in summer averaged 42% of the night grazing [17], and the four watched in winter spent 31%.

Twenty-four Hour Grazing Budgets

Twenty-four-hour activity budgets calculated from these data showed that, with the exception of milk hinds, the animals spent more time grazing in February and March than in July and August (see table 11.3), a result that parallels previous results for red deer (Georgii and Schröder 1978). The total grazing time of stags in summer was intermediate between that of milk and yeld hinds, while in February and March stags grazed about 2 h longer than hinds.

TABLE 11.3   Twenty-four-Hour Grazing Budgets in Summer and Winter (h/grazing/24 h day)

	Summer	Winter
Milk hinds	11.76	11.08
Yeld hinds	9.84	> 10.00[a]
Stags	10.40	12.88

[a] No measures were available for the proportion of nighttime spent grazing by yeld hinds in winter. However, 9.92 h/day were spent grazing during daytime, and incidental observations of yeld hinds at night showed that feeding was not restricted to the day. Extrapolating from the weight of their stomach contents compared with that of milk hinds suggests that they spent between 10 and 11 h grazing per 24-h day in winter.

Why was the total grazing time longer in winter than in summer? The most likely explanation was that food biomass was low in winter and that this was associated with a reduction in ingestion rates. Several lines of evidence showed that, within seasons, there was a close relationship between grazing time and the biomass of the plant communities used. Grazing bouts were longer on the short greens (where standing crops were always low) than on other communities, and day-to-day variation in grazing time within seasons was related to the biomass of the plant communities used (see above). Moreover, studies of domestic herbivores have shown that, as the abundance of vegetation is reduced, bite size and the rate of food intake decline and the animals compensate by increasing the amount of time spent feeding (Hancock 1953; Arnold 1964; Allden and Whittaker 1970; Stobbs 1973a,b; Charcon and Stobbs 1976; Hodgson and Milne 1978).

11.4   USE OF DIFFERENT ALTITUDES

At all times of year, both stags and hinds selectively used the lower parts of the study area. Proportions of animals seen below 30 m were lowest from February to April (fig. 11.5), when standing crops on the short and long greens were low. In May, shortly after the onset of spring growth, both sexes began to use the ground below 30 m to an increased extent, while the stags did so from June onward.

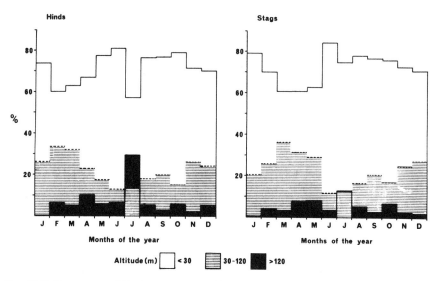

FIG. 11.5. Percentage of hinds one year or older and stags more than three years old seen at different altitudes in each month of the year in 1975 (based on census data).

Particularly among hinds, there was a marked decline in the use of low ground in July. This coincided with a peak in the numbers of bloodsucking tabanid flies, of which four species *(Tabanus bisignatus, T. montanus, Haematopota crassicornis,* and *Crysops relicta)* are common on the island. On sunny days attack rate by tabanids was high, but it declined with increasing altitude (fig. 11.6), while on cloudy days attack rate was reduced. As would be expected if tabanid activity affected the distribution of the deer, the proportion of animals seen over 120 m on different days in July was negatively correlated with cloud cover [18]. Similar patterns of altitude use have been found in other populations of Scottish red deer (see Colquhoun 1971; Staines 1977; Mitchell, Staines, and Welch 1977).

In contrast to red deer populations in eastern Scotland, where stags tend to use lower ground than hinds in winter but higher ground in summer (Staines 1969), stags in our study area used the lowest areas more and the highest levels less than hinds in most months of the year (fig. 11.4). In addition, the timing of altitude use varied between the sexes. In winter, the hinds moved down earlier in the day: for example, in the winter of 1975, 73.4% of all hinds seen between 0800 and 1200 were on ground below 30 m, while only 41.1% of stags were seen below the same level; after

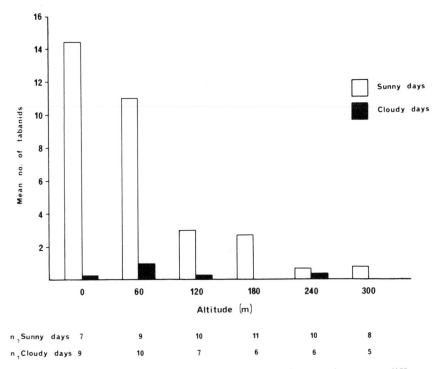

FIG. 11.6. Attack rate by tabanid flies (all species) on a human observer at different altitudes on predominantly sunny and predominantly cloudy days. In each sample the observer stood still for 5 min on *Calluneatum* and caught all tabanids that landed on her. (Attack rate on a human observer had previously been shown to be correlated with attack rate in an enclosure herd of deer.) The figure shows means and ranges of the numbers of tabanids collected in samples at each altitude level (Holt, n.d.).

1400, 76.2% of hinds were and 80.0% of stags were seen below 30 m. During the evening, hinds left the glen bottom and moved onto the lower slopes of the hills earlier than stags.

11.5 USE OF DIFFERENT PLANT COMMUNITIES
Seasonal Changes

Annual, seasonal, and individual differences in the extent to which the deer used the different plant communities were all pronounced. In the following subsections, we describe how the use of different plant communities varied in the population, only returning to a direct comparison between the sexes at the end of the section.

Both stags and hinds fed relatively little on the short greens from November until April or May, using instead the long greens,

FIG. 11.7. A ten-year-old stag feeding in winter on *Calluna* moorland at Kilmory.

the *Juncus* marsh, the *Molinia* grasslands, and the *Calluna/
Trichophorum*-dominated communities of the middle hill slopes
(fig. 11.7, Appendix 19) as well as feeding on seaweed (fig. 11.8).
Use of the short greens (see fig. 11.9) peaked in early summer
and then declined as standing crop and productivity fell. The
long greens and *Juncus* marsh were heavily used by hinds in the
late winter and early spring and by stags in midsummer (June–
August), and the *Molinia* grasslands on the lower hill slopes provided the main winter staple of both sexes.

   Both in winter and in summer, the deer showed strong selection for grazing on the greens and selection against *Calluna/
Trichophorum* moorland (fig. 11.10). As in other herbivores
(Sinclair 1977), the extent to which different communities were
selected was the product of an interaction between the quality and
the abundance of the food they offered. The *Agrostis/Festuca*
swards on the short greens provided the highest proportion of
easily digestible dry matter and the largest fraction of available
protein in all months (see Mitchell, Staines, and Welch 1977; van
de Veen 1979; Kay and Staines 1981). However, in late winter the
standing crop on the short greens was so low (see chap. 2) that the

Fig. 11.8. An eight-year-old hind feeding on seaweed at Samhnan Insir.

deer were probably unable to satisfy their total nutritional requirements by feeding on them extensively and moved to other
communities where higher standing crops compensated for inferior quality. This explanation of the reduction in the use of the
short greens during the winter assumed that the deer could not
have increased their feeding time enough to compensate for the
decline in bite size. Studies of other wild ruminants show that
grazing budgets of 11–12 h/day are unusually high for ruminants
(Owen Smith, unpublished data), and experiments with domestic
sheep, which presumably have fewer other demands on their time
than most wild animals, have demonstrated that they will graze
for over 11 h/day only under conditions of extreme food shortage
(Allden and Whittaker 1970). An additional reason for avoiding
the short greens during periods of food shortage may have been
that the deer were liable to consume loose sand. This was probably
both uncomfortable and damaging: research on sheep and deer
has shown that there is a close relationship between the amount of

FIG. 11.9. A hind feeds close to her yearling son and two-year-old daughter on the short greens at Samhnan Insir.

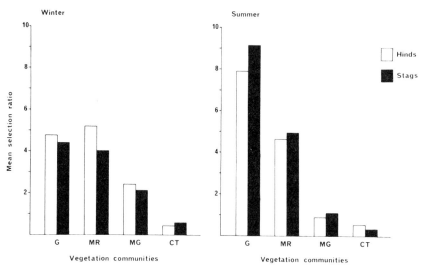

G greens, MR Juncus marsh, MG · Molinia grasslands, CT · Calluna Trichophorum moorland

FIG. 11.10. Mean selection ratios for each plant community calculated across years (1973–75) by dividing the proportion of feeding records on each community by the area (ha) of the community in the study area.

soil ingested while grazing and tooth wear (Healy and Ludwig 1965; Severinghaus and Cheatum 1956), which is known to have important effects on condition and survival in deer.

The extent to which the deer selected particular plants within communities evidently varied widely. When they were grazing on the short greens, they showed little apparent selectivity, and the sward was evenly cropped to within 2 cm of ground level for much of the year. In contrast, when feeding on the long greens and *Juncus* marsh, they selected *Agrostis, Festuca,* and *Carex* species and rarely ate from the stands of *Juncus effusus* and *J. articulatus* (though *J. squarrosus,* which is common in heather moorland, was frequently eaten, particularly in early spring). On the *Molinia* grassland, *Molinia* itself was eaten only during spring; at other times the animals selected broad-leaved grasses growing between the *Molinia* tussocks. *Eriophorum* was usually ignored, and, unlike sheep (Grant et al. 1976), the deer rarely fed on the considerable areas of floating bog. When feeding on the moorland communities, they took both *Calluna* and *Molinia,* seldom eating *Trichophorum* except in spring.

A previous study on Rhum that examined food selection by analysis of fecal particles produced broadly similar results (Charles, McCowan, and East 1977, unpublished data). About half the identified plant particles were from grassland swards, and regular peaks occurred in the use of *Molinia* and *Trichophorum* during the early summer and of *Calluna* during the winter months, with broad-leaved grasses, *Festuca,* and *Carex* being used throughout the year (see fig. 11.11). The study differed from ours in that *Agrostis*-dominated grasslands were used extensively in winter but this probably reflected the fact that the study was carried out in the western part of Rhum, where large areas of long *Agrostis* and *Festuca*-dominated grassland grow on the upper slopes of the hills. When the deer using these areas have reduced the standing crops on the herb-rich greens growing on the lower slopes, they turn progressively to these longer grasslands. In contrast, the area of long *Agrostis/Festuca* greens in our study area is more limited, and, when the herb-rich greens were eaten out, the animals were forced to use the *Molinia* and *Calluna*-dominated communities on the hill slopes. Detailed descriptions of the use of different plants by red deer are available in several other studies and provide broadly similar results (Colquhoun 1971; van de Veen 1979; Jensen 1968; Dzieciolowski 1969, 1970a,b).

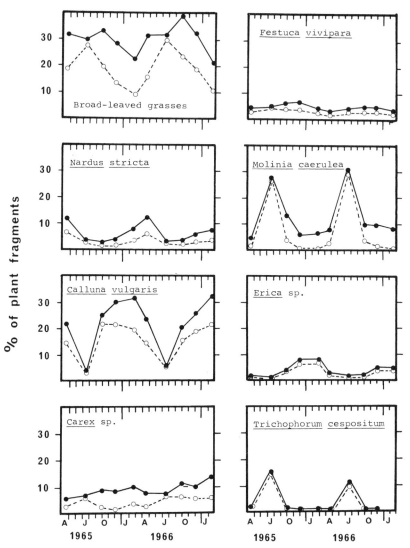

FIG. 11.11. Percentages of different types of plant fragments in feces collected between April 1965 and February 1967 in the western part of Rhum. The open circles show the percentage of fragments from living plants only, and the solid circles show those from living plus dead plants (from Charles, McCowan, and East, unpublished data).

Like many other herbivores (e.g., Sekulig and Estes 1977; Nesbit-Evans 1970; Western 1971; Wyatt 1971), both hinds and stags chewed bones and cast antlers (see fig. 11.12) when they found them, particularly in spring and early summer.

Fig. 11.12. A mature stag chewing a cast antler in early spring, shortly before he cast his own antlers.

Differences between Years

The extent to which the deer used the different plant communities in particular months varied widely from year to year as a result of differences in weather conditions. For example, in 1975 little rain fell in February and March, temperatures were low and spring growth was delayed, with the result that both sexes spent the early spring feeding principally on the *Molinia* grasslands rather than the short greens as in other years.

Among milk hinds, variation in habitat use in summer was

closely related to annual rainfall. In years when rainfall between May and July was heavy, they used the short greens and the heather moorland more and the long greens and *Juncus* marsh, and *Molinia* grassland less than in years when rainfall was comparatively low [19] (see fig. 11.13). In contrast, the extent to which yeld hinds used the short greens and heather moorland in summer was not correlated with rainfall, though there was a tendency for use of the long greens, *Juncus* marsh, and *Molinia* grassland to decline in wet years [20]. Neither among milk hinds nor among yelds was variation in the extent to which different communities were used correlated with summer temperature or with changes in population density (see chap. 12). Stags, like milk hinds, tended to feed more on the short greens and less on *Molinia* grassland in wet summers [21], though there was no evidence that their use of heather moorland increased in wet years.

Use of the short greens by milk hinds probably declined in dry summers because grass growth was reduced and the animals used the long greens, *Juncus* marsh, and *Molinia* grassland to a greater extent instead. In wet years standing crops on the short greens were probably greater, while growth on the long greens and *Juncus* marsh may have been reduced by waterlogging. Both milk hinds and stags may have been more sensitive to changes in standing crop on the short greens than yeld hinds because their protein requirements were greater (see below).

Although habitat use between January and March varied between years as widely as in summer, changes were not correlated with rainfall, temperature, or population density. However, there was a weak tendency for hinds to use the short greens less in colder winters, though this did not approach statistical significance.

As a result of variation between years, there was little concordance across years in the extent to which the animals used particular communities in the same months [22]. This raised important problems when comparisons were made between years, and to avoid this difficulty we restricted most comparisons involving data from several years to the three winter months (January, February, March) and the three summer months (May, June, July) when the animals' behavior varied least between years.

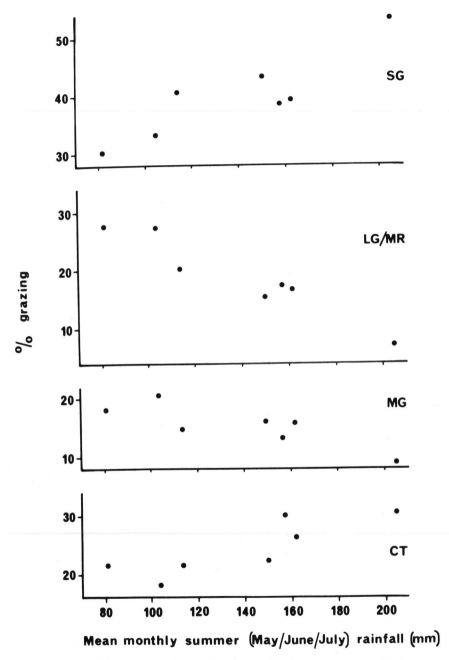

Fig. 11.13. Proportion of grazing milk hinds seen on different plant communities between May and July in different years, 1974–80, plotted on mean rainfall during these months in each year.

Individual Variation

There were also pronounced individual differences in the extent to which hinds and stags used different plant communities. For example, some individuals were seldom seen grazing on the short greens in either summer or winter, while others were recorded on this community in over 50% of cases in both seasons. Among hinds, these differences were stable across years [23] and were related to the abundance of different plant communities in the animal's core area [24]. Since hinds shared their mothers' ranges, they normally showed feeding behavior similar to that of both their mothers [25] and their sisters, particularly during the winter months.

Consistent differences in the extent to which particular plant communities were used also existed among stags. However, their feeding behavior appeared to be less closely constrained by the abundance of greens in their core area than that of hinds, and there was no significant relationship between the frequency with which they were seen feeding on short greens and the abundance of short greens in their core areas [26], nor was their feeding behavior significantly correlated with that of their mothers [27].

Differences between Milk and Yeld Hinds

Since both nutritional requirements and activity budgets differed between milk and yeld hinds, we were not surprised to find that their use of plant communities also varied. However, comparisons of habitat use were complicated by a tendency for hinds using the Upper Glen and Samhnan Insir both to spend less time feeding on the greens and to show lower fecundity than those using Kilmory (see chap. 12). To avoid the possibility that spatial variation in habitat use and fecundity might confound comparisons between milk and yeld hinds, we extracted data only for individuals that had been observed both in years when they were milk hinds and in years when they were yeld.

In summer, hinds tended to feed more on the short greens in years when they were milk than in years when they were yeld [28], and they used the *Molinia* grassland and heather moorland to a lesser extent (see table 11.5), though these differences were not pronounced. The tendency for hinds to use the short greens more and the heather moorland less when they were milk than when they were yeld persisted through the winter [29], when it was greater than in summer (see table 11.5).

A similar tendency for lactating hinds to feed more on grasses and less on heather than yeld hinds has been noted in other studies (Ineson and Mohun, unpublished data quoted in Blaxter et al. 1974). There are at least two plausible explanations for this difference. Milk hinds may feed more on the short greens because they require a higher proportion of available protein in their diet to cover the costs of lactation. Alternatively, their calves (which begin to graze within their first month of life) may require higher-quality forage than adults as a consequence of their smaller body size (see Bell 1971; Clutton-Brock and Harvey 1982), and this may constrain their mothers' habitat use. That the difference is most pronounced in late winter, when most calves have been weaned, favors the latter explanation, though it is possible that milk hinds may need to feed more on protein-rich foods after the end of lactation to replace lost reserves.

Sex Differences

Previous studies of red deer in eastern Scotland suggest that stags tend to use habitats inferior to those used by hinds. Hinds are more likely to be found in areas overlying base-rich rock (Watson and Staines 1978) or offering large areas of flushed grassland, while stags occur on ground where heather predominates (Lowe 1966; Staines and Crisp 1978; Jackes 1973; Charles, McCowan, and East 1977). Examination of rumen contents confirms that the diets of stags include a higher fiber content and a lower fraction of available protein than those of hinds in the winter months: rumen contents of hinds show higher nitrogen levels, and food particles tend to be smaller (see table 11.4). Both differences are reduced or absent in summer.

In our study area on Rhum, the areas used by stags and hinds were not widely separated as in many parts of the Scottish mainland. Areas used by stag groups lay within a few hundred meters of localities heavily used by hinds, and, though the two sexes were usually found in separate parties, their ranges overlapped extensively. In these circumstances, did hinds and stags differ in their use of the various plant communities?

Comparisons of the frequency with which hinds and stags were seen feeding on different plant communities provided no simple

Table 11.4   Sex Differences in Habitat Use and Rumen Nitrogen Levels of
Red Deer in Aberdeenshire in Winter

	Hinds	Stags
Proportion of home range area of hind and stag groups overlying base-rich rock	32.9	6.1
Proportion covered by well-drained grassland	76.3	20.2
Rumen nitrogen (g/100 g dry matter)		
July–September	3.08	3.03
November–March	2.40	2.15

Source: Watson and Staines (1978); Staines and Crisp (1978).

contrasts. In summer, throughout the study area as a whole, stags
fed slightly less on the short greens and more on *Molinia* grassland
than either milk or yeld hinds during the summer months (see
table 11.5). In winter, stags fed slightly more on the short greens
than yeld hinds but less than milk hinds. They also used the long
greens and *Juncus* marsh less and *Calluna* moorland and seaweed
more than either category of hinds. Thus, although some differences in feeding behavior existed between hinds and stags [30],
there was no clear evidence that hinds fed consistently on foods
qualitatively superior to those taken by stags.

Table 11.5   Mean Proportion of Grazing Hinds and Stags over Three Years
Old Seen on Different Plant Communities in Summer and Winter, 1974–79

	Short Greens	Long Greens and *Juncus* Marsh	*Molinia* Grassland	*Calluna* Moorland	Seaweed
Summer[a]					
Milk hinds	38.2	20.2	16.4	24.4	0.8
Yeld hinds	35.0	19.6	17.4	28.6	1.2
Stags	33.9	21.7	24.3	18.8	1.2
Winter[b]					
Milk hinds	23.0	19.8	24.2	25.7	8.2
Yeld hinds	13.7	20.5	21.6	29.0	14.4
Stags	20.8	15.8	23.4	21.6	18.3

Note: Figures based on census data. Figures for hinds show means calculated across 29
individuals in years when they were milk versus yeld.

[a] May–July.

[b] January–March.

TABLE 11.6   Mean Proportion of Feeding Hinds and Stags Seen on Different
Plant Communities at Samhnan Insir and Kilmory

	Kilmory		Samhnan Insir	
	Summer	Winter	Summer	Winter
Short greens				
Hinds	43.3	27.2	52.5	23.9
Stags	63.1	28.9	22.8	9.5
Long greens and				
***Juncus* Marsh**				
Hinds	25.8	28.4	7.9	9.5
Stags	22.5	13.7	1.0	2.7
***Molinia* Grassland**				
Hinds	10.6	20.8	9.7	25.5
Stags	8.3	21.7	16.0	16.2
***Calluna* moorland**				
Hinds	18.1	14.3	28.8	26.5
Stags	5.0	17.5	58.6	61.1
Seaweed and littoral				
debris				
Hinds	2.2	9.2	1.1	14.6
Stags	10.6	18.3	1.9	1.6

Note: Hinds were one year old or older; stags were over three years old. Figures are from
census data, 1973–76.

However, pronounced differences between the sexes occurred
in some parts of the study area (see table 11.6). Whereas at Kilmory (where population density was relatively low and the reproductive performance of hinds was high [see chap. 12]) stags
used the short greens more than hinds in summer and to approximately the same extent in winter, stags at Samhnan Insir (where
population density was high and the greens were very heavily
grazed) used the short greens, long greens, and *Juncus* marsh
considerably less than hinds at all times of year, compensating for
this by an increase in their use of *Calluna* moorland. In this part of
the study area, at least, it seems likely that the average quality of
the diet of stags may have been inferior to that of hinds, particularly during the winter months.

11.6   RANGING BEHAVIOR
  Day Range Length

The extent of daily movement was similar in hinds and stags (see table 11.7). In both sexes it was lower at night than during the day and lower in winter than in summer. When differences in day length were taken into account, our data indicated that the deer covered about 1.8 km per 24-h day in winter compared with about 3.0 km in summer. This estimate falls within the limits of day range length recorded in other studies (Darling 1937; Craighead et al. 1973).

TABLE 11.7   Mean Rate of Movement (quadrats/h) in Hinds and Stags in Summer and Winter

| | Hinds | | Stags |
	Day	Night	Day
Summer	1.68	0.67	1.51
N	9	5	20
Winter	1.40	0.33	1.42
N	11	5	11

Note: Figures based on continuous watches of individuals carried out 1974–76.

In both seasons there was also a marked diurnal pattern of ranging behavior: animals moved consistently farther during the morning and evening than in the middle of the day [31] and moved little at night (see fig. 11.14). High rates of movement occurred in both seasons on *Molinia* and *Calluna*-dominated plant communities [32] but were lower at times when hinds aggregated in large groups [33], particularly on low ground.

A similar decline in daily movement in late winter and early spring occurs in elk (Craighead et al. 1973) and in other deer species (Moen 1973). The reduction of movement among animals that were feeding in large groups probably occurred because large groups formed in areas when feeding conditions were optimal.

  Home Range and Core Area Size

Our analysis of home range size was based on records for forty-three breeding hinds over three years of age and twenty-nine stags over five years of age and employed census data collected

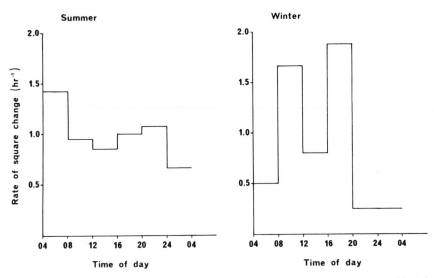

FIG. 11.14. Rate of movement (median number of quadrats changed per hour) in hinds in winter and summer. Estimates for daytime based on samples of twenty-three days' observation of one hind in February and March 1977 and twenty-seven days' observation of the same animal in July and August 1976. Estimates for nighttime based on five night watches of the same animal. (Estimates for nighttime are means, not medians, since the median value was zero in both cases.)

between 1973 and 1975. We considered both summer (May–July), winter (January–March), and annual (April–April) range sizes (the total size of the area used by the method described in Appendix 13) as well as the sizes of 65% core areas (the smallest area that accounts for 65% of sightings of that individual). Measures of range and core area size proved to be closely correlated with each other across individuals, though it is important to note that our figures represent minimal estimates, since animals straying outside the study area were not recorded: for example, neither range nor core areas included the rutting areas of stags that rutted outside the study area.

Both hinds and stags had larger ranges and core areas in summer than in winter [34] (fig. 11.15), a change that was associated with a reduction in the extent to which the higher altitudes were used and an increase in the amount of time spent grazing on the longer greens, *Juncus* marsh, and *Molinia* grasslands on the glen bottoms and on the lower hill slopes. Similar reductions in the sizes of winter and early spring ranges have been found in several other studies (Darling 1937; Murie 1956; Craighead et al. 1973; Staines 1974; Georgii 1980). In both seasons hinds had larger

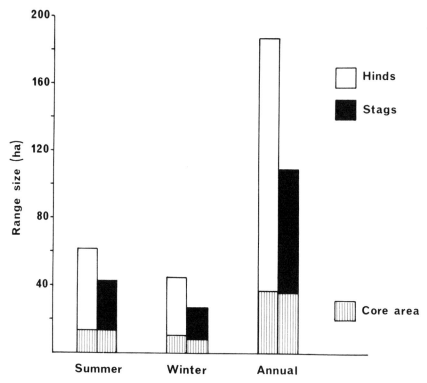

FIG. 11.15. Range and 65% core area size in hinds and stags in summer and winter. (Analysis based on records for forty-three breeding hinds and twenty-nine stags over five years old between 1973 and 1975.)

ranges than stags [35, 36], and their annual ranges (excluding records collected during the rut, when stags dispersed widely) were also larger [37]. In contrast, the size of their core areas differed little, though in winter stags had slightly smaller core areas than hinds [38].

Individual differences in ranging behavior among hinds were related both to their reproductive status and to the food available in their range. Hinds tended to have larger summer core areas when they were yeld than when they were milk [39]. No difference was evident in core area size in winter [40] or in total range size in either season.

Individual differences in range size and core area size among hinds were correlated negatively with the proportion of the range covered by greens [41, 42] and positively with the proportion covered by *Calluna* moorland [43]. This probably helped explain why both range size and core area size differed between parts of the study area (see chap. 12), though the negative correlation between the area of greens in an animal's range and the size of the range remained within most areas [44].

Individual differences in range and core area size among stags were as large as those among hinds. However, they bore little relationship to the distribution of greens, and individual core area sizes among stags were not correlated with the proportion of greens they contained [45, 46].

The ranging behavior of red deer and elk populations is evidently highly variable. In nonmigratory populations of elk, individuals can use seasonal home ranges as small as 5–6 ha (Yanushko 1957), while migratory populations can move between summer and winter ranges more than 50 km apart (e.g., Knight 1970). Our estimates of range size for red deer on Rhum are considerably smaller than estimates of home range size in a heavily culled population in northeastern Scotland, which exceed 600 ha (Staines 1970, 1974), perhaps because the small scale of the topography on Rhum is associated both with the increase in the availability of shelter (see Staines 1974, 1976) and with more even distribution of the plant communities preferred by the deer. A variety of studies of other mammalian species have shown that where food supplies are aggregated in large, irregularly distributed clumps, range size tends to be large. Where clump size is smaller and clumps are more evenly distributed, range size is reduced (Clutton-Brock and Harvey 1978).

Sex differences in home range size also appear to vary between red deer populations: in some studies (e.g., Darling 1937) stags are reported to have larger ranges than hinds, while others have found little difference in home range size between the sexes (Knight 1970). These differences may be partly a product of variation in home range definitions and seasonal differences in the pattern of home range use between the sexes: stags often travel considerable distances to their rutting grounds (e.g., Lincoln, Youngson, and Short 1970), and if these areas are included within their home ranges they are likely to show larger ranges than hinds. It is not yet clear whether the summer and winter ranges of stags are generally larger or smaller than those of hinds in other populations.

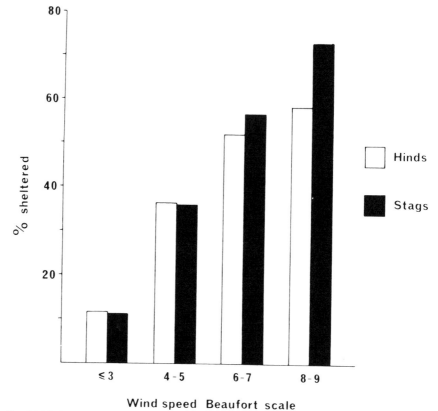

FIG. 11.16. Percentage of grazing hinds (over one year old) and stags (over three years old) seen on sheltered ground in winter on days of different wind speed. (Analysis based on census data, 1975–76.)

11.7   USE OF SHELTER

While the site of food resources has an important influence on the usual distribution of hinds and stags (Lowe 1966), their location on particular days was often strongly influenced by wind speed and direction (see Staines 1976, 1977). The proportion of animals of both sexes seen in sheltered positions increased when wind speed exceeded force 5 [47] (see fig. 11.16). Windy days were associated with an increase in the number of deer seen on *Molinia* grassland and heather moorland (where shelter was usually available) and a reduction in the number seen on the short greens (table 11.8).

Though the reactions of both sexes to windy conditions were similar, stags were more often seen in sheltered positions than

TABLE 11.8   Percentage Change in Proportions of Hinds and Stags seen Feeding in Winter on Different Plant Communities on Windy Days versus Calm Ones

	Short Greens	Long Greens and *Juncus* Marsh	*Molinia* Grassland	*Calluna* Moorland	Seaweed
Hinds	−17.0	−2.9	− 0.1	+18.6	−1.9
Stags	−12.7	+3.6	+10.6	+ 1.2	−4.1

Note: Calm = wind speed ≤ force 5, Beaufort scale; windy = wind speed ≥ force 6, Beaufort scale. Figures based on census data, winter 1975.

were hinds (fig. 11.16). This was not a consequence of sex differences in the plant communities used, for similar differences were found within communities [48], and our data thus supported previous suggestions that, despite their larger body size, stags may be more vulnerable to exposure than hinds (Jackes 1973). However, research at Glenfeshie has produced contrary results (B. W. Staines, pers. comm.): in very cold weather hinds were found in *more* sheltered positions than stags. One possibility is that at Glenfeshie stags and hinds were more widely segregated than on Rhum, and stags were generally found in areas where shelter was less accessible.

11.8   ORIGINS OF SEX DIFFERENCES IN FEEDING AND RANGING BEHAVIOR

Thus, compared with hinds, stags spent more time grazing (particularly in winter) and grazed more at night (see table 11.9). They also used the lowest levels of the study area more than hinds throughout the year, though in winter they moved down to the glen bottoms later in the day. Across the study area as a whole, sex differences in habitat use were slight, though stags ate more seaweed than hinds, but in areas heavily used by hinds, stags used the short greens, long greens, and *Juncus* marsh less than hinds. Individual differences in feeding behavior between hinds were more closely related to the abundance of greens in their core area than among stags. Finally, stags spent more time feeding in sheltered positions than hinds.

Sex differences in activity budgets were probably consequences of variation in nutritional requirements and body size. The greater nutritional requirements of stags resulting from their larger size, perhaps accentuated by differences in heat loss (see

TABLE 11.9   Sex Differences in Feeding and Ranging Behavior

Activity budgets: In summer, stags spent more time grazing than yeld hinds but less than milk hinds. In winter, stags spent more time grazing than hinds.

Diurnal variation in activity: Both in summer and in winter, stags grazed more at night than hinds.

Use of altitude levels: Stags used lower ground more than hinds throughout the year. In winter, stags moved down to the glen bottoms later than hinds.

Use of plant communities: In summer, stags fed less on the short greens and heather moorland than hinds. In winter, stags used the long greens and heather moorland less than hinds and ate more seaweed. Differences in habitat use varied between parts of the study area: Where hind density was highest, stags used the short greens less than hinds throughout the year.

Range area: In both seasons, stags had smaller ranges than hinds, though core area size did not differ.

Individual differences in feeding and ranging: Differences in feeding behavior and core area size were related to abundance of greens in their core areas among hinds but not among stags.

Use of shelter: Stags spent more time feeding in shelter than hinds.

Jackes 1973), probably explained why they spent more time grazing than nonlactating hinds in both summer and winter. The rate of ingestion probably varied little between the sexes when they fed on similar swards, and an increase in the amount of time spent grazing by stags was consequently to be expected. The greater amount of time needed to collect sufficient food may also have required stags to spend a larger proportion of the nighttime grazing: particularly in winter, temperatures were substantially lower at night, and the deer apparently minimized nocturnal grazing.

The tendency for stags to feed less on the short greens and more on heather moorland in areas heavily used by hinds may also have been related to their greater body size and energy requirements. At Samhnan Insir, standing crops on the short greens were very low (see chap. 2), and if, as we have argued, the rate of food intake by both sexes was approximately similar, stags may have been unable to afford to spend as much time feeding on this community as hinds and may have been forced to move to areas where vegetation was of lower quality but standing crops were greater. In addition, like yeld hinds, they may not have needed such high levels of available protein in their diet as milk hinds, and reduction in the amount of time spent on the short greens may have allowed them to minimize energy expenditure. It is consistent with this explanation that sex differences in habitat use at Samhnan Insir were most pronounced during the winter months when standing crops were lowest. In addition, studies of feeding behavior on the Scottish mainland (Osborne 1980) indicate that the presence of sheep (which typically select for areas of *Agrostis/Festuca* even more strongly than deer) depresses the use of greens by stags more than by hinds. A similar tendency for larger herbivore *species* to move down the food quality spectrum ahead of smaller ones as food abundance declines has also been observed among African antelopes (Bell 1971; Jarman 1974).

The increased amount of time stags spent in shelter may have been related to increased heat loss as Jackes (1973) has suggested. There is suggestive evidence from a variety of vertebrates that males lose heat faster than females (Glucksman 1974; Slee 1970, 1972), perhaps partly because they have lower fat reserves and higher metabolic rates, and partly because they apparently have a reduced capacity to vary their metabolic rate in response to heat loss or to regulate it by nonshivering thermogenesis (Slee 1972). An additional reason for increased heat loss in red deer stags in winter is that their poor body condition after the rut delays the growth of their winter coat, which is often thinner than that of hinds.

11.9 SUMMARY

1. During the peak of lactation hinds compensated for their increased nutritional requirements by increasing their grazing time. Stags grazed more at night than hinds in both summer and winter.
2. Grazing time increased when the animals fed to a greater extent on swards where standing crops were low, and it showed a general increase in winter compared with summer.
3. Both hinds and stags used the lowest parts of the study area most heavily, but throughout the year stags were more often seen on low ground than hinds. However, in winter, they moved down into the glen bottoms later in the mornings than hinds.
4. The deer strongly selected the short greens throughout the year but used these comparatively little from November until April, using instead the *Juncus* marsh, *Molinia* grasslands, and *Calluna*-dominated communities.
5. Throughout the study area as a whole, sex differences in the

use of different plant communities were slight. However, in areas heavily used by hinds, stags fed little on the short greens.
6. Hinds had larger ranges than stags but similar-sized core areas. Core area size among hinds (but not among stags) was negatively correlated with the proportion of the core area covered by greens.
7. Wind speed and direction were important determinants of the distribution of deer on particular days. On windy days, stags were more often seen in sheltered positions than hinds.

Statistical Tests

1. Comparison of daytime grazing bout length in hinds in July and August 1975 versus February and March 1976.
    Mann-Whitney U test: $z = 1.669$, $N = 70$, $.1 > p > .05$.
2. Comparison of daytime grazing bout length in stags in July and August 1975 versus February and March 1976.
    Mann-Whitney U test: $z = 3.239$, $N = 118$, $p < .002$.
3. Comparison of nighttime grazing bout length in July and August in hinds versus stags.
    Mann-Whitney U test: $U = 25$, $n_1 = 10$, $n_2 = 13$, $p < .02$.
4. Comparison of nighttime grazing bout length in February and March in hinds versus daytime bout length.
    Mann-Whitney U test: $z = 2.762$, $N = 43$, $p < .01$.
5. Comparison of nighttime grazing bout length in stags in February and March versus daytime bout length.
    Mann-Whitney U test: $z = 3.382$, $N = 38$, $p < .001$.
6. Comparison of percentage of daytime spent grazing in summer by milk versus yeld hinds, controlling for area differences.
    Mann-Whitney U test: $z = 2.86$, $N = 33$, $p < .005$.
7. Comparison of percentage of daytime spent grazing in hinds in summer 1975 versus winter 1976.
    Mann-Whitney U test: $U = 16$, $n_1 = 9$, $n_2 = 11$, $p < .02$.
8. Comparison of percentage of daytime spent ruminating in hinds in summer 1975 versus winter 1976.
    Mann-Whitney U test: $U = 23$, $n_1 = 9$, $n_2 = 11$, $p < .05$.
9. Comparison of percentage of daytime spent lying inactive in hinds in summer 1975 versus winter 1976.
    Mann-Whitney U test: $U = 21$, $n_1 = 9$, $n_2 = 11$, $p < .05$.
10. Comparison of percentage of daytime spent standing inactive in hinds in summer 1975 versus winter 1976.

Mann-Whitney U test: $U = 12$, $n_1 = 9$, $n_2 = 11$, $p < .02$.
11. Comparison of percentage daytime spent grazing in winter by barren versus pregnant hinds.
    Mann-Whitney U test: $U = 24$, $n_1 = 8$, $n_2 = 11$, $.1 > p > .05$.
12. Comparison of percentage daytime spent grazing in hinds versus stags in summer 1975.
    Mann-Whitney U test: $U = 28$, $n_1 = 9$, $n_2 = 18$, $p < .02$.
13. Comparison of percentage daytime spent inactive in hinds versus stags in summer 1975.
    Mann-Whitney U test: $U = 41$, $n_1 = 9$, $n_2 = 18$, $p < .05$.
14. Comparison of percentage daytime spent grazing in stags in summer 1975 versus winter 1976.
    Mann-Whitney U test: $U = 22$, $n_1 = 11$, $n_2 = 18$, $p < .002$.
15. Correlation between percentage daytime spent grazing and percentage grazing time spent on short greens in one yeld hind across twenty-four days.
    Pearson product-moment correlation coefficient: $r = .594$, $t = 3.46$, d.f. $= 22$, $p < .01$.
16. Correlation between percentage of time spent ruminating divided by percentage time spent grazing and percentage grazing time spent on short greens in a single yeld hind observed for twenty-four days in summer.
    Pearson product-moment correlation coefficient: $r = -.587$, $t = 3.40$, d.f. $= 22$, $p < .01$.
17. Comparison of percentage nighttime spent grazing by hinds versus stags in summer.
    Mann-Whitney U test: $U = 9$, $n_1 = 5$, $n_2 = 15$, $p < .01$.
18. Correlation between (a) percentage of hinds and (b) percentage of stags seen above 120 m on different days in July and cloud cover (census data 1974–76).
    (a) Spearman rank correlation coefficient: $r_s = -.538$, $t = 2.83$, d.f. $= 21$, $p < .01$.
    (b) Spearman rank correlation coefficient: $r_s = -.560$, $t = 3.10$, d.f. $= 21$, $p < .01$.
    (From M. Holt, unpublished data.)
19. Correlation between mean rainfall in May to July in different years, 1974–80, and the percentage of grazing milk hinds seen on different plant communities.
    Short greens, Pearson product-moment correlation coefficient: $r = .854$, $t = 3.670$, d.f. $= 5$, $p < .02$.

Long greens/*Juncus* marsh, Pearson correlation coefficient:
$r = -.960, t = 7.666$, d.f. $= 5, p < .01$.

Molinia grassland, Pearson product-moment correlation
coefficient: $r = -.837, t = 3.420$, d.f. $= 5, p < .02$.

Calluna moorland, Pearson product-moment correlation
coefficient: $r = .808, t = 3.066$, d.f. $= 5, p < .05$.

20. Correlation between mean rainfall in May to July in different
years, 1974–80, and the percentage of yeld hinds over three
years old seen feeding on different plant communities.

Short greens, Pearson product-moment correlation
coefficient: $r = .16, t = .362$, d.f. $= 5$, n.s.

Long greens/*Juncus* marsh Pearson correlation coefficient:
$r = -.506, t = 1.312$, d.f. $= 5$, n.s.

Molinia grassland, Pearson product-moment correlation
coefficient: $r = -.489, t = 1.254$, d.f. $= 5$, n.s.

Calluna moorland, Pearson product-moment correlation
coefficient: $r = .260, t = .602$, d.f. $= 5$, n.s.

21. Correlation between the percentage of grazing stags over
three years old seen on different plant communities and mean
rainfall in May to July in different years, 1974–80.

Short greens, Pearson product-moment correlation
coefficient: $r = .714, t = 2.280$, d.f. $= 5, .1 > p > .05$.

Long greens/*Juncus* marsh, Pearson correlation coefficient:
$r = -.357, t = .855$, d.f. $= 5$, n.s.

Molinia grassland, Pearson product-moment correlation
coefficient: $4 = -.844, t = 3.519$, d.f. $= 5, p < .02$.

Calluna moorland, Pearson product-moment correlation
coefficient: $r = -.144, t = 0.051$, d.f. $= 5$, n.s.

22. Concordance in the percentage of observations of (*a*) hinds
and (*b*) stags seen in each month on different plant communities across 1973 to 1975.

(*a*) Short greens, Kendall coefficient of concordance:
$W = .530, \chi^2 = 17.48$, d.f. $= 11, .1 > p > .05$.

Long greens/*Juncus* marsh, Kendall coefficient of concordance: $W = .510, \chi^2 = 16.84$, d.f. $= 11$, n.s.

Molinia grassland, Kendall coefficient of concordance:
$W = .645, \chi^2 = 21.59$, d.f. $= 11, p < .05$.

Calluna moorland, Kendall coefficient of concordance:
$W = .450, \chi^2 = 14.86$, d.f. $= 11$, n.s.

Seaweed, Kendall coefficient of concordance: $W = .488$,
$\chi^2 = 16.10$, d.f. $= 11$, n.s.

(b) Short greens, Kendall coefficient of concordance,
$W = .476, \chi^2 = 15.62$, d.f. $= 11$, n.s.
Long greens/*Juncus* marsh, Kendall coefficient of concordance: $W = .420, \chi^2 = 13.85$, d.f. $= 11$, n.s.
Molinia grassland, Kendall coefficient of concordance:
$W = .415, \chi^2 = 13.69$, d.f. $= 11$, n.s.
Calluna moorland, Kendall coefficient of concordance:
$W = .631, \chi^2 = 20.82$, d.f. $= 11$, n.s.
Seaweed, Kendall coefficient of concordance: $W = .641$,
$\chi^2 = 21.15$, d.f. $= 11, p < .05$.

23. Concordance of percentage feeding records in which individual hinds were seen on different vegetation communities in (a) winter and (b) summer across years (1974–77).
(a) Short greens, Kendall coefficient of concordance:
$W = .462, \chi^2 = 58.35$, d.f. $= 42, p < .01$.
Long greens/*Juncus* marsh, Kendall coefficient of concordance: $W = .648, \chi^2 = 81.64$, d.f. $= 42, p < .001$.
Molinia grassland, Kendall coefficient of concordance:
$W = .448, \chi^2 = 56.425$, d.f. $= 42, p < .02$.
Calluna moorland, Kendall coefficient of concordance:
$W = .619, \chi^2 = 78.01$, d.f. $= 42, p < .001$.
(b) Short greens, Kendall coefficient of concordance:
$W = .497, \chi^2 = 44.37$, d.f. $= 29, p < .05$.
Long greens/*Juncus* marsh, Kendall coefficient of concordance: $W = .389, \chi^2 = 35.05$, d.f. $= 29$, n.s.
Molinia grassland, Kendall coefficient of concordance:
$W = .373, \chi^2 = 32.42$, d.f. $= 29$, n.s.
Calluna moorland, Kendall coefficient of concordance:
$W = .574, \chi^2 = 49.914$, d.f. $= 29, p < .01$.

24. Correlation and regression between the percentage of feeding observations of individual hinds in which they were seen feeding in winter on short or long greens and the abundance of greens in their core areas.
Pearson product-moment correlation coefficient: $r = .450$,
$t = 3.38$, d.f. $= 45, p < .01; y = 13.1 + 1.3x$.

25. Correlation between the proportion of feeding observations in which individual hinds were seen feeding on short greens in winter and summer and the proportion of observations in which their mothers were seen on the same community.
Winter, Pearson product-moment correlation coefficient:
$r = .545, t = 3.12$, d.f. $= 23, p < .02$.

Summer, Pearson product-moment correlation coefficient:
$r = .400$, $t = 2.05$, d.f. $= 22$, $.1 > p > .05$.

26. Correlation and regression between the percentage of feeding observations in which individual stags over five years old were seen feeding on short or long greens in winter and the abundance of greens in their core areas.

Pearson product-moment correlation coefficient: $r = .072$, $t = 0.290$, d.f. $= 16$, n.s.

27. Correlation between the percentage of feeding observations in which individual stags over five years old were seen feeding on short greens in winter and summer and the proportion of observations in which their mothers were seen on the same community.

Winter, Pearson product-moment correlation coefficient:
$r = .10$, n.s.

Summer, Pearson product-moment correlation coefficient:
$r = .393$, $t = 1.76$, d.f. $= 17$, n.s.

28. Comparison of the frequency with which individual hinds were seen feeding on short greens in summer in years when they were milk versus years when they were yeld (census data).

Wilcoxon matched-pairs test: $z = 1.806$, d.f. $= 28$, $.1 > p > .05$.

29. Comparison of the percentage frequency with which individual hinds were seen feeding on short greens in winter in years when they were milk versus years when they were yeld.

Wilcoxon matched-pairs test: $T = 19.5$, $N = 16$, $p < .01$.

30. Comparison of the percentage frequency with which hinds and stags were observed feeding on different plant communities in winter and in summer (census data 1973–76).

Winter, G test: $G = 271.70$, d.f. $= 4$, $p < .001$. $N = 1059$.
Summer, G test: $G = 52.07$, d.f. $= 4$, $p < .001$. $N = 1104$.

31. Comparison of rate of quadrat change by a single hind between three periods of the day (0800–2000) in winter, and four periods (0400–2000) in summer.

Winter, Friedman two-way analysis of variance: $\chi_r^2 = 8.08$, d.f. $= 2$, $p < .02$. $N = 21$ days.

Summer, Friedman two-way analysis of variance, $\chi_r^2 = 22.37$, d.f. $= 3$, $p < .001$. $N = 24$ days.

32. Comparison of rates of quadrat change by a single hind in both winter and summer when grazing different plant communities.

Summer: $\chi^2 = 23.85$, d.f. $= 3, p < .001. N$ 24 days.
Winter: $\chi^2 = 58.69$, d.f. $= 3, p < .001. N = 21$ days.

33. Comparison of rate of quadrat change by a single hind in winter and summer when she was a member of a group of fewer than ten animals compared with times when she belonged to a group of ten animals or more.

Summer, overall: $\chi^2 = 24.88$, d.f. $= 1, p < .001. N = 24$ days.
Short greens: $\chi^2 = 15.45$, d.f. $= 1, p < .001$.
Long greens and *Juncus* marsh: $\chi^2 = 7.26$, d.f. $= 1, p < .01$
Calluna moorland: $\chi^2 = 0.04$, d.f. $= 1$, n.s.
Winter, overall: $\chi^2 = 31.83$, d.f. $= 1, p < .001. N = 21$ days.
Long greens and *Juncus* marsh: $\chi^2 = 6.23$, d.f. $= 1, p < .02$.
Calluna moorland, $\chi^2 = 0.17$, d.f. $= 1$, n.s.

34. *(a)* Comparison of mean range size (1973–75) in hinds in summer versus winter.

Wilcoxon matched-pairs test: $z = -5.68, N = 43, p < .001$.

(b) Comparison of mean core area size (1973–75) in hinds in summer versus winter.

Wilcoxon matched-pairs test: $z = -5.1, N = 43, p < .001$.

(c) Comparison of mean range size in stags in summer versus winter.

Wilcoxon matched-pairs test: $T = 95.5, N = 23$, n.s.

(d) Comparison of mean core area size in stags in summer versus winter.

Wilcoxon matched-pairs test: $T = 39.5, N = 23, p < .01$.

35. Comparison of mean range size in stags versus hinds in summer 1974.

Mann-Whitney U test: $z = -2.729, N = 53, p < .01$.

36. Comparison of mean range size in stags versus hinds in winter 1975.

Man-Whitney U test: $z = -3.336, N = 53, p < .001$.

37. Comparison of annual range size in stags versus hinds (April 1974–March 1975).

Mann-Whitney U test: $z = 2.79, N = 58, p < .01$.

38. Comparison of mean core area size in stags versus hinds in winter 1975.

Mann-Whitney U test: $z = -2.653, N = 53, p < .01$.

39. Comparison of summer core area size in twenty-seven hinds between years in which they were yeld and years in which they were rearing a calf.

Wilcoxon matched-pairs test: $z = 2.04, N = 27, p < .05$.

40. Comparison of winter core area size in thirty hinds between years in which they were yeld and years in which they were rearing a calf.

Wilcoxon matched-pairs test: $z = 0.504, N = 30$, n.s.

41. (a) Correlation between the size of ranges of individual hinds in summer and the proportion of their ranges covered by greens.

Pearson product-moment correlation coefficient: $r = -.543$, $t = 3.36$, d.f. $= 27, p < .01$.

(b) Correlation between the size of ranges of individual hinds in winter and the proportion of their ranges covered by greens.

Pearson product-moment correlation coefficient: $r = -.491$, $t = 2.93$, d.f. $= 27, p < .01$.

42. (a) Correlation between the size of core areas of individual hinds in summer and the proportion of their core areas covered by greens.

Pearson product-moment correlation coefficient: $r = -.562$, $t = 3.53$, d.f. $= 27, p < .01$.

(b) Correlation between the size of core areas of individual hinds in winter and the proportion of their core areas covered by greens.

Pearson product-moment correlation coefficient: $r = -.569$, $t = 3.59$, d.f. $= 27, p < .01$.

43. (a) Correlation between the size of core areas in individual hinds in summer and the proportion of their core areas covered by *Calluna* moorland.

Pearson product-moment correlation coefficient: $r = .577$, $t = 3.67$, d.f. $= 27, p < .01$.

(b) Correlation between the size of core areas in individual hinds in winter and the proportion of their core areas covered by *Calluna* moorland.

Pearson product-moment correlation coefficient: $r = .536$, $t = 3.30$, d.f. $= 27, p < .01$.

44. (a) Correlation between the size of core areas of individual hinds in different parts of the study area in summer and the proportion of their ranges covered by greens.

Upper Glen: sample size inadequate.

Intermediate Area, Pearson product-moment correlation coefficient: $r = -.25, t = 0.8$, d.f. $= 12$, n.s.

Kilmory, Pearson product-moment correlation coefficient:
$r = -.57, t = 3.02$, d.f. $= 19, p < .01$.

Samhnan Insir, Pearson product-moment correlation
coefficient: $r = -.86, t = 5.85$, d.f. $= 12, p < .001$.

(b) Correlation between the size of core areas of individual
hinds in different parts of the study area in winter and the
proportion of their ranges covered by greens.

Intermediate, Pearson product-moment correlation
coefficient: $r = -.07, t = 0.24$, d.f. $= 12$, n.s.

Kilmory, Pearson product-moment correlation coefficient:
$r = -.39, t = 1.87$, d.f. $= 19, .1 > p > .05$.

Samhnan Insir, Pearson product-moment correlation
coefficient: $r = -.67, t = 3.16$, d.f. $= 12, p < .01$.

45. Correlation between the size of core areas of individual stags
in summer and the proportion of their core areas covered by
greens.

Pearson product-moment correlation coefficient: $r = -.31$,
$t = 1.695$, d.f. $= 27$, n.s.

46. Correlation between the size of core areas of individual stags
in winter and the proportion of their core areas covered by
greens.

Pearson product-moment correlation coefficient: $r = -.007$,
$t = 0.007$, d.f. $= 27$, n.s.

47. Comparison of the frequency of hinds and stags seen feeding
in sheltered positions in winter on days of different wind speed.

Hinds, G test: $G = 50.12$, d.f. $= 3, p < .001. N = 1200$.

Stags, G test: $G = 71.36$, d.f. $= 3, p < .001. N = 463$.

48. Comparison of the frequency of stags versus hinds seen feeding on different plant communities that were in sheltered positions.

G test: Interaction, $G = 0.781$, d.f. $= 3$, n.s. $N = 581$.

12 Population Dynamics

The males of some species kill one another by fighting; or they drive one another about until they become greatly emaciated. They must also be exposed to various dangers, whilst wandering about in eager search for a female.

Charles Darwin (1871)

12.1 INTRODUCTION

As we have described in the previous chapter, both nutritional requirements and feeding behavior differ between the sexes in red deer. Consequently, environmental factors that influence food availability or energy expenditure might be expected to have different effects—at least in degree—on males and females. So far, studies of population dynamics in large mammals have paid little attention to the comparative effects of either density-dependent or density-independent variables on males and females. Even in studies of small mammals, whose population dynamics have been more extensively investigated (see Krebs and Myers 1974; Southern 1979), research has only recently focused on the contrasting effects of population density on the two sexes (Redfield, Taitt, and Krebs 1978a,b; Boonstra and Krebs 1979).

In this chapter we examine changes in habitat use, reproductive performance, and mortality among hinds and stags using our study area between 1971 and 1980. Over this period, population density in our study area has increased steadily since the cessation of culling in 1972–73—from 19.8 deer/km² in 1971 to 30.5 deer/km² in 1980. The number of hinds using the study area has increased more than the number of stags: hinds have risen from 60 animals one year old or older in 1971 to 149 in 1979, whereas stags have increased from 124 to 135 over the same period (see fig. 12.1).

As in many other temperate ungulates (Klein 1968; Grubb 1974), population density and climatic factors interact in their effects on the deer: when population density is high, cold winters or dry summers have a greater influence on reproduction or survival than when density is low. Consequently, it is often difficult to separate the effects of density from those of weather variation unless data samples are available spanning many years. Nevertheless, our results show that some variables are closely correlated with weather variation but not with population density, whereas others are correlated with density changes but not with climatic variation.

An additional way of separating the effects of density from the effects of weather variation is to investigate whether similar correlations between population density and reproduction or survival apply across populations at different densities. Variation in the density of deer within our study area (see Appendix 6) allowed us

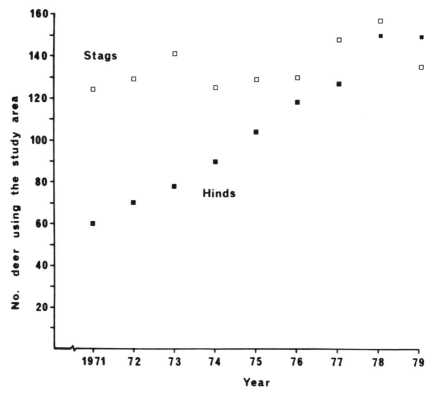

FIG. 12.1. Total number of hinds and stags one year old or older that regularly used the Rhum study area in different years.

to compare the reproductive performance of Samhnan Insir hinds with that of those using Kilmory to see whether differences between those two areas resembled temporal changes in the study area as a whole.

Section 12.2 describes changes in the habitat use and reproductive performance of hinds, and section 12.3 examines the same changes in stags. In section 12.4 we compare these results with studies of population dynamics in other ungulates, and discuss the reasons why harsh weather and high population density affect growth and survival in males more than in females.

12.2 CHANGES IN THE REPRODUCTIVE PERFORMANCE OF HINDS
Calf Birth Weight

Variation in the mean birth weight of calves caught in different years showed no obvious correlation with population size. Since

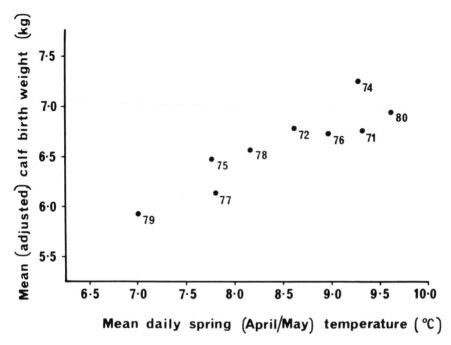

FIG. 12.2. Average birth weight of calves caught in our study area in different years (adjusted for the effects of mother's age and mother's range area) plotted against mean daily temperature in April and May of each year.

food availability during the last months of gestation is known to have an important effect on birth weight (Sadleir 1969), we examined correlations between mean birth weight in different years and weather variation in April and May. Mean birth weight proved to be closely correlated with mean daily temperatures in April and May [1], presumably because grass growth was advanced in warm springs (see fig. 12.2). In fact, the date on which the cuckoo was first heard at Kilmory each spring was a reliable predictor of calf birth weight in June (fig. 12.3) [2]! Presumably the association arose because spring temperatures affected both the arrival of summer migrants and the rate of vegetation growth. The relationship between spring temperatures and calf birth weights in our study area was supported by analysis of the mean weights of calves caught throughout Rhum between 1961 and 1970 by the Nature Conservancy stalkers [3].

Although there was no significant correlation between temporal changes in birth weight and population density, there was a tendency for birth weights to be lower at Samhnan Insir than at

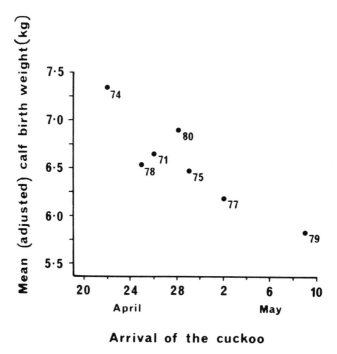

FIG. 12.3. Mean calf birth weight in each year (adjusted for the effects of mother's age and mother's range area) plotted against the date on which the cuckoo was first heard or seen at Kilmory in each year.

Kilmory [4], suggesting that population density can affect birth weight, but that climatic factors usually exert a stronger influence.

Conception Date

Unlike birth weight, the average conception date in different years was not correlated with any climatic variable that we examined but became progressively later as population size increased [5] (fig. 12.4). In addition, hinds using Samhnan Insir conceived, on average, six days later than those using Kilmory [6].

Fecundity

Like changes in conception dates, variation in fecundity was not related to any of the climatic variables we examined but was significantly correlated with changes in population size. Between 1971 and 1980 the proportion of both three-year-old hinds [7] and milk hinds [8] that calved each year fell as population size increased (see fig. 12.5).

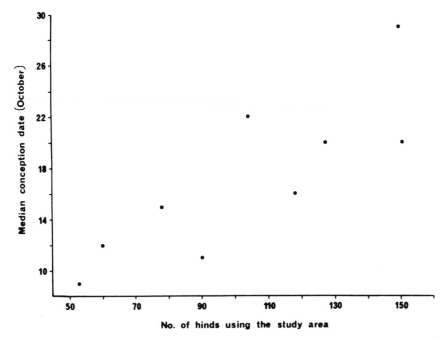

FIG. 12.4. Median conception date for calves born in each year to milk hinds plotted against the number of hinds resident in the study area.

Comparisons of pooled data for 1971–79 showed that fecundity was lower among hinds occupying home ranges at Samhnan Insir than among those based at Kilmory (fig. 12.6), especially three-year-olds [9]. In addition, hinds at Samhnan Insir that calved as three-year-olds and succeeded in raising their calves into the winter never calved again the following year, whereas they commonly did so at Kilmory [10]. Fecundity in hinds of five years old or more did not differ significantly between the two areas. A similar relationship between population density and fecundity is found across other populations of red deer in Scotland (Mitchell, Staines, and Welch 1977).

Calf Mortality in Winter

Like fecundity and conception date, the proportion of calves that died in winter grew as population size increased [11] (fig. 12.7) and was not significantly correlated with changes in temperature or rainfall in either summer or winter. In contrast, calf mortality in summer was not related to population size.

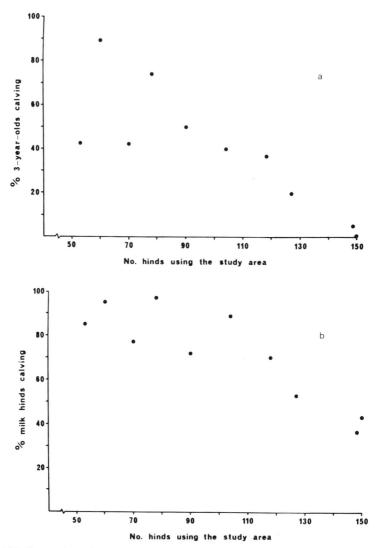

FIG. 12.5. Proportion of *(a)* three-year-old hinds, and *(b)* milk hinds calving each year plotted against the total number of hinds resident in the study area.

Both summer and winter calf mortality tended to be greater at Samhnan Insir than at Kilmory, and total calf mortality was significantly higher at Samhnan Insir [12].

Increasing calf mortality in the study area may have been caused partly by changes in the milk yields of hinds and in the growth of their calves: between 1975 and 1978, the mean duration of suckling bouts among three-month-old calves declined

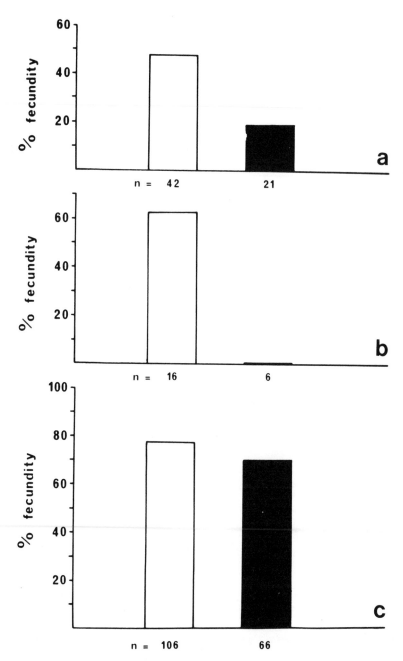

Fig. 12.6. Comparison of fecundity (percentage of hinds that calved) among hinds using Kilmory versus Samhnan Insir: (a) three-year-olds; (b) four-year-old milk hinds; (c) all hinds of five to ten years old. Open histograms, Kilmory; solid histograms, Samhnan Insir.

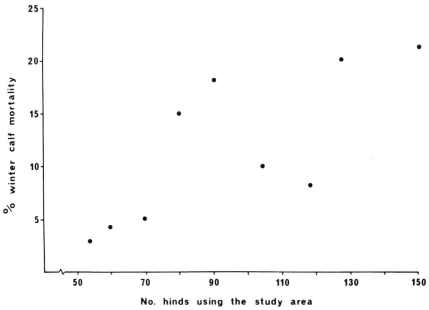

Fɪɢ. 12.7. Percentage of calves dying in winter (October–April) plotted against the number of hinds resident in the study area.

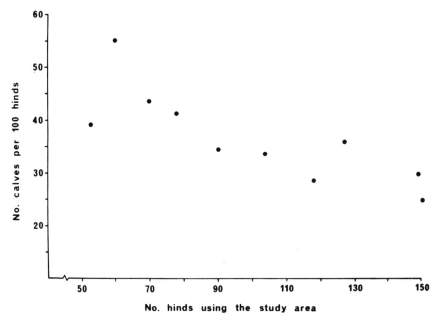

Fɪɢ. 12.8. Calf/hind ratios in April plotted against the number of hinds resident in the study area.

from 78 sec to 54 sec [13], and a similar association between high population density and short suckling bouts has been recorded in mountain sheep (Geist 1971*b*). In other deer species, as well as in domestic mammals, low food availability or quality is associated with reduced milk yields and retarded growth in juveniles (Kitts et al. 1956; Verme 1963; Thomson and Thomson 1953; Allden 1970; Grubb 1974).

Calf/Hind Ratios

As fecundity fell and calf mortality rose, the ratio of calves to hinds in spring declined from 53.7 calves per 100 hinds one year old or older in 1971 to 26.9 in 1979 [14] (fig. 12.8). Variation in calf/hind ratios was not correlated with any weather variable examined, nor did calf/hind ratios change in the rest of the island during the same period (see Appendix 2).

A similar relationship between calf/hind ratios and population density is found across deer populations on the Scottish mainland. Considering all areas over 100 km² counted by the Red Deer Commission between 1973 and 1977, calf/hind ratios fall as population density increases [15] (fig. 12.9*a*), and a similar relationship applies across smaller population units within South Ross [16] (fig. 12.9*b*). It has been suggested that this relationship occurs because high-density populations happened to be counted in years when breeding succes was low. However, this seems unlikely, since, when values of calf/hind ratios are adjusted for differences between years, a greater proportion of the overall variance is explained by density [15].

Coat Change

As population size in our study area increased, there was a gradual decline in the proportion of hinds that had already changed into winter coat by 1 November [17] (see fig. 12.10). Though the approximate timing of coat change is determined by changes in photoperiod, factors affecting female body condition also have some influence: deer wintered indoors change earlier than those wintered outdoors (Ryder 1977), and in our study area milk hinds changed later than yelds (see chap. 4).

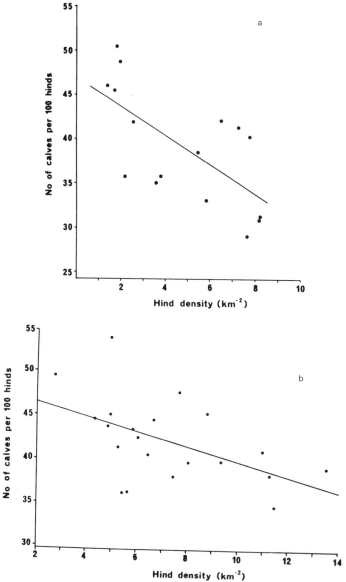

Fig. 12.9. (a) Calf/hind ratios (calves per 100 hinds) plotted against hind density for all areas of Scotland counted by the Red Deer Commission 1973–77. (b) Similar figures for different deer clans within South Ross (data from Mutch, Lockie, and Cooper 1976).

Mortality and Emigration

Our sample size is as yet too small to allow us to investigate corre-lations between population and adult mortality. Nevertheless,

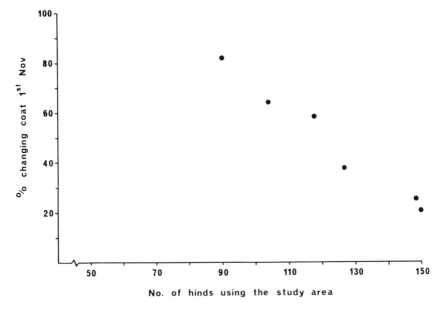

F_IG. 12.10. Proportion of milk hinds changing into winter coat by 1 November in different years. Animals were scored as changing coat if more than half their summer coat had already been shed.

mortality appears to have increased during the later years of the study, especially after cold winters. In particular, mortality of female yearlings appears to have risen: for example, between 1971 and 1975, less than 2% of female yearlings died each year, whereas between 1975 and 1980, 12% died. Yearling deaths typically followed cold or prolonged winters: after the hard winter of 1978–79, 25% of all female yearlings in the study area died, whereas after the mild winter of 1979–80, only 11% died.

Increased population density had no obvious effects on the number of animals that emigrated to other parts of the island. Between 1971 and 1974, two hinds left the study area, while between 1975 and 1979 only one left.

Habitat Use

Though variation in habitat use between years was not correlated with population size, the feeding behavior of hinds using Samhnan Insir differed in a variety of ways from that of animals using Kilmory. Hinds at Samhnan Insir fed less on the short greens in winter (table 11.6) and spent less time grazing [18] and more time ruminating: ruminating/grazing ratios for hinds observed were

0.21 at Kilmory compared with 0.48 at Samhnan Insir. When they fed on short greens, threat rates were higher at Samhnan Insir [19], and mother/offspring association was reduced (Cockerill, n.d.). In addition, both ranges and core areas were smaller at Samhnan Insir [20].

12.3 CHANGES IN GROWTH AND REPRODUCTIVE PERFORMANCE AMONG STAGS

Antler Growth

During the course of the study, the mean size of antlers of sixteen-month-old stags showed a dramatic decline that was significantly correlated with population size in the study area [21] (fig. 12.11), though not with either rainfall or temperature in summer or winter. The antlers of yearlings reared at Samhnan Insir were consistently smaller than those of animals raised at Kilmory [22]. Changes in the size of antlers among yearlings presumably reflected variation in body weight (Hyvarinen, Kay, and Hamilton 1977): it is interesting that the mean duration of suckling bouts involving three-month-old males fell from 84 sec in 1975–76 to 59 in 1977–78, whereas duration of suckling bouts involving females of the same age fell from 69 sec in 1975–76 to 57 in 1977–78.

Antler weight among adult stags wintering in the study area was also negatively correlated with population size [23], and declined during the course of the study from 800 g per antler between 1971 and 1973 to less than 700 g in 1979 (fig. 12.12).

Both antler casting and cleaning dates became later as population size increased. The median date of antler casting in seven-year-old stags changed by nearly two weeks from 29 March in 1971 to 8 April in 1979 [24] (fig. 12.13). In addition, median cleaning dates tended to become later.

Mortality

Although differences were not significant, mortality tended to be higher in stags than hinds, except between the ages of three and six, when fatalities occurred among young milk hinds (see table 12.1). Differential juvenile mortality was pronounced in some years and reduced or absent in others. For example, following the hard winter of 1978–79, 34.8% of male calves and 34.5% of male yearlings died, compared with 15.8% of female calves and 25% of

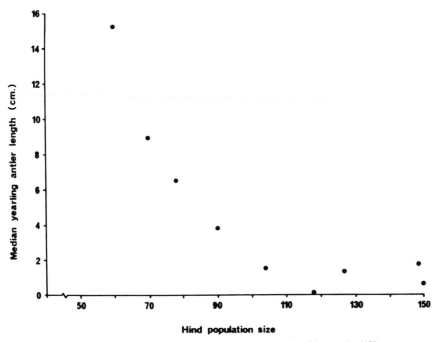

FIG. 12.11. Median length of cleaned antlers of sixteen-month-old stags in different years.

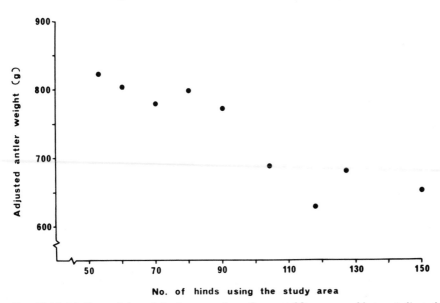

FIG. 12.12. Median weights of single cast antlers of two- to fifteen-year-old stags (adjusted for differences in age) plotted against the number of hinds regularly using the study area in different years. Weights shown are those of an average seven-year-old.

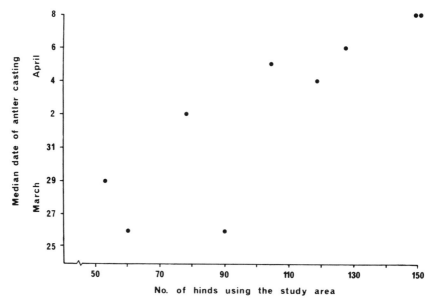

FIG. 12.13. Median casting dates of seven-year-old stags plotted against the number of hinds using the study area from 1971 to 1979 (excluding 1973).

female yearlings. Male mortality was particularly pronounced at Samhnan Insir, where 50% of male yearlings died, compared with only 11% at Kilmory (see p. 90).

Increase in population size was also associated with higher mortality rates in adult stags. As population density increased, the mean age at death of stags over three years old that died in each year fell from fourteen in 1974 (the first year in which adequate data were available) to nine in 1980 [25]. No such decline was evident among hinds.

Adult Sex Ratio

As density rose, there was a substantial change in adult sex ratios [26] (see fig. 12.1). This was not a consequence of any increase in the proportion of stags that emigrated from the study area, since, contrary to our initial expectations, dispersal rates of stags tended to decline as population density rose. For example, while 37.3% of stags born between 1970 and 1972 dispersed from the study area before their fifth winter, only 13.3% of those born between 1973 and 1975 did so [27].

The change in adult sex ratios occurred for three reasons. First, the number of stags that immigrated into the study area remained

TABLE 12.1   Mortality in Stags and Hinds of Different Ages, 1974–79

| Age | Percentage Dying per Year | |
	Females	Males
0–6 months	12.4 (137)	12.4 (178)
6–12 months	15.0 (120)	20.8 (154)
Yearlings	7.4 (95)	13.0 (115)
2 years	1.1 (94)	1.8 (110)
3–4 years	3.6 (166)	1.7 (181)
5–6 years	3.8 (132)	2.2 (136)
7–8 years	2.3 (86)	6.1 (115)
9–10 years	2.8 (71)	16.3 (80)
11–12 years	8.7 (46)	37.0 (46)

Note: Sample size is shown in parentheses.

approximately constant throughout the study and was typically lower than the number that emigrated: on average, four to five animals immigrated each year while seven to eight emigrated. Second, stags were more likely to wander out of the study area in the autumn and be shot: between 1974 and 1979, thirty-five stags usually resident in the study area were shot in peripheral parts of the island, whereas only two hinds were lost during the same period. Finally, the tendency for stag mortality to increase to a greater extent than mortality among hinds also contributed to changes in adult sex ratios.

Similar differences in sex ratio existed between Samhnan Insir, which was used by relatively few stags (adult sex ratio averaged about 1.4 hinds per stag), and Kilmory, which was used by a larger number (about 0.6 hinds per stag). However, in this case differences in the rate of emigration were also involved: 88.9% of males born at Samhnan Insir subsequently either moved to some other part of the study area or left it altogether, whereas only 30% of those born at Kilmory did so [28].

Rutting Behavior

Changes in population size and adult sex ratio were correlated with a decline in breeding competition between stags during the rut. As the number of hinds in the study area rose, the size of harems increased [29] (fig. 12.14*a*), harem stability tended to decline [30], the rut became progressively later [31], and the frequency of fights tended to fall [32] (fig. 12.14*b*).

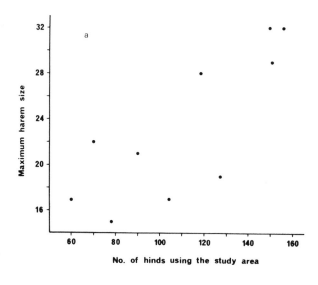

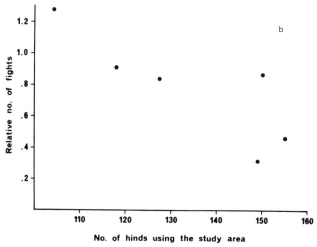

FIG. 12.14. (a) Maximum harem size (calves excluded) observed in each year. (b) Relative frequency of fights in different years plotted against the total number of hinds using the study area in different years (the relative frequency of fights was the total number seen by all observers during the peak rut divided by the number of different stags that held harems in the study area during this period).

To examine changes in rutting behavior, we replicated in the 1980 rut the observation samples originally collected in October 1976. Comparisons showed that, compared with harem-holders in 1976, those in 1980 spent more time feeding [33] and less moving [34], and that the frequencies of all the rutting activities we measured had declined (table 12.2).

TABLE 12.2   Mean Frequency of Different Rutting Activities in Six Harem-Holding Stags Observed in 1976, versus Nine Observed in 1980

Year	Percentage of Time Spent Feeding	Chivies/h	Herds/h	Displacements/h	Chases stag/h	Roars/min
1976	4.49	5.0	4.23	1.08	1.76	2.92
1980	9.81	4.28	3.33	0.70	0.75	2.39

12.4   DENSITY EFFECTS IN OTHER UNGULATES
Effects among Females

Similar relations between population density and growth, fecundity, and calf mortality have been found in a wide variety of other studies of cervids (e.g., Grimble 1901; Banwell 1968; Klein 1968, 1970; Challies 1973, 1974; Caughley 1970; Staines 1970, 1978; Miers 1962; Mitchell, Staines, and Welch 1977; Dasmann and Blaisdell 1954; Marburger and Thomas 1965) as well as in other ungulates (Sinclair 1977; Grubb 1974). It is commonly the case that the first consequences of increasing population density are a rise in juvenile mortality and that effects on fecundity and adult mortality are initially less pronounced. Of particular interest was the absence of any increase in the rate of emigration of hinds, even though by the end of the study reproductive success was lower in the study area than in adjoining areas of the island. This supports the suggestion that association with relatives probably has important advantages and suggests that the rate of emigration in hinds may remain low until whole matrilineal groups begin to disperse. It is not yet clear why the proportion of stags that disperse from the study area has declined.

Population Density and Sexual Dimorphism

Of all the changes in our study population, the reduction in the length of the antlers of yearling stags was the most dramatic (see fig. 12.11). Other studies of cervids have shown that food shortage has a strong effect on antler growth (Richie 1970; Roseberry and Klimstra 1974; McCullough 1979), and in other animals, too, it is commonly the case that the "secondary sexual" characters of males have low growth priorities and are particularly strongly influenced by food shortage (Huxley 1932). The decline in antler size was probably also associated with a greater reduction in the

TABLE 12.3   Sex Differences in Reactions to Increased Population Density

---

Reproduction: Increased population density associated with reduced reproductive success in both sexes.

Emigration: Approximately 30% of stags born in the study area emigrated to other parts of the island before the age of five; dispersal rate declined with increasing population density. Dispersal negligible in hinds throughout the study.

Growth: Male size and development of secondary sexual characters strongly influenced by population density. Sexual dimorphism among adults reduced in populations showing low reproductive performance.

Mortality: Male calves, yearlings, and two-year-olds more likely to die than females of the same age, especially after prolonged winters. Life expectancy of stags, but not hinds, declined with increasing population density.

Adult sex ratio: Usually biased toward females in high-density populations.

---

overall growth of males compared with females, an effect of food shortage that has been documented in other cervids (red deer: Challies 1974; Wegge 1975; reindeer: McEwan 1968; Reimers 1972; Leader-Williams and Ricketts 1982; white-tailed deer: Ullrey et al. 1967; McCullough 1979; mule deer: Julander, Robinette, and Jones 1961) as well as in domestic animals (Widdowson 1976). Across Scottish red deer populations, size dimorphism is positively related to measures of the reproductive performance of hinds that are sensitive to changes in food availability (see fig. 12.15), and similar effects have been found in other deer (e.g., Leader-Williams 1980).

Why is the growth rate of males usually more strongly affected by food shortage than that of females? In part this is probably because the nutritional requirements of adolescent males are greater than those of females (owing to differences in growth rates and metabolism), whereas the rate at which they can ingest and digest food is similar. However, other factors probably also contribute to this effect. Experiments with adult and subadult rats subjected to complete starvation for six days showed that males used relatively more body protein and less fat than females (Widdowson and McCance 1956; Widdowson 1976): of the total energy expended by males, 33% came from body protein and 67% from fat, whereas among females only 8% of energy expended came from body protein and 92% came from fat. This was partly because of the smaller fat reserves of males (170 g/kg) compared with females (210 g/kg), but males also used a lower proportion of their fat reserves before catabolizing muscle, and the fat content of the two sexes was similar at the end of the study (160 g/kg). As a

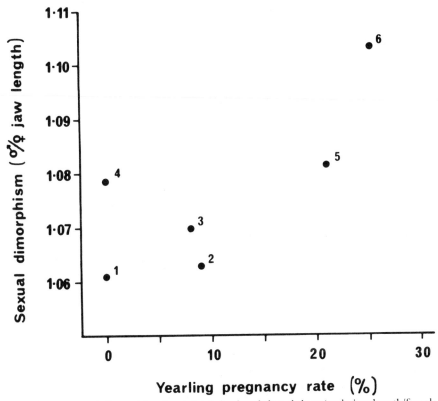

FIG. 12.15. Sexual dimorphism in skeletal size in adult red deer (male jaw length/female jaw length) from different Scottish estates plotted against the proportion of yearling females in the population that were pregnant. (Mitchell 1973; Staines 1978; Mitchell and Crisp 1981. Data from Glen Dye were excluded, since the population was reduced by over 100% immediately before Staines's study and this was likely to have affected pregnancy rates but not sexual dimorphism among adults). 1, Glenfeshie; 2, Rhum; 3, Fiunary; 4, Scarba; 5, Invermark; 6, Glen Fiddich.

result of their increased mobilization of protein, starvation had permanent effects on the size of males but not on that of females (Widdowson, Mavor, and McCance 1964). Widdowson (1976) suggests that the increased rate of protein breakdown in males is a consequence of the effects of androgens on catabolism.

While it is possible that the greater sensitivity of male growth to food reduction is an inevitable consequence of these physiological mechanisms, it may also be an adaptive strategy. As we have already described, several studies indicate that, under conditions of food shortage, large body size increases the risk of starvation (see below). Moreover, the tendency for intermale competition to fall when food availability is low or population density is high (see

Geist 1971*b*) suggests that the benefits of large size will be reduced in the same circumstances. This argument may explain why it is that secondary sexual characteristics used in competitive inter- actions, like antlers and horns, tend to have low growth priority and to decline in relative size in high density populations. Future studies could usefully examine the effects of food shortage on males and females in monogamous and polyandrous species.

Population Density and Male Mortality

Increased male mortality under conditions of environmental stress or food shortage has been observed in previous studies of red deer in Europe (Ahlén 1965). For example, an early onset of winter combined with a late spring in 1976–77 produced a par- ticularly high mortality differential between stags and hinds in Siberia (Sobanskii 1979). In Scottish populations Darling (1937) noticed that: "there is an undoubtedly higher mortality among stags than hinds. It is the common experience to find the popula- tion of deer in a given area composed of females to males in the proportion of two to one, though among the calf crop, the ratio is almost equal or, if anything in favour of males." However, carcass counts have so far failed to find any evidence of differential mortality (see Mitchell, McCowan, and Parish 1973; Mitchell, Staines, and Welch 1977), probably either because most mortality surveys have been carried out where populations are maintained below carrying capacity by shooting or because differential mor- tality occurs mostly in particular years.

There is also extensive evidence of increased mortality among males in other ungulates. Though differential mortality during the first year of life is typically slight and both male and female biases have been reported (Robinette et al. 1957; Woodgerd 1964; Nievergelt 1966; Gossow 1973; Grubb 1974; Geist 1971*b;* Woolf and Harder 1979), many studies have observed increased mortal- ity in yearling or two-year-old males, or both, compared with females. These include studies of mule deer (Taber and Dasmann 1954; Longhurst and Douglas 1953; Klein and Olson 1960; Robinette et al. 1957), white-tailed deer (Hesselton, Severinghaus, and Tanck 1965; Woolf and Harder 1979), wapiti (Flook 1970), caribou (Bergerud 1971), wildebeest (Child 1972), and Soay sheep (Grubb 1974). In addition, evidence that mature males show in- creased mortality and reduced life spans compared with females in high-density populations or in situations where food abundance and quality are low has been found in elk (Anderson 1958; Peek, Lovaas, and Rouse 1967; Flook 1970), mule deer (Robinette et al. 1957; Klein and Olson 1960), reindeer and caribou (Banfield 1954; Klein 1968; Leader-Williams 1980), musk oxen (Tener 1954), wildebeest (Talbot and Talbot 1963) and tsessebe (Child, Robbel, and Hepburn 1972) as well as in macropods (Newsome 1977), rodents (White 1914; Myers and Krebs 1971; Boonstra and Krebs 1979), carnivores (Schaller 1972), domestic animals, and humans (Widdowson 1976).

Why are yearling and two-year-old males more likely to die than females of the same age in many ungulate populations? There are many possible explanations, but several can be excluded with some confidence. It seems unlikely that higher mortality rates among young males are a consequence of their "unguarded" X chromosome since increased male mortality under conditions of environmental stress is found among the young of a number of dimorphic birds even though the *female* is the heterogametic sex (Latham 1947; Lack 1954). Higher mortality rates among young males cannot be accounted for purely by the fact that males commonly disperse from their natal areas whereas females do not: in many cases, differential mortality occurs before the age of dispersal, while higher mortality rates among males are also found in domestic and laboratory animals subjected to food shortage or other forms of environmental stress (Widdowson 1976). Nor does it seem likely that increased predation on young males explains the difference, for it is found in areas where predators are absent (see Grubb 1974). The most likely explanation appears to be that young males are more likely to die from starvation on account of their greater nutritional requirements, lower fat reserves (table 7.2), and increased protein catabolism, perhaps accentuated by higher rates of heat loss.

The principal cause of increased mortality among adult males may be their heavy energy expenditure during the rut and their poor condition at the onset of winter (see Klein and Olson 1960; Robinette et al. 1957; Heptner, Nasimovitsch, and Bannikov 1961; Flook 1970), which may render them more liable to starvation, disease (see Sinclair 1977), and predation (see Schaller 1972; Scrimshaw, Taylor, and Gordon 1968) than adult females, but their larger body size may also contribute to this effect.

It is sometimes suggested that higher mortality rates among males must have evolved for the benefit of the population (e.g., Klein and Olson 1960). No such explanation is necessary. The ultimate cause of many of these differences is probably that selection pressures favor the greater development in males of traits that promote early growth rates, large body size, and success in competitive interactions, and that these have energetic costs that reduce male survival in times of food shortage. That a close association exists between differential mortality, imbalanced adult sex ratios, and traits that favor the competitive ability of one sex was well appreciated by Fisher (1930):

> any great and persistent inequality between the sexes at maturity should be found to be accompanied by sexual differentiations having a very decided bionomic value.

12.5 Summary

1. During the course of the study, population density increased substantially. Habitat use showed no consistent change over this period.
2. As population density increased, conception dates and the timing of coat change became later, calf mortality in winter increased, and fecundity and calf/hind ratios declined. Only three hinds emigrated from the study area during the course of the study.
3. Among stags, antler cleaning and casting dates became later, while antler growth, breeding competition, and emigration rates declined. Mortality rates increased, both in yearlings and in adult animals.
4. Reduced sex differences in growth and increased differences in mortality under conditions of food shortage have been recorded in many other ungulate populations. Both are probably a consequence of the increased nutritional requirements of males compared with females.

Statistical Tests

1. Correlation between the mean birth weight (extrapolated by the method described on p. 46) of calves caught in our study area and mean daily temperature in April and May 1971–80 (1973 excluded, since sample size < 10).

Pearson product-moment correlation coefficient: $r = .875$, $t = 4.783$, d.f. $= 7$, $p < .01$.

2. Correlation between the mean birth weight of calves caught in our study area and the date on which the cuckoo was first heard in Kilmory Glen each year.

Spearman rank correlation coefficient: $r_s = -.857$, $N = 7$, $p < .05$.

3. Correlation between the mean birth weight of calves caught by Nature Conservancy stalkers on Rhum and mean daily temperature in April/May 1961–70. Sample restricted to calves caught when less than eight days old.

Pearson product-moment correlation coefficient: $r = .603$, $t = 2.143$, d.f. $= 8$, $.1 > p > .05$.

4. Comparison of mean birth weights (adjusted for sex of calf and mother's age) between hinds using Kilmory versus Samhnan Insir 1971–79. (Adjusted means: Kilmory 6.70 kg; Samhnan Insir 6.29 kg.)

Analysis of variance: $F_{1,181} = 3.070$, $.1 > p > .5$.

5. Correlation between median conception date in each year (calculated by backdating from calf birth dates using gestation lengths of 234 days for female calves and 236 days for male calves) and the number of hinds using the North Block (1971–80—excluding 1973 when sample size < 10).

Spearman rank correlation coefficient: $r_s = .846$, $N = 9$, $p < .02$.

6. Comparison of median conception dates of hinds using Kilmory versus Samhnan Insir, 1970–78. $N = 139, 53$.

G test: $G = 4.06$, d.f. $= 1$, $p < .05$.

7. Correlation between the proportion of three-year-old hinds that calved each year and the number of hinds regularly using the North Block of Rhum in different years, 1971–80.

Spearman rank correlation coefficient: $r_s = -.852$, $t = 4.593$, d.f. $= 8$, $p < .01$.

8. Correlation between the proportion of milk hinds of four to ten years that calved each year (1971–80) and the number of hinds regularly using the North Block of Rhum in different years.

Spearman rank correlation coefficient, $r_s = -.794$, $t = 3.693$, d.f. $= 8$, $p < .01$.

9. Comparison of numbers of hinds calving as three-year-olds (i.e., conceiving as two-year-olds) between Kilmory and Samhnan Insir (1971–79). $N = 42, 21$.

   G test: $G = 5.15$, d.f. $= 1$, $p < .05$.

10. Comparison of numbers of hinds that reared calves as three-
    year-olds and calved again the following year between Kilmory
    and Samhnan Insir (1971–79).
       Fisher's exact probability test: $p < .01$. $N = 12, 10$.

11. Correlation between the percentage of calves (both sexes)
    dying in their first winter and the number of hinds using the
    study area.
       Spearman rank correlation coefficient: $r_s = .80$, $N = 9$,
       $p < .02$.

12. Comparison of total calf mortality at Kilmory versus Samhnan
    Insir calculated across years (means $= 26\%$ and 36% re-
    spectively).
       Wilcoxon matched-pairs test: $T = 2$, $N = 7$, $p < .05$.

13. Comparison of duration of suckling bouts to calves of three
    months between years, 1975–78.
       Analysis of variance: $F_{3.105} = 2.663$, $p < .05$.
       This result does not appear to be a product of any associa-
          tions between year and changes in the sex of calves, the
          ages of mothers, or their reproductive status: none of
          these variables interacted significantly with year of obser-
          vation.

14. Correlation between calf/hind ratios in April in the study area
    and the number of hinds using the study area in different
    years.
       Spearman rank correlation coefficient: $r_s = -.80$, $N = 9$,
       $p < .02$.

15. Regression of calf/hind ratios on hind population density in
    Red Deer Commission counts of areas 100 km², 1973–77. Data
    were extracted from published commission reports, and den-
    sity was calculated by dividing the number of deer seen by the
    area counted.
       (a) Simple regression of calf/hind ratios on hind density.
       $y = 46.89 - 1.64x$.
       Pearson product-moment correlation coefficient: $r = -.663$,
          $F_{1.14} = 10.956$, $p < .01$.
       (b) Regression of calf hind ratios adjusted for interyear dif-
          ferences using ANOVA (see p. 141).
       Pearson product-moment correlation coefficient: $r = -.958$,
          $F_{1.10} = 53.50$, $p < .001$.

16. Regression of calf/hind ratios on hind population density in data collected in South Ross (extracted from Mutch, Lockie, and Cooper 1976).

$y = 48.2 - 0.83x$.

Pearson product-moment correlation coefficient: $r = -.492$, $F_{1.19} = 6.08$, $p < .02$.

17. Correlation between the percentage of milk hinds that had completed at least half their molt into winter coat before 1 November and the number of hinds regularly using the study area. (Data on coat change were only collected between 1974 and 1980, excluding 1976).

Spearman rank correlation coefficient: $r_s = -1.00$, $N = 6$, $p < .02$.

18. Comparison of percentage of daytime spent grazing in winter by milk hinds at Kilmory versus Samhnan Insir.

Mann-Whitney U test: $U = 18$, $n_1 = 9$, $n_2 = 11$, $p < .05$.

19. Comparison of frequency of threats between hinds feeding on short greens at Kilmory versus Samhnan Insir in winter.

$\chi^2 = 5.60$, d.f. $= 1$, $p < 0.02$.

20. Comparison of (a) mean range size, (b) mean core area size between hinds using different parts of the study area in summer and winter.

(a) Summer, Mann-Whitney U test: $U = 58$, $n_1 = 13$, $n_2 = 16$, $p < .05$.

Winter, Mann-Whitney U test: $U = 49$, $n_1 = 13$, $n_2 = 16$, $p < .02$.

(b) Summer, Mann-Whitney U test: $U = 0$, $n_1 = 13$, $n_2 = 16$, $p < .001$.

Winter, Mann-Whitney U test: $U = 35$, $n_1 = 13$, $n_2 = 16$, $p < .001$.

21. Correlation between the median length of antlers of sixteen-month-old stags and the number of hinds using the North Block in different years.

Spearman rank correlation coefficient: $r_s = -.833$, $N = 9$, $p < .02$.

22. Comparison of frequency of yearlings with antlers greater and less than the median for the population that were reared at Kilmory and Samhnan Insir.

G test: $G = 9.36$, d.f. $= 1$, $p < .01$. $N = 39$, 25.

23. Correlation between median antler weight (adjusted for age

differences) of stags wintering in the study area and the number of hinds regularly using the study area in different years (1971–79).

  Spearman rank correlation coefficient: $r_s = -.967$, $N = 9$, $p < .001$.

24. Correlation between median antler casting dates of seven-year-old stags, 1971–80 (excluding 1973), and the number of hinds regularly using the study area.

  Spearman rank correlation coefficient: $r_s = .883$, $N = 9$, $p < .02$.

25. Correlation between the mean age of stags over three years old that died each year and the number of hinds using the study area, 1974–80.

  Spearman rank correlation coefficient: $r_s = -.750$, $N = 7$, $.1 > p > .05$.

26. Correlation between adult sex ratio (hinds per stag) in the study area and the number of hinds regularly using the study area.

  Spearman rank correlation coefficient: $r_s = .967$, $N = 9$, $p < .001$.

27. Comparison of the number of stags born in the study area 1970–73 that had dispersed before their fifth winter with the number born 1974–76 that did so. $N = 59, 46$.

  G test: $G = 6.587$, d.f. $= 1$, $p < .02$.

28. Comparison of emigration frequency of males born at Kilmory and Samhnan Insir.

  G test: $G = 38.1$, d.f. $= 1$, $p = .001$. $N < 38, 23$.

29. Correlation between (a) the mean, (b) the maximum size of harems in different years, 1971–80, and the number of hinds resident in the study area.

  (a) Spearman rank correlation coefficient: $r_s = .650$, $N = 9$, $.1 > p > .05$. (Data for 1973 not available.)

  (b) Spearman rank correlation coefficient: $r_s = .739$, $t = 3.106$, $N = 10$, $p < .02$.

30. Correlation between the highest level of harem stability reached on any day during the peak period of the rut and the number of hinds using the study area, 1972–80 (excluding 1973, when data were incomplete).

  Spearman rank correlation coefficient: $r_s = -.607$, $N = 8$, $.1 > p > .05$.

31. Correlation of timing of start of rut (= timing of first turning point in the cumulative sum plot of harem stability over time) and the number of hinds regularly using the study area.

Spearman rank correlation coefficient: $r_s = .900$, $N = 9$, $p < .02$.

32. Correlation between the total number of fights seen during the peak rut by all observers in each year, divided by the number of stags that held harems in the study area, and the number of hinds using the study area, 1975–80.

Spearman rank correlation coefficient, $r_s = .714$, $N = 6$, n.s.

33. Comparison of percentage of time spent feeding by harem-holding stags in 1976 versus 1980.

Mann-Whitney U test: $U = 8$, $n_1 = 6$, $n_2 = 9$, $p < .05$.

34. Comparison of percentage of time spent moving by harem-holding stags in 1976 versus 1980.

Mann-Whitney U test: $U = 8$, $n_1 = 6$, $n_2 = 9$, $p < .05$.

13 The Evolutionary Ecology of Males and Females

For it is difference in feeding habits that make some animals live in herds and others scattered about; some are carnivorous, some vegetarian, others will eat anything. So, in order to make it easier for them to get these nutrients, nature has given them different ways of life.

Aristotle, *The Politics*

13.1 SEX DIFFERENCES IN RED DEER

As we have argued in previous chapters, the origins of most of the anatomical, physiological, behavioral, and ecological differences between hinds and stags probably lie in the contrasting factors that affect their reproductive success. Because paternal assistance in rearing calves is unnecessary and hinds occur in groups that an individual stag can defend, direct competition between stags is intense, and variance in reproductive success is greater among stags than among hinds. Consequently, selection pressures favor heavy energy expenditure by stags in the autumn rut as well as the development of traits that enhance fighting ability—including antlers and increased body size. The relationship between body size and reproductive success leads to selection favoring rapid early growth rates and reduced fat deposition in males, as well as to increased early investment by hinds in their male calves. These sex differences have a variety of ecological consequences: as a result of their greater size and faster growth rates, and perhaps also because of increased heat loss, stags require absolutely more energy than hinds except when the latter are lactating. Consequently, stags have to spend more time feeding than hinds during the winter months, particularly at night. The smaller fat reserves of young stags and the poor condition of mature stags at the onset of winter, perhaps combined with increased energy requirements, higher rates of heat loss, and the catabolic effects of androgens, probably explain why stags show higher mortality than hinds, especially after prolonged winters. Increased mortality in stags contributes to the tendency for populations close to carrying capacity to show female-biased adult sex ratios and reduced competition between breeding males. The high costs and lower benefits of large body size in these populations provide a functional explanation for the reduction in male secondary sexual characteristics including size dimorphism and antler development.

Selection pressures operating on hinds also contribute to differences between the sexes. The proximate factors affecting reproductive success in hinds are differences in life span, calf survival, and fecundity. Hinds' heavy investment of energy and protein in calves has probably arisen through selection pressures favoring the reduction of calf mortality and requires lactating hinds to feed on qualitatively superior foods and to occupy habitats of superior quality to those that can be tolerated by yeld hinds and stags. The relation between the breeding success of hinds and their access to high-quality food patches may help explain why hinds remain in their natal area and why mothers tolerate the proximity of their mature daughters. As a result, hinds usually associate with relatives, and the rate of agonistic interactions between hinds is low compared with that between stags.

This reconstruction of the adaptive significance of sex differences in red deer assumes that our observations on Rhum reflect similar differences in other red deer populations. Although Rhum is not typical of Highland estates and Scottish red deer populations differ in many ways from European ones, similar sex differences in breeding behavior, parental investment, group structure, feeding behavior, and mortality patterns have been found in other populations of red deer and elk. This does not say that sex differences do not vary in degree between red deer populations—it seems likely that they do, and different traits may even vary in opposite directions. For example, among red deer populations living on abundant food supplies, sex differences in growth rates and breeding behavior may be augmented, while differences in feeding behavior and mortality may be reduced (see chap. 12). Unfortunately, except in the case of adult size (see fig. 12.15), it is not yet possible to examine correlations across populations.

13.2   INTERSPECIFIC COMPARISONS

Our explanation of the functional significance of sex differences between red deer hinds and stags carries the implication that sex differences that have arisen as a consequence of selection for fighting ability in males should be most highly developed in species where intermale competition is intense and less developed where competition is reduced. Is this the case? Unfortunately, little information is available on the distribution of reproductive success among males in other cervids. Nevertheless, it is reasonable to assume that, in species where females form large parties during the breeding season, individual males are able to monopolize access to a larger number of breeding females than in species where females are widely dispersed in small parties or are solitary. Consequently, variance in reproductive success, intensity of breeding competition between males, and sexual dimorphism should all be more pronounced in the former species than in the latter.

Weight Dimorphism

Comparisons between European cervids support this prediction. The two species that form the largest breeding parties, red deer and reindeer, also show the greatest degree of size dimorphism: in red deer, average male weights in autumn are 1.7 times female weight, while in reindeer they are 1.6 times female weight. In contrast, moose, roe deer, and muntjac, none of which form large breeding aggregations, show reduced size dimorphism (male weights in summer are about 1.2, 1.1, and 1.0 times female weight respectively).

To test this prediction more systematically, we extracted from the literature measures of body weight for adult males and females of fifteen different cervid species (fig. 13.1), allocating species to three categories based on the reported size of breeding parties: A, party size two animals or fewer; B, party size three to five; and C, party size 6 or more. As figure 13.1 shows, the average degree of dimorphism is greatest among species allocated to group C and least among those allocated to group A.

Antlers

In several mammalian groups, male weaponry is also more highly developed in species where males have the opportunity to monopolize access to considerable numbers of females. Among primates (whose principal weapons are their teeth), the relative size of male canines is greater in polygynous species than in monogamous ones (Harvey, Kavanagh, and Clutton-Brock 1978). This suggests that antler size should be greatest in deer species that form large breeding parties and smallest in species where breeding females are widely dispersed. Since antler size across species increases with body size (Huxley 1932), it is necessary to remove the effects of size by comparing antler size relative to body size. When this is achieved by plotting antler length on shoulder height for different cervid species, those allocated to group A have the shortest antlers for their body size, while those allocated to group C have the largest ones (see fig. 13.2).

Obviously this comparison does not account for many differences in antler shape and structure between species. Though the

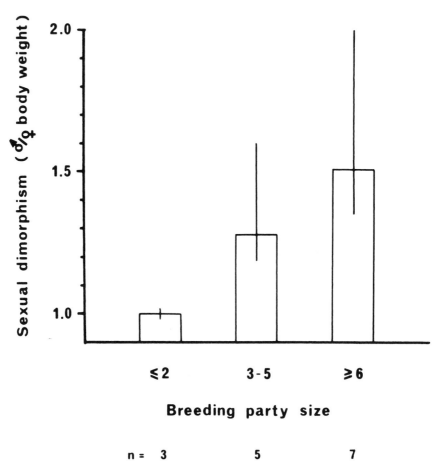

FIG. 13.1. Mean body weight dimorphism (male/female weight) for different species of cervids (see table 13.1) allocated to three categories on the basis of the size of parties in the breeding season. Party size ≤ 2; party size 3–5; party size ≥ 6. Extending lines show ranges of sexual dimorphism in each category.

conformation of antlers has been widely discussed (Geist 1966), many questions remain unanswered. Why do some species have palmated antlers? Why is the overall shape of the antlers of *Odocoileus* species fundamentally different from that of the Old World Cervinae? And why do antler casting times vary so widely? In none of these cases is it easy to find an obvious ecological correlate for the variation.

One other aspect of weapon development must be mentioned. Of all deer species, only among reindeer and caribou do females habitually develop antlers. Most explanations of the occurrence of

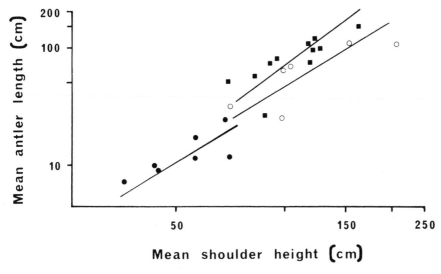

F_{IG}. 13.2. Relationship between shoulder height and antler length in different cervid species (from Clutton-Brock, Albon, and Harvey 1980). *Solid circles,* group A: *Muntiacus muntjak, M. reevesi, Axis kuhlii, Mazama americana, M. gouazoubira, M. rufina, Pudu pudu; open circles:* group B: *Cervus unicolor, Odocoileus hemionus, O. virginianus, Capreolus capreolus, Alces alces, Hippocamelus bisulcus;* group C: *Dama dama, Axis axis, A. porcinus, Cervus duvauceli, C. elaphus, C. canadensis, C. eldi, C. nippon, Elaphurus davidianus, Rangifer tarandus, Hippocamelus antisensis.*

antlers in female reindeer stress the hard conditions they live under and the intensity of competition for food during the long winter (e.g., Espmark 1964; Henshaw 1968, 1969). Espmark (1964) suggests that their antlers allow female reindeer to dominate bulls during the late winter (after the males have cast their antlers) and thus to ensure adequate food supplies for their calves. Henshaw (1968) argues against this on the dubious grounds that females should allow the viability of their calves to be tested by competition for food during their first winter in order to avoid wasting food on animals unlikely to attain maturity. Henshaw suggests instead that antlers alow females to dominate males and ensure that pregnant females obtain adequate food supplies. In contrast, Stonehouse (1968) suggests that antlers are necessary in female reindeer but not females of other species because reindeer possess unusually thick insulating fat and fur and consequently have a greater need of heat-radiating mechanisms in summer. However, as Henshaw (1969) has pointed out, most female reindeer carry little body fat and are shedding their coats during the early summer period when their antlers are growing.

Their effects in protecting calves and in ensuring adequate food supplies for the cows themselves both appear likely to be important advantages of antlers in female reindeer. However, we doubt that the severity of the climate and the related shortage of food supplies explains why antlered females occur only in *Rangifer,* since severe conditions and prolonged food shortage are commonly faced by other cervids. A more striking difference between reindeer and caribou and other species is the size and composition of late winter herds, which are large and contain a considerable proportion of bulls (Bergerud 1974*a,b*). Competition both with males and with unrelated females is likely to be more intense than in other deer and may select for the development of weapons in females. An alternative explanation is that antlers have evolved in female reindeer since, unlike other cervids, they do not hide their calves and thus must defend them against predators (P. Packer, n.d.).

Testis Size

Competition between stags for access to breeding hinds does not necessarily end at copulation; a wide variety of studies show that competition between sperm in the uterus is common. Recent work on primates (Harcourt et al. 1981) suggests that one way males can ensure that they are successful in fertilizing females they have copulated with is to produce a relatively large volume of ejaculate: and that, in species where sperm competition is likely, males have relatively larger testes than in species where it is unlikely as a result of the structure of breeding systems. In deer, sperm competition is likely to be most frequent in species that form large breeding groups that vary in membership from day to day. In species where males defend breeding territories that overlap female ranges (as in muntjac and roe deer), or where males guard a single estrous female for several days on end (as is probably the case in moose as well as in white-tailed deer), sperm competition is probably less likely.

Is relative testis size related to variation in breeding systems in the Cervidae? Data extracted from the literature suggest that this is the case: for their body size, fallow deer, reindeer, red deer, sika, and wapiti have considerably larger testes that muntjac, roe deer, white-tailed deer, mule deer, and moose (see fig. 13.3).

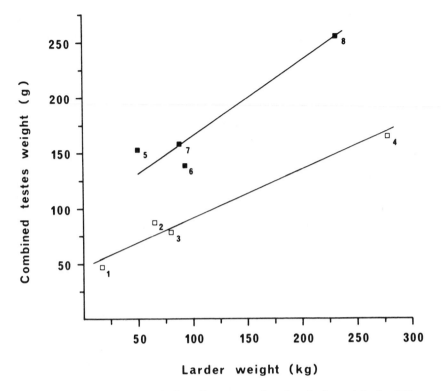

FIG. 13.3. Testis weight during the breeding season plotted on larder weight for different cervids. 1, *Capreolus capreolus* (Short and Mann 1966; Bramley 1970); 2, *Odocoileus virginianus* (Mirarchi, Scanlon, and Kirkpatrick 1977); 3, *O. hemionus* (Anderson, Medin, and Bowden 1974); 4, *Alces alces* (Peek 1962); 5, *Dama dama* (Chaplin and White 1972); 6, *Rangifer tarandus* (Leader-Williams 1979); 7, *Cervus elaphus* (Mitchell, McCowan, and Nicholson 1976); 8, *Cervus canadensis* (Flook 1970). Testis weight includes epididymides; larder weight is approximately 70–75% of live weight.

Whether this also indicates that average testosterone levels in the blood are higher in the first group of species is not yet known.

Growth

Like dimorphism in adults, sex differences during growth should be most pronounced in species showing a high degree of polygyny. Unfortunately, it is not yet possible to make systematic comparisons of growth parameters between deer species. As in red deer, males tend to be born heavier than females in wapiti (Johnson 1951), reindeer and caribou (Leader-Williams 1980), mule deer (Cowan and Wood 1955; Moen 1973), and white-tailed deer (McCullough 1979). Comparative data on growth and fat deposition are unavailable, though we would predict that species allocated to groups A and B would show less sexual dimorphism in growth rates and fat deposition than those in group C. It is of particular interest that the sex difference in fat deposition is reversed in some vertebrate species where females are larger than males—for example, in European sparrow hawks, male nestlings grow more slowly than females but have larger fat reserves (I. Newton 1979, pers. comm.). If our explanation of the sex difference in body fat among growing red deer is correct, we should predict that in sparrow hawks and other species where female juveniles allocate a smaller proportion of their available resources to the development of fat reserves, the reproductive success of females will prove to be more strongly influenced by their early growth rates than will that of males.

Mortality

Our explanation of differential mortality also predicts that sex differences in mortality should be most pronounced in species where polygyny and sexual dimorphism are highly developed. However, many factors contribute to mortality that do not necessarily vary with the degree of polygyny. In particular, increased intolerance between males and male-biased dispersal are found in both sexually monomorphic and sexually dimorphic species (see Greenwood 1980), and this probably explains why mortality differentials are pronounced in some monogamous species as well as in species showing reversed sexual dimorphism (Dubost 1978; Ralls, Brownwell, and Ballou 1980). Consequently, when considering the effects of sexual dimorphism on survival, it is important to exclude as far as possible differences in mortality due to dispersal.

The available literature provides some indication that mortality among males is increased in highly polygynous mammals showing pronounced dimorphism compared with less dimorphic species. Stirling (1975) has argued that, among seals, male mortality rates are higher, relative to female rates, in strongly dimorphic, polygynous species, whereas studies of several monogamous canids (coyote, gray fox, silver-backed jackal) show that sex differences in adult mortality are slight or absent (Ralls, Brownwell, and Ballou 1980). Moreover, in spotted hyenas (where females are larger than males), females have shorter life spans (Kruuk 1972). However, in all these cases, differential mortality may be a consequence of increased aggression between members of the larger sex rather than a cost of dimorphism per se.

Stronger evidence that sexual dimorphism can cause increased mortality in the larger sex in times of environmental stress is provided by comparisons of adult mortality during periods of rapid population decline. When the reindeer population of Saint Matthew Island crashed from 6,000 to 42, only one of the survivors was a male (Klein 1968), whereas in die-offs of wildebeest and reedbuck (both considerably less dimorphic than reindeer) sex differences in mortality were slight (Child 1972; Ferrar and Kerr 1971). In addition, research on game birds shows that sex differences in juvenile mortality are greater in polygynous, dimorphic species than in monogamous, monomorphic ones. Latham (1947) compared the effects of abnormally high and low temperatures on the young of polygynous game birds (pheasant, turkey, mallard) versus monogamous ones (quail, partridge). He found that, in the polygynous species in his sample, females showed improved survival compared with males, whereas this difference was reversed or absent in the monogamous ones. These results are of particular significance, for in this case differences in mortality cannot be attributed to chromosomal differences between the sexes (see p. 279). Finally, in hamsters, the only common laboratory mammal that shows reversed sexual dimorphism, females have shorter life spans in caged populations than males (Festing 1972).

Though several lines of evidence suggest that sex differences in mortality are most pronounced in species where polygyny is highly developed, the relative contribution of different traits to mortality differences is still far from clear. Are mortality differentials principally a consequence of sex differences in size or growth rate or metabolic rate or heat loss? Are they more closely related to the degree of polygyny than to size dimorphism? Do species showing reversed sexual dimorphism show female-biased mortality in times of food shortage? These questions will be answered only by systematic comparisons of mortality differentials in carefully selected groups of species.

Finally, though both juvenile mortality and adult mortality are commonly male-biased among mammals, this is not universal, and other factors can influence mortality differentials. For example, in African elephants, adult cows are more likely than bulls to die during periods of food shortage, though they are the smaller sex (Corfield 1973). This is apparently because elephant calves are more dependent on water sources than adults, and mothers of dependent offspring thus are restricted to the proximity of water and, during periods of food shortage, are more likely to die than bulls, which can wander farther afield (Corfield 1973; D. Western, pers. comm.). In toque macaques, female juveniles are more likely to die from starvation than males (Dittus 1977, 1979). In this case increased mortality among female juveniles apparently occurs both because male juveniles, being larger, are socially dominant over females and can gain priority of food access and because adult females are less tolerant of female juveniles than of males.

Feeding Behavior and Dispersal

In contrast to sex differences arising principally from selection favoring fighting ability in males, differences associated with the nutritional demands of lactation on females should occur in both monogamous and polygynous species. Is this the case? Data on sex differences in feeding behavior in monogamous, monomorphic species are sparse, and no studies have yet distinguished between lactating and nonlactating females. However, in at least some primates that show little or no sexual dimorphism, female populations are clumped in areas of superior habitat, while the intervening areas are used by nonbreeding males (Charles-Dominique and Martin 1972; Clark 1978). In addition, sex differences in feeding behavior occur in some monogamous species (Pollock 1977), though whether they are as pronounced as in polygynous ones is not known. Increased male dispersal at adolescence is widespread among mammals and occurs in a variety of mammals showing little or no sexual dimorphism (Greenwood 1980) as well as in at least one species where females are the larger sex (Dubost 1978).

13.3 IMPLICATIONS OF SEX DIFFERENCES
  For Population Dynamics

The existence of regular differences between males and females in behavior, physiology, and anatomy, as well as in their reactions to environmental stress and to population density, has far-reaching implications in other areas of ecology. Our study suggests that, like competition between species (Lawton and Hassell 1981), competition between the sexes may frequently be one-sided. Models of population dynamics will need to take this into account, and the fact that they have not done so in the past may help to explain the comparatively poor fit between theoretical predictions and observed trends (see Caughley 1977).

Moreover, almost all studies of the consequences of population density have examined correlations between measures of reproductive performance and the combined density of both sexes, making the implicit assumption that the same density-limiting factors apply to males and to females. This is not necessarily so. Where there are pronounced differences in feeding behavior between the sexes, the growth, survival, or reproductive performance of members of one sex may be more closely related to the density of members of the same sex than to the combined density of both sexes. And, in cases where competition between the sexes is one-sided, it is even possible that the performance of one sex may be constrained principally by the density of members of the opposite sex. For example, we should not be surprised if, in red deer, the growth and survival of stags was best predicted by the density of hinds alone.

Last, it is quite possible that interspecific competition has differential effects on males and females. In fact, some evidence for red deer already points in this direction. In an area of Argyll resembling Rhum in altitude and topography but, unlike Rhum, carrying 13.5 sheep/km², niche separation between the sexes was reduced, and the sex ratio was considerably more strongly biased toward hinds (1:1.9) (Osborne 1980).

For Management

Sex differences in behavior and ecology also have implications for the management of ungulate populations. Since male secondary sexual characters have low growth priority and males are more susceptible to starvation than females, managers of ungulate populations cannot hope to maximize both the size and number of shootable males and total meat yield from the same population.

Sex differences in dispersal also have profound implications for culling policy. Since local populations of females are commonly stable and individuals are reluctant to move from their natal area, it is important to take local variation in population density and reproductive success into account when culling females of species that form matrilineal groups, for if they are overculled in one locality and underculled in another they are likely to be slow to redistribute themselves, and the total population will decline. An example of this effect is provided by Osborne's study of red deer on a 15,000 ha estate in Argyll (Osborne 1980). When the number of hinds taken from different parts of the estate was compared with the number usually seen in each area, it was found that the percentage culled each year varied between areas from 0 to 77.1 and that the density of hinds in different areas of the estate was negatively correlated with the proportion that were taken each year, indicating that culling was depressing population density. By allowing the areas that had been overculled to recover and by distributing the cull in relationship to the distribution of hinds, Osborne calculated that the potential sustainable yield could be increased by over 50%. In ungulate species where males disperse, overculling local populations of males is unlikely to have as pronounced effects as overculling females. However, since the size of the sustainable yield that can be taken from populations of males in these species depends extensively on rates of immigration and emigration, and since virtually nothing is known of these rates or of the factors that affect them, it is impossible to calculate the maximum sustainable yield of local populations of males.

Finally, since the nutritional requirements of males and females differ, many common land management practices may have different effects on males and females. Little is known of the effects of land management on wild ungulate populations in general or on red deer in particular (see de Nahlik 1974), but there is some indication that both competition with domestic stock and moorburning can have different effects on stags and hinds. As we have already described, Osborne's study suggests that competition with sheep may affect the habitat use of stags more than that of hinds (Osborne 1980), while a study of the effects of moor-burning on red deer in Jura during the previous century (Evans 1890) indicates that this may influence hind numbers more than stag numbers. Evans leased an estate for twenty-one years and shot the deer lightly each year. During the first six years of his study, little moor burning was carried out, and the deer population remained approximately stable. Subsequently, regular burning was instituted, and the hind population began to increase rapidly whereas the increase in stag numbers was more gradual (see Lowe 1971).

For Studies of Evolution

An understanding of the distribution of sex differences can also offer important insights into the behavior of fossil species. We shall never know for certain how the giant Irish elk *(Megaceros giganteus)* lived in the bleak hills of Pleistocene Europe or what caused it to disappear quite suddenly about eleven thousand years ago, but we can make one prediction with some certainty: the relative size of its antlers and the difference in body size between the sexes indicate that it was polygynous. The possibility of reconstructing the social organization of fossil species from relationships between sexual dimorphism and breeding systems in contemporary species is now widely recognized in studies of primate evolution (e.g., Martin 1980) and holds considerable promise for studies of ungulate phylogeny.

A more doubtful extrapolation from contemporary ecology is the suggestion that, because the costs of secondary sexual characteristics to males may be high, dimorphic species are especially likely to become extinct. The Irish elk is the type example, and the suggested disadvantages of growing and supporting a pair of antlers that often weighed over 30 kg have ranged from epilepsy and sterility to miring in ponds and tangling in trees (Gould 1974). The elk's antlers presumably played such an important role in breeding competition that their benefits offset their costs. Though the costs of growing antlers of this size were presumably high, it seems most unlikely that they could ever have been sufficient to endanger the species—for, before they reached this level, selection would have favored individuals with smaller antlers and reduced antler size in the population.

13.4   Variation in Breeding Systems among Cervids

As we have described, gross differences in the form of breeding systems apparently affect competition between males and the extent of sexual dimorphism. But why have different breeding systems evolved? In other mammals, the degree of polygyny that evolves appears to depend principally on the distribution of females (Emlen and Oring 1977; Clutton-Brock and Harvey

1978). Where females are distributed singly or in small groups, males cannot monopolize large numbers of them, and extreme polygyny cannot develop, whereas in species where females occur in large herds, the potential for polygyny is high.

But what determines the distribution of females? Comparative evidence from several species suggests that variation in the size of female groups is closely related to the distribution of food (Jarman 1974; Estes 1974; Altmann 1974; Clutton-Brock 1974; Owen-Smith 1977; Clutton-Brock and Harvey 1977b; Emlen and Oring 1977). Where food supplies can be defended, females typically occupy feeding territories and live solitarily or in small groups. Where food supplies are too widely distributed in time and space to be defensible, females may live in sizable groups, though they do not always do so. Whether this is the case appears to depend partly on the microdistribution of food. If food is distributed in small, discrete pockets that are widely separated from each other, the presence of other animals feeding in the immediate vicinity is likely to interfere with the rate at which an animal can collect food (see, for example, Wrangham 1977), and females tend to feed alone or in small groups. In contrast, if the microdistribution of food is more even and feeding interference less intense, females commonly feed in large herds (Jarman 1974). There are probably many advantages to aggregating in open country (Bertram 1978), of which, in ungulates, the detection and avoidance of predators may be the most important (see chap. 9).

Can differences in female group size among deer be explained within this general framework? Though adequate studies of the behavior and ecology of cervids are sparse, this appears to be the case (see table 13.1). Both in temperate regions and in the tropics, species that commonly colonize closed-canopy forests and are predominantly browsers occur singly or in small groups both in the breeding season and throughout the rest of the year. These include roe deer and moose in the temperate regions, muntjac and Bawean deer in Asia, and the brocket deer of South America. In some of these species males defend territories overlapping female home ranges, as in muntjac, while in others, like moose, males range widely, associating with individual females for the duration of their estrus.

Among species principally found in broken-canopy forest, forest fringe, or deciduous woodland, which mostly both browse and graze extensively, breeding systems vary widely. In white-tailed deer, bucks associate with small groups of does until they locate a doe in estrus and then form a close "tending bond" with her (Hirth 1977). In hog deer, males and females collect in sizable breeding groups, and males again form close bonds with receptive does and do not attempt to collect harems (Schaller 1967). And, in sambar, males defend rutting territories in traditional mating areas, excluding their rivals and attempting to attract female groups by vocal and olfactory displays (Schaller 1967).

The largest breeding groups are found in species occupying open environments, which rely extensively on grazing but may browse as well. In sika, wapiti, and Père David's deer, males typically collect and defend harems of females that can consist of twenty animals or more (Struhsaker 1967; Horwood and Masters 1970; Schaller and Hamer 1978), and in reindeer, caribou, and barasingh, breeding herds are commonly even larger (Schaller 1967; Espmark 1964; Lent 1965; Schaller and Hamer 1978). In the three latter species dominant males commonly tolerate the proximity of subordinates, and access to receptive females is related to dominance rank. Breeding behavior is highly variable in these species: in forest or in areas where they are heavily hunted, males of red deer and wapiti may tend single females rather than defend harems (Burckhardt 1958; Altmann 1956), while, in some populations, bull reindeer and caribou typically form harems from which subordinate males are excluded (Bergerud 1974a,b). Moreover, not all deer living in open country form large groups: the marsh-living Chinese water deer may be monogamous, and both marsh deer and huemul are typically seen in small parties (Whitehead 1972; Schaller and Hamer 1978).

Another way the distribution of resources can influence the extent of intermale competition is through its effect on breeding synchrony among females (Emlen and Oring 1977). In deer species occupying temperate environments, where food quality and abundance varies widely throughout the year and is highly predictable, reproductive synchrony among females is typically close. Though birth peaks probably occur in most tropical cervids, they are less pronounced than in temperate species, and births occur throughout the year (Schaller 1967; Whitehead 1972). This may help to explain why, in tropical deer, the defense of exclusive harems is less common and why males continue to feed during the

TABLE 13.1 Breeding Group Size and Habitat Type in Deer

Species	Breeding Group Size	Closed-Canopy Forests	Broken-Canopy Forest, Woodland, Forest Fringe	Open Country: Marshes, Moorland, Tundra, Montane Grassland	References
TEMPERATE AND ARCTIC SPECIES					
Capreolus capreolus (roe deer)	B	——	——		Anderson (1953); Prior (1968); Bramley (1970); Klein and Strandgaard (1972)
Alces alces (moose)	B	——	——		Peterson (1955); Geist (1963); de Vos, Brokx, and Geist (1967); Houston (1974)
Odocoileus hemionus (mule deer)	B		——		Dasmann and Tabor (1956); Cowan (1956); de Vos, Brokx, and Geist (1967)
Odocoileus virginianus (white-tailed deer)	B		——		Cowan (1956); de Vos, Brokx, and Geist (1967); Hirth (1977)
Dama dama (fallow deer)	C		————		Chapman and Chapman (1975)
Cervus nippon (sika deer)	C		————		Horwood and Masters (1970); Lowe (1977)
Cervus elaphus (red deer)	C		—————————		Darling (1937); Mitchell, Staines, and Welch (1977)
Cervus canadensis (wapiti)	C		—————————		McCullough (1969); Flook (1970)
Rangifer tarandus (reindeer, caribou)	C		—————————————		Henshaw (1970); Bergerud (1974a)
ASIAN TROPICS					
Elaphodus cephalophus (tufted deer)	?	——			
Muntiacus muntjak (Indian muntjac)	A	——			Barette (1977)

Species		Reference
M. reevesi (Reeve's muntjac)	A	McCullough (1974)
Axis kuhlii (Bawean deer)	A	Blouch and Atmosoedirdjo (1978)
Cervus unicolor (sambar)	B	Boonsong and McNeely (1977); Kurt (1978)
Axis porcinus (hog deer)	C	Schaller (1967); Boonsong and McNeely (1977)
Cervus timorensis (rusa)	?C	Kurt (1978)
Cervus eldi (brow-antlered deer)	C	Sinh (1975)
Hydropotes inermis (Chinese water deer)	A	Whitehead (1972)
Cervus duvauceli (barasingh)	C	Schaller (1967); Boonsong and McNeely (1977)
NEOTROPICS		
Mazama gouazoubira (brown brocket)	A	Whitehead (1972)
M. rufina (little red brocket)	A	Whitehead (1972)
Pudu pudu (pudu)	?A	Roe (1975)
Hippocamelus bisulcus (Chilean huemul)	?C	Povilitis (1978)
Hippocamelus antisensis (Peruvian huemul)	?C	Roe and Rees (1976)
Blastocerus dichotomus (marsh deer)	?B	G. Schaller, pers. comm.
Ozotoceros bezoarticus (pampas deer)	?C	G. Schaller, pers. comm.

breeding season: an individual's ability to remain in breeding condition for several months may be a more important determinant of his reproductive success than the number of hinds he can defend.

Though gross differences in breeding systems among cervids are apparently related to variation in habitat and climate, the study of cervid breeding systems is in its infancy, and many important questions remain unanswered. Why do male sambar and fallow deer defend territories while white-tailed and mule deer bucks tend individual does? Why are Chinese water deer and marsh deer usually seen in pairs or small parties, in contrast to other species occupying open environments? And why does the seasonal pattern of antler growth differ so widely in timing and regularity between species? We hope this study will encourage other fieldworkers to examine the adaptations of the males and females of other species.

13.5 SUMMARY

1. Sex differences that are a consequence of selection for fighting ability in males should be more pronounced in polygynous deer than in territorial or monogamous species. Comparative data indicate that the distribution of sex differences in body weight, growth, weapon development, and mortality follow the predicted pattern.

2. The existence of pronounced sex differences in behavior and ecology among deer has important implications for studies of their population dynamics and evolution as well as for management policies.

3. These comparisons raise the underlying question why breeding systems differ between deer species. Though there are many gaps in our knowledge, current evidence suggests that variation in breeding systems is related to the distribution and density of food supplies, as it is in other groups of vertebrates.

Appendixes

Appendix 1 Taxonomy of Contemporary Cervidae

The taxonomy of existing cervids is confused, and the number of species and subspecies recognized varies widely (see, for example, Corbet 1978; Whitehead 1972). For convenience, we have followed Ellerman and Morrison-Scott's (1951) classification of cervid species (with the exception that we have distinguished between *Cervus elaphus* and *C. canadensis*), dividing them into five sub-families, seventeen genera, and thirty-seven species.

MOSCHINAE
Moschus moschiferus Linnaeus 1758 (musk deer), central and northern Asia, Far East.

HYDROPOTINAE
Hydropotes inermis Swinhoe, 1870 (Chinese water deer), eastern Asia

MUNTIACINAE
Muntiacus muntjak Zimmerman, 1780 (Indian muntjac), India, Asia, and Far East.

Muntiacus rooseveltorum Osgood, 1932, Indo-China

Muntiacus reevesi Ogilby, 1839 (Reeve's muntjac), China

Muntiacus crinifrons Sclater, 1885 (Black muntjac), China

Muntiacus feae Thomas and Doria, 1889 (Fea's muntjac), Tenasserim and Thailand

CERVINAE
Elaphodus cephalophus Milne-Edwards, 1872 (tufted deer), Burma and China

Dama dama Linnaeus, 1758 (Fallow deer), Europe, Asia Minor

Dama mesopotamica Brooke, 1875 (Persian fallow deer), Iran

Axis axis Erxleben, 1777 (chital or axis deer), India, Sri Lanka

Axis porcinus Zimmerman, 1780 (hog deer), northern India to Far East

Axis kuhlii Müller and Schlegel, 1844 (Kuhl's, Bawean deer), Bawean Island

Axis calamianensis Heude, 1888 (Calamian deer), Calamian Islands

Cervus unicolor Kerr, 1792 (sambar), India, Far East

Cervus duvauceli Cuvier, 1823 (barasingh, swamp deer), India

Cervus eldi McClelland, 1842 (Eld's deer), Far East

Cervus nippon Temminck 1838 (sika, Japanese deer), Far East, Japan

Cervus albirostris Przewalski 1883 (Thorold's deer), Tibet

Cervus elaphus Linnaeus 1758 (red deer, shou, hangul, etc.), Europe, North Africa, Asia

Cervus canadensis Erxleben 1777 (wapiti), Asia, North America

Elaphurus davidianus Mine-Edwards 1866 (Père David's deer), formerly China

ODOCOILEINAE
Odocoileus hemionus Rafinesque, 1817 (mule, black-tailed deer), western and central North America

Odocoileus virginianus Zimmerman, 1780 (white-tailed deer), North America, northern South America

Capreolus capreolus Linnaeus, 1758 (roe deer), Europe, central and northern Asia

Alces alces Linnaeus, 1758 (moose, elk), northern Europe, North America

Rangifer tarandus Linnaeus, 1758 (reindeer, caribou) northern Europe, North America

Blastocerus dichotomus Illiger, 1815 (marsh deer), central Brazil, northern Argentina

Ozotoceros bezoarticus Linnaeus, 1766 (pampas deer), Brazil, Argentina

Hippocamelus bisulcus Molina, 1782 (Chilean huemul), western South America

Hippocamelus antisensis d'Orbigny, 1834 (Peruvian huemul), western South America

Mazama americana Erxleben 1777 (red

307

brocket), Central and South America
Mazama gouzaoubira Fischer 1814
(brown brocket), Central and South
America
Mazama rufina Bourcier and Pucheron
1852 (little red brocket), Central and
South America

Mazama chunyi Hershkovitz 1959
(dwarf brocket), Central and South
America
Pudu pudu Molina 1782 (Pudu), Chile,
Argentina
Pudu mephistophiles de Winton 1896
(Pudu), Columbia, Ecuador, Peru

Appendix 2 Red Deer Population of Rhum, 1957–80

Year	Island Population				North Block		
	Stags	Hinds	Calves	Total	Stags	Hinds	Calves
1957	564	741	274	1,584	116	158	66
1958	602	818	295	1,715	116	157	67
1959	626	871	236	1,733	94	196	58
1960	624	862	314	1,800	72	151	67
1961	774	831	325	1,830	97	168	65
1962	721	737	295	1,753	65	83	33
1963	718	725	248	1,691	59	87	27
1964	737	752	353	1,842	68	93	49
1965	703	758	299	1,760	59	80	33
1966	630	676	300	1,606	49	69	31
1967	604	686	280	1,570	51	73	33
1968	598	721	309	1,628	58	69	33
1969	No count						
1970	609	539	210	1,358	171	60	26
1971	492	509	174	1,175	141	66	29
1972	537	624	231	1,392	135	73	35
1973	No count						
1974	585	604	249	1,438	103	82	42
1975	649	700	280	1,629	94	113	49
1976	626	676	260	1,562	128	127	54
1977	596	736	246	1,578	150	139	46
1978	588	650	271	1,509	99	149	61
1979	551	640	239	1,430	79	120	52
1980	502	645	259	1,406	76	148	62

Note: Total numbers of stags, hinds, and calves on Rhum in early spring. Counts by the Nature Conservancy (1957–65) and the Red Deer Commission (1966–80).

Appendix 3 Climatological Summary for Rhum

	Jan.	Feb.	Mar.	Apr.	May	June	July	Aug.	Sep.	Oct.	Nov.	Dec.	Year
Mean maximum temperature (°C)	7.3	7.1	8.4	10.6	13.6	15.9	17.0	17.4	14.8	12.6	9.3	8.3	11.9
Mean minimum temperature (°C)	2.1	1.7	2.3	3.2	5.7	8.4	10.4	10.1	8.8	7.4	3.7	3.5	5.6
Mean daily temperature (°C)	4.6	4.6	5.5	7.1	9.8	11.9	13.6	13.7	11.8	10.0	6.4	5.9	8.7
Mean monthly rainfall (mm)	291	172	221	132	107	161	165	162	260	241	342	257	2511
Mean monthly rainfall, Kilmory (mm)	255	134	171	87	75	119	125	109	197	179	258	204	1913
Number of wet days (> 1.0 mm. of rainfall)	22	17	18	15	15	15	16	15	20	20	23	25	221
Mornings with snow lie (at sea level)	4	3	1	0	0	0	0	0	0	0	1	2	11
Mean wind speed, 1958–68 (kph)	19.7	17.9	18.1	15.2	14.2	14.9	13.6	14.6	14.6	17.1	15.7	21.3	16.5

Note: The averages are for the period 1971–80 at Kinloch Castle except where indicated.

Appendix 4 Characteristics of Soils Connected with Different Plant Associations on Rhum

Number	Grid Reference	pH	loss at 550°C % dry weight	Extractable					Total		Origin of Soil
				Na mg/100g dry weight	K mg/100g dry weight	Ca mg/100g dry weight	Mg mg/100g dry weight	Mn mg/100g dry weight	P mg/100g dry weight	N %	
1	362025	4.8	31.2	11	22	24	19	6.9	0.9	0.89	Torridonian sandstone
2	394005	4.3	71.0	28	51	69	80	0.4	2.2	1.72	Peat on Torridonian
3	307987	4.6	66.3	56	48	125	98	0.2	1.0	1.57	Peat on granophyre
4	332960	4.7	89.1	73	73	139	153	1.1	6.9	2.42	Granophyre
5	347955	5.5	72.7	45	47	182	208	3.6	1.6	2.45	Ultrabasic rock
6	333018	6.8	13.9	10	9	274	16	0.3	0.6	0.59	Trias limestone
7	386005	5.5	44.6	25	33	209	72	2.9	1.7	1.55	Torridian
8	330026	5.8	13.4	26	20	98	52	3.9	0.6	0.36	Trias limestone
9	345950	6.4	15.8	14	11	75	114	Negligible	0.2	0.55	Ultrabasic rock
10	386005	5.5	70.8	37	37	758	101	3.0	1.9	2.87	Torridian
11	320976	6.2	29.6	27	24	225	67	0.5	0.7	1.27	Granophyre
12	353962	5.6	22.9	10	14	58	133	0.4	0.6	0.90	Ultrabasic rock
13	340956	5.1	43.7	31	54	76	79	—	2.6	1.40	Raised beach

Source: Nature Conservancy Council.

Key: 1. *Calluna* heath.
2. *Calluna/Trichophorum/Molinia* wet heath.
3. *Eriophorum/Calluna* bog.
4. *Nardus/Calluna* heath.
5. *Schoenus/Molinia* fen.
6. *Schoenus* flush.
7. *Molinia* grassland.
8. Herb-rich *Calluna* heath (faces on Ca-rich soils).
9. Herb-rich *Calluna* heath (faces on Mg-rich soils).
10. *Juncus acutiflorus* marsh.
11. Species-rich *Agrostis/Festuca* grassland (Ca-rich soil).
12. Species-rich *Agrostis/Festuca* grassland (Mg-rich soil).
13. Species-poor *Agrostis/Festuca* grassland.

Na, K, Ca, and Mg extracted with nitrogen ammonium acetate solution (pH 7.0).
P extracted with Truog's reagent (0.002$_N$–H_2SO_4 buffered at pH 3.0).
pH by glass electrode 1 : 2.5 soil : water dilution.

Appendix 5   Digestibilities of Common Upland Plants

	Spring	Summer	Autumn	Winter
Grasses				
Deschampsia flexuosa	75	57–68	77	62–69
Festuca rubra	75	57–68	72–73	48–63
Holcus molis	76	62–72	69–73	42–53
Agrostis tenuis	76	62–72	64–70	35–41
Molinia caerulea	67	47–65	50	—
Agrostis/Festuca				
grassland	32–44	47–65	50	—
Sedges				
Eriophorum vaginatum	62	37–49	39	43
Trichophorum				
caespitosus	67	58–64	40	—
Carex spp.	67	58–64	56	54
Dwarf shrubs and trees				
Calluna vulgaris	60	44–52	—	—
(Mature)	—	41–42	—	—
(Mature)	29–31	—	24	27
(Mature)	—	—	45	—
(Pioneer)	—	47	—	—
(Building)	—	46–48	46	42
(Mature)	—	49	—	—
Juniperus communis	41–43	—	38	43
Betula spp.	27–34	—	25	29

Note: In vitro percentage of dry matter digestibilities, usually using sheep liquor (from Kay and Staines 1981).

Appendix 6 Number of Hinds and Stags One Year Old or Older Resident in the Study Area in Different Years.

| Year | Stags | Hinds | | | | |
		Upper Glen	Kilmory	Intermediate	Samhnan Insir	Total
Total area (km²)		3.0	2.9	2.0	1.4	
Area of greens (ha)		5.1	47.8	2.5	18.0	
1970	—	11	13	15	14	53
1971	124	11	14	18	17	60
1972	129	12	19	20	19	70
1973	141	14	23	22	19	78
1974	125	15	29	25	21	90
1975	129	16	37	28	23	104
1976	130	20	40	29	29	118
1977	148	25	46	29	28	127
1978	157	29	52	33	36	150
1979	135	25	57	35	32	149
1980	124	—	—	—	—	157

Note: Between 1974 and 1979 thirty-five stags and two hinds with core areas in our study area were shot while in peripheral localities.

Individuals were defined as resident if they were seen in at least 10% of census in at least four months of the year. Hinds were allocated to the four main parts of the study area depending on the location of their core areas. These figures are lower than in Appendix 2, since island counts inevitably include a proportion of animals not resident in the study area.

Appendix 7   Altitude and Vegetation in the Study Area

1.   Proportion of Ground within the Study Area at Different Altitudes (m)

0–15	16–30	29–60	61–120	121–240	> 240 m
8.0	16	18	31	24	7

2.   Proportion of Ground at Each Level Covered by Different Plant Communities

	Altitude (m)				
Community	0–15	16–30	31–60	61–120	> 120
Heather moorland	7.3	26.9	43.0	53.4	81.5
Eriophorum bog	18.3	30.0	27.9	26.2	15.7
Molinia grassland	19.2	26.9	21.8	18.1	2.8
Agrostis/Festuca greens	43.9	12.3	4.2	1.0	—
Juncus marsh	11.0	3.9	3.0	1.3	—

Appendix 8 Estimates of Lifetime Reproductive Success in Hinds

Since red deer hinds on Rhum can live as long as nineteen years (Guinness 1979) and our study covers only ten years, measures of the variance in lifetime reproductive success were not available for any cohort. Nevertheless, our data span a large proportion of the breeding life span of a considerable number of hinds and can be used to estimate the distribution of lifetime breeding success in the population. We wish to emphasize that these estimates are only approximate, for they are necessarily based on age-specific fecundity and mortality curves calculated on an expanding population (see Caughley 1977), and our methods will tend to underestimate the variance between individuals.

The sample of hinds we followed throughout their entire breeding life span was biased toward animals that had died young. To obtain an estimate of variance in lifetime reproductive success that was free from bias, we extracted records for all forty-three hinds alive in 1979 that were born between 1963 and 1971. The reproductive performance of all these animals had been monitored for at least six years and in some cases for as many as eleven years. Each hind was allocated an age at "death" at random from the observed distribution of life expectancy for hinds of her age between 1971 and 1978. To estimate the lifetime success of these animals we then:

1. Counted the number of calves each hind had reared to one year old during the period when she had been observed (her observed reproductive success, ORS).

2. Determined the mean deviation for each hind from the average ORS for hinds of the same age in the population over the same period.

3. Estimated the reproductive success of each animal during the period of her life span for which we had no records from the mean reproductive success for hinds of the same age, corrected by adding/subtracting the individual's mean deviation from this average.

4. Repeated this process ten times for each hind to avoid any random effects of step 1 and calculated the hind's mean estimated reproductive success (ERS) over this period, adding this to her ORS. These values were then cast in a frequncy distribution.

5. This still ignored hinds born in the 1963–71 cohorts that died before 1979. From our knowledge of age-specific mortality between 1971 and 1978, we estimated the number of hinds born in these cohorts that would have died before 1979: our data indicated that thirty-five would have died before reaching breeding age (three years) and twenty-one would have died between three and fifteen years old. Of the latter, we estimated that one animal would have died at each of the following ages: 3, 4, 5, 6, and 15, and two would have died at 7, 8, 9, 10, 11, 12, 13, and 14 years of age. Each of these animals was allocated the mean reproductive success of individuals "dying" at the same age in the sample described in step 4 and these figures were added to the estimates described in step 4. This distribution (which approximated a normal distribution) was smoothed by reiteration using the same mean and variance.

6. The estimated number of individuals that died before age three was added to produce figure 5.1.

Appendix 9 Reproductive Value

A female's reproductive value, as originally defined by Fisher (1930), is her age-specific expectation of producing future offspring (Pianka 1978). Reproductive value is defined as

$$\frac{V_x}{V_o} = \frac{e^{rx}}{l_x} \int_x^\infty e^{-rt} l_t m_t dt \ ,$$

where V_o is the expectation of future offspring at birth;

V_x is the expectation of future offspring from age x until the postreproductive stage is reached;

e^{rx} and e^{-rt} are the instantaneous rate of increase per individual at age x and age t respectively (these weight offspring according to the direction in which the population is changing).

l_x and l_t are the probability of surviving to age x and age t respectively;

m_x and m_t are age-specific fecundity rates, usually calculated in terms of the number of *female* offspring that will be produced (Caughley 1977).

Reproductive value is commonly used by population biologists to calculate the contribution of different age groups to future generations. In this study we were interested in the general form of the curve in order to predict how parental investment might change with age (see chaps. 5 and 7). Unlike most data sets, which provide measures of age-specific fecundity at a single point in time, ours allowed us to calculate the fecundity of the same cohort at different ages. This permitted the use of Pianka's simplified definition of reproductive value,

$$V_x = \sum_{t=x}^{\infty} \frac{l_t}{l_x} mt \ .$$

Taking the 1972 cohort, for which measures of fecundity were available in each year from 1975 to 1980, we estimated the average fecundity of its members beyond 1980. This was achieved by calculating their mean deviation between the ages of three and eight from the age-specific fecundity curve for all animals in the study population and extrapolating this until the postreproductive state was reached. Curves calculated for other cohorts followed the same pattern, though at different elevations, as a result of density-dependent changes in fecundity.

One important point concerning reproductive value should be noted. It ignores any relation between the mother's age and calf survival. If, as our data indicate, calf survival is related to maternal age, this will affect both the individual's contribution to the next generation and selection pressures affecting parental investment. Since we could recognize individuals, we were able to modify these for different age categories by the age-specific probability of calf mortality (see fig. 5.5).

Reproductive value normally is calculated only for females. However, a male's investment in reproduction should also be sensitive to changes in his future expectancy of breeding success. Our data allowed us to estimate reproductive value for stags using the same methods as among hinds. Figure 7.5 compares estimates of the reproductive value of stags born in 1972 with estimates of that of hinds born in the same year.

316

Appendix 10   Fighting Success in Stags

Measuring the fighting success of individual stags posed several problems. Though it was clear that large individual differences existed, it was not possible to establish a rank order between stags within particular breeding seasons: not only did many stags never fight each other, but dominance relations between individuals changed frequently during the course of the rut as stags exhausted their energy reserves or were wounded. To rank stags on the number of fights won or the number of different individuals beaten would have yielded unrealistic results, since the status of those beaten was of obvious importance. Our way of solving this problem was to weight an individual's rank according to the ranks of his opponents (see Sade 1972): for each stag in each year between 1974 and 1976, we calculated the number of stags he beat (B), plus the total number they beat excluding the subject (Σb), the number of stags he lost to (L), plus the total number they lost to, excluding the subject (Σl). Our index of fighting success was

$$\frac{B + \Sigma b + 1}{L + \Sigma l + 1}$$

(adding one to each side of the ratio reduced the chance of an anomalous result in cases where individuals were either never observed to beat other animals or never observed to be beaten).

To check that this measure gave as realistic an estimate of social rank as traditional dominance measures (see Brantas 1968; Richards 1974; Syme 1974), we calculated dominance ranks using our method for published data on reindeer (Espmark 1964), horses (Clutton-Brock, Greenwood, and Powell 1976), and bison (McHugh 1958). We then compared these with the authors' own measures, based either on the technique of minimizing irregularities in the hierarchy (see Brantes 1968) or on the ratio of the number of individuals that each animal dominated to the number that dominated him. In all cases our estimates gave results very similar to the original rank orders ($r_s = .929$ [reindeer]; .980 [Highland ponies]; .992 [American Buffalo]; $p < .001$).

Like most other measures of dominance, ours assumed that an individual that beat another individual could beat all those animals the latter beat: though this was generally true, it was not always the case. It was also likely to have given unrealistic estimates of fighting success for stags that rutted in outlying areas and held few hinds within the study area, as well as for stags that were never seen either to beat or to be beaten and consequently scored the relatively high index of 1. However, such cases should have tended only to obscure relationships between fighting success and other variables.

In our calculations of fighting success we included all twenty-five stags over five years old that rutted in the study area between 1974 and 1976, but we excluded three animals that rutted on the periphery of the area and were consequently never seen involved in fights and two whose ages were uncertain.

Appendix 11 Estimates of Lifetime Reproductive Success in Stags

As we describe in chapter 7, stags rarely held harems before they were five years old, and most individuals did not continue to do so beyond the age of eleven. Since 1971, thirteen stags have rutted in the study area each year between their fifth year and their death. Since mean age at death in this sample (11.1) did not differ from that in the population as a whole (11.0), we assumed that it was not biased toward animals that had died young.

Reproductive success for each stag in each year was calculated by the method described on p. 146, and these values were added to provide a measure of the success of each animal over the whole period. These values were then cast in a frequency distribution that was again smoothed by reiteration.

Since approximately 35% of stags died before the age of five, we added five zero values to this distribution, to produce figure 7.4.

318

Appendix 12 Definition of Party Size

Where animals aggregate in parties of unstable membership within which nearest-neighbor distances are highly variable, there is often a problem in deciding where one party stops and another begins. In practice, it is usually necessary to employ a fixed nearest-neighbor distance below which two individuals are regarded as belonging to the same party and above which they are allocated to different parties. The observer can then use this criterion to delineate parties, stringing together nearest neighbors not separated by more than this distance. For example, see figure A.1a.

At what absolute value should the nearest neighbor distance be set to

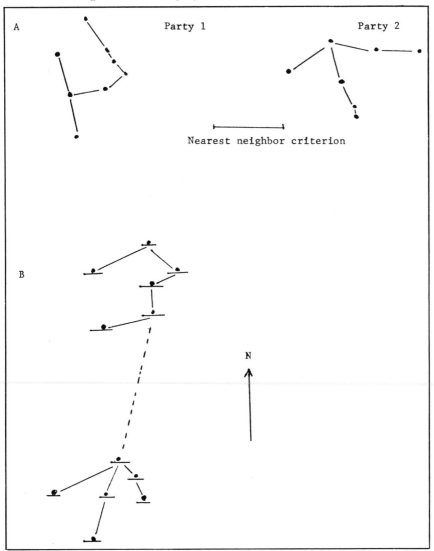

Fig. A.1. Methods used for determining party membership.

provide a reasonable definition of parties? If animals are aggregated in parties, the distribution of distances between *all* dyads in a population should be bimodal, with the lower mode representing intraparty distances and the higher mode representing interparty distances. Though such data are often impossible to collect, the form of the distribution can be identified by measuring the distance between each individual and its nearest neighbor in a constant direction. For example, in figure A.1*b* the nearest neighbor distances are measured only in a northerly direction. A plot of the distribution of nearest-neighbor distances calculated in this way will also show a bimodal distribution of animals aggregated in parties.

On one day in July and on one in January, we estimated the distances between all deer in Kilmory Glen and their nearest neighbors in a given direction (as in fig. A.1*b*). On both days, a distinct bimodality was present, with the discontinuity falling at between 40 and 60 m. We consequently used a 50 m distance criterion to define parties.

To check whether this provided a meaningful definition of grouping, we also examined the proportion of individuals that were engaged in the same activity as their nearest neighbor, predicting that the degree of activity synchrony should be higher within parties than between them. Among nearest neighbors allocated to the same party, 90% of individuals were engaged in the same activity, whereas, among those allocated to different parties, only 56% were engaged in the same activity, a level similar to that found throughout the population as a whole.

Appendix 13 Range and Core Area Size

Range Size

Each individual deer recorded in each census was allocated to a 100 m by 100 m quadrat. A computer program was developed to plot these sightings for the period in question (e.g., winter) and draw a concave boundary around the outermost locations in the four steps described below.

1. The algorithm began at the most northerly location and moved south, examining each row of quadrats in turn, searching for the most westernly observation at each latitude (fig. A.2). If the sightings in row 3 occurred east of the western boundary of row 2, they were ignored and the computer proceeded to the next row until it found an observation to the west of the rows above. Rows that did not have sightings had an interpolated boundary averaged between the previous row of observations and the next. This procedure continued south until the most extreme westerly location was found.

2. Having found the most westerly location, the computer returned to the northernmost sighting and drew the eastern boundary in the same way as step 1.

3. Once the most easterly observation was located, the computer moved to the most southerly sighting and then proceeded northwards row by row, interpolating the western boundary until the most westerly location, identified in step 1 was reached.

4. The boundary was closed by returning to the most southerly location and linking the eastern observations as described above.

The number of grid squares within the boundary were then

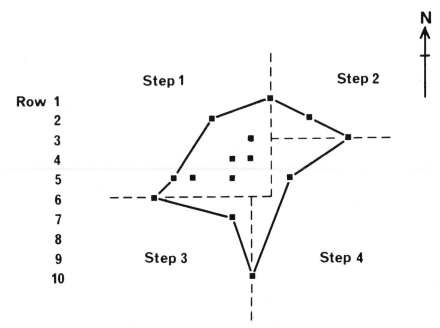

FIG. A.2. Schematic representation of the method used to define range boundaries.

counted to provide a minimum estimate of the area (in hectares) of each individual.

Core Area Size

Although most deer ranged widely, individuals usually concentrated their time in a small proportion of their range. To identify in a nonarbitrary way the part of each animal's range that was heavily used, we developed an algorithm that began by calculating the centroid of the range and then excluded the most distant sightings until only a specified percentage of observations remained. The algorithm then identified the quadrats included, drew a new boundary around them using the technique described above, and calculated the area within this boundary.

Though this provided a nonarbitrary method of defining the center of the range, it did not itself provide any indication of the proportion of sightings that should be excluded to provide a meaningful definition of each individual's "core area." Should this be 10% or 80%? One way of answering this question is to test the proportion of each animal's range that is included within the boundary when the percentage of observations included is progressively reduced (see fig. A.3). If

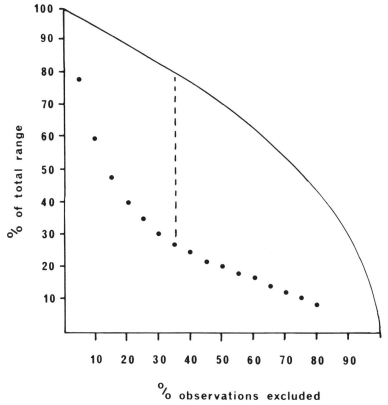

FIG. A.3. Schematic representation of the change in the proportion of an animal's range that is included within the concave boundary as the percentage of outlying observations that are excluded is increased.

the animal used part of its range consistently heavily and the remainder occasionally, a change in the gradient of this curve should be apparent at the point at which successive increases in the percentage of sightings excluded begin to erode the heavily used portion of the range.

This analysis was performed on sightings for twenty hinds. Though not all analyses showed a discontinuity, the majority did so when 30–40% of sightings had been excluded. We consequently defined an animal's core area in the number of adjacent quadrats accounting for 65% of all sightings.

Vegetation Distribution

The plant community occupying the midpoint of each quadrat in the study area was determined from Ferreira's (1970) vegetation map. This allowed us to compare the proportion of ground within ranges and core areas that was covered by different plant communities.

Range Overlap

A final subroutine of the ranging algorithm calculated the number of quadrats common to the core areas of pairs of individuals and expressed these as a proportion of the total number of quadrats in each animal's core area.

Appendix 14   Grazing Bouts

While it is obvious that grazing is not distributed randomly throughout the day but occurs in bouts, defining such bouts in a useful way presents a variety of problems. All activities, particularly grazing, are frequently interrupted. For example, a hind may be grazing intensively but will, from time to time, raise her head and stare around her (see Clutton-Brock and Guinness 1975). The observer, recording her behavior at the end of each minute, will record her as "inactive" at such times. If the length of grazing "bouts" is calculated from the number of consecutive minute records of a particular activity, results are unlikely to be biologically meaningful (see Slater 1974).

There are various ways of avoiding this problem. Field observers commonly claim that the distributions of such feeding "bouts" are bimodal, with the lower mode representing interrupted bout sections and the upper mode representing full bouts. However, there is no published quantitative data on this topic collected on free-ranging populations, and no such bimodality is found among deer.

Alternatively, bouts may be defined according to critical time limits, by disregarding gaps of less than a specified length. In this case the problem arises as to what length of "gap" should be included within a bout (Slater 1974). In some cases arbitrary criteria are used (e.g., Rowell 1961), while in others, critical gap length has been chosen after identifying some discontinuity in the frequency distribution of gaps (see Slater 1974). A useful way of distinguishing irregularities in the distribution of gap length is to plot the log survivor function of gap frequency (Slater 1974). If events of a given type are distributed at random (i.e., forming a Poisson process), the histogram of intervals between them will follow a negative exponential distribution (Slater 1974). When gap frequency is cumulated backward and plotted on a semilogarithmic scale, the distribution of gaps between independent events yields a straight line, the slope of which is proportional to

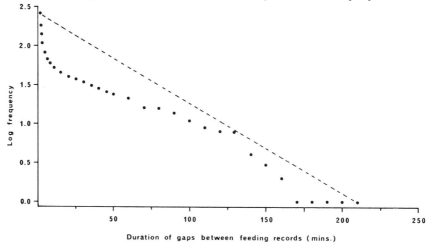

Fig. A.4. Log survivorship curves of gaps between periods of grazing for one milk hind observed in July 1974. Observations of other individuals in winter and summer produced similar results.

324

the probability of an event at any given time after the last event. By plotting frequencies in this way, it is relatively easy to compare the observed distribution with a negative exponential one and to identify changes in probability, which appear as changes in slope.

When the data collected in continuous watches of hinds were treated in this way (fig. A.4), it was apparent that interfeeding gap length differed markedly from a negative exponential distribution. Consequently, it was reasonable to suppose that very short interfeeding gaps represented temporary interruptions within feeding bouts (see Slater 1974). The observed distributions diverged maximally from the negative exponential distribution at 10 min, indicating that all gaps between feeding of less than this length should be included within feeding. When gaps less than 10 min long were excluded, the numbers of bouts occurring in the four quartiles of the expected negative exponential distributions did not differ (see Slater 1974).

When watching individual deer throughout the day, we often did not observe the start of their first grazing bout in the morning or the end of their last bout in the evening. This was comparatively infrequent in summer, and only complete bouts were included in our analysis. In winter, day lenght was reduced while feeding bout length increased, and as a result the proportion of bouts that either began or terminated outside the duration of our watches was high. The exclusion of incomplete bouts from the sample would have produced a strong bias against long bouts, so incomplete bouts longer than 20 min were included, and our figures underestimate bout lengths. Despite this, as we describe in chapter 11, grazing bout lengths were longer in winter than in summer.

Appendix 15 Milk Composition

			Gross Composition		
Stage of Lactation (days)	Total Solids	Fat	$N \times 6.38$	Lactose	Gross Energy (kcal/100 g milk)
3–30	21.1	8.5	7.14	4.45	130
31–100	23.5	10.3	7.63	4.45	156
100	27.1	13.1	8.59	4.46	185

			Mineral Composition				
Stage of Lactation (days)	Ash	Ca	P	Mg	Na	K	Cl
3–30	1.18	0.22	0.22	0.018	0.033	0.12	0.070
31–100	1.11	0.22	0.18	0.018	0.037	0.13	0.071
100	1.10	0.25	0.19	0.022	0.035	0.12	0.080

Source: Arman et al. (1974).

Note: Mean composition (g/100 g milk) of milk of five red deer hinds.

Appendix 16 Suckling Bout Duration

							Calf Age in Weeks							
	0–2	3–4	5–6	7–8	9–10	11–12	13–14	15–16	17–18	19–20	21–22	23–24	25–26	27–28
Mean (sec)	177.0	105.8	88.2	100.1	61.2	75.0	74.6	45.6	38.1	37.2	43.6	33.8	28.4	39.1
SD	110.8	70.3	42.0	64.8	6.9	80.1	65.8	26.9	23.7	14.0	34.2	13.8	9.3	25.1
N	53	63	37	16	6	14	25	31	31	25	25	12	14	9

Source: Cockerill (n.d.).

Note: Mean duration (seconds) of suckling bouts involving calves of different ages. Estimates were based on records of bouts collected from a large sample of calves during censuses from 1975 to 1978. Rejections (bouts < 1 sec) were excluded.

Appendix 17 Frequency of Rutting Activities

	Period of Rut		
Activity	Early	Middle	Late
Roar bouts	15.9	53.3	11.50
Grunt	1.87	3.74	1.00
Thrash	0.26	0.98	0.39
Wallow	0.00	0.13	0.11
Flehmen	0.44	0.92	0.34
Lick hind	1.0	3.90	1.9
Herd hind	0.50	4.89	1.56
Herd calf	0.00	0.65	0.00
Chivy hind	4.44	9.22	1.56
Mount hind	0.00	0.99	0.33
Ejaculate	0.05	0.24	0.06
Hours of Observation	39	61	18

Note: Mean frequencies (per hour) of different rutting activities for one nine-year-old stag (SAGY) when holding a harem of approximately twenty hinds in 1973. Separate values are shown for the early rut (20 September–1 October), the middle rut (1 October–15 October), and the late rut (15 October–25 October).

Appendix 18 Grouping Criteria

Year	Group Threshold	Number of Groups of ≥ 2 Animals	Sample Size	Percentage of Deer Allocated to a Group of ≥ 2	Distribution of Group Size					
					2	3	4	5	6	7+
1973	22.0	17	82	64.6	9	4	2	0	1	1
1974	25.5	25	108	61.1	16	5	2	1	1	0
1975	24.0	32	115	80.9	19	8	0	1	2	2
1976	22.0	33	126	76.9	20	6	1	2	3	1

Note: Results of a single-link cluster analysis on association frequency between hinds one year old or older for 1973–76.

Appendix 19 Habitat Use by Month in Hinds and Stags

Community	Jan.	Feb.	Mar.	Apr.	May	June	July	Aug.	Sep.	Oct.	Nov.	Dec.
Hinds												
Short greens	13.4	22.9	25.9	25.3	28.0	32.9	30.4	32.9	34.5	34.8	18.0	11.2
Long greens and												
Juncus marsh	16.2	17.3	28.6	20.8	21.7	20.1	12.4	11.9	9.1	11.3	10.8	15.5
Molinia grassland	27.1	28.8	23.8	21.8	14.5	8.6	12.7	10.2	15.2	22.3	27.6	25.7
Calluna moorland	28.1	23.0	19.3	25.0	21.9	30.5	40.8	39.1	31.5	20.0	27.7	27.8
Seaweed and littoral debris	14.4	5.0	3.3	7.9	0	1.2	0.7	4.2	6.5	6.1	7.5	7.0
N (feeding records) =	(219)	(265)	(153)	(293)	(308)	(186)	(206)	(243)	(249)	(336)	(199)	(248)
Stags												
Short greens	13.7	13.5	32.9	25.3	29.3	40.8	41.4	38.3	23.6	33.2	35.8	18.4
Long greens and												
Juncus marsh	7.3	9.3	5.6	16.8	23.6	22.1	19.2	14.2	15.4	23.6	10.8	11.4
Molinia grassland	25.9	19.1	26.3	26.8	25.9	13.0	13.6	16.0	9.0	13.0	25.7	24.2
Calluna moorland	26.6	31.2	21.7	19.7	13.4	15.9	23.8	26.8	46.3	17.2	9.6	31.8
Seaweed and littoral debris	21.4	15.7	6.4	4.8	0.9	0	0.3	1.4	1.6	10.5	10.1	8.5
N =	(89)	(109)	(63)	(133)	(179)	(115)	(97)	(119)	(96)	(63)	(58)	(79)

Source: Figures based on census data.

Note: Mean proportion of grazing hinds one year old or older and stags more than three years old seen on different plant communities in different months, 1973–75. Mean sample size per month per year is given in parentheses.

References

Ahlén, I. 1965. Studies on the red deer, *Cervus elaphus* L., in Scandinavia. III. Ecological investigations. *Viltrevy* 3:177–376.

Alexander, G. 1956. Influence of nutrition upon duration of gestation in sheep. *Nature* 178:1058–59.

Alexander, G., and Davis, H. L. 1959. Relationship of milk production to number of lambs born or suckled. *Austr. J. Agric. Res.* 10:720–24.

Alexander, R. D. 1974. The evolution of social behaviour. *Ann. Rev. Ecol. Syst.* 5:325–83.

Alexander, R. D.; Hoogland, J. L.; Howard, R. D.; Noonan, M.; and Sherman, P. W. 1979. Sexual dimorphism and breeding systems in pinnipeds, ungulates, primates and humans. In *Evolutionary biology and human social behavior: An anthropological perspective,* ed. N. A. Chagnon and W. Irons, pp. 402–35. North Scituate, Mass.: Duxbury Press.

Allden, W. G. 1970. The effects of nutritional deprivation on the subsequent productivity of sheep and cattle. *Nutr. Abstr. Rev.* 40:1167–84.

Allden, W. G., and Whittaker, I. A. MacD. 1970. The determinants of herbage intake by grazing sheep: The interrelationship of factors influencing herbage intake and availability. *Austr. J. Agric. Res.,* 21:755–66.

Altmann, M. 1952. Social behaviour of elk *(Cervus canadensis nelsoni)* in the Jackson Hole area of Wyoming. *Behaviour* 4:116–43.

———. 1956. Patterns of social behaviour in free-ranging elk of Wyoming. *Zoologica* 41:65–71.

———. 1963. Naturalistic studies of maternal care in moose and elk. In *Maternal behavior in mammals,* ed. L. H. Rheingold, pp. 233–53. New York: Wiley.

Altmann, S. A. 1974. Baboons, space, time and energy. *Am. Zool.* 14:221–48.

Anderson, A. E.; Medin, D.E.; and Bowden, D. C. 1974. Growth and morphometry of the carcass, selected bones, organs and glands of mule deer. *Wildl. Monogr.* 39:1–122.

Anderson, C. C. 1958. The elk of Jackson Hole. *Wyoming Game Fish Comm. Bull.* 10:1–184.

Anderson, J. 1953. Analysis of Danish roe deer population (*Capreolus capreolus* L.) based on extermination of the total stock. *Dan. Rev. Game Biol.* 2:127–55.

Anderson, J. E. M. 1976. Food energy requirements of wild Scottish red deer. In *The red deer of South Ross,* eds. W. E. S. Mutch, J. D. Lockie, and A. B. Cooper. Edinburgh: Department of Forestry and Natural Resources, University of Edinburgh.

Appleby, M. C. 1980. Social dominance and food access in red deer stags. *Behaviour* 74:294–309.

————. 1981. Social dominance: Functional aspects in red deer stags. Ph.D. diss., University of Cambridge.

Arbuthnott, J. 1710. An argument for divine providence, taken from the constant regularity observed in the births of both sexes. *Phil. Trans. Roy. Soc.* 27:186–90.

Arman, P. 1974. A note on parturition and maternal behaviour in captive red deer (*Cervus elaphus* L.). *J. Reprod. Fert.* 37:87–90.

Arman, P.; Hamilton, W. J.; and Sharman, G. A. M. 1978. Observations on the calving of free-ranging tame red deer (*Cervus elaphus*). *J. Reprod. Fert.* 54:279–83.

Arman, P.; Kay, R. N. B.; Goodall, E. D.; and Sharman, G. A. M. 1974. The composition and yield of milk from captive red deer (*Cervus elaphus* L.). *J. Reprod. Fert.* 37:67–84.

Arnold, G. W. 1964. Factors within plant associations affecting the behaviour and performance of grazing animals. In *Grazing in terrestrial and marine environments,* ed. D. J. Crisp, pp. 133–54. Oxford: Blackwell.

Arnold, G. W., and Dudzinski, M. L. 1969. The effect of pasture density and structure on what the grazing animal eats and animal productivity. In *Intensive utilisation of pastures,* ed. B. J. F. James, pp. 42–48. Sydney: Angus and Robertson.

————. 1978. *Ethology of free-ranging domestic animals.* Amsterdam: Elsevier.

Aschaffenburg, R.; Gregory, M. E.; Kon, S. K.; Rowlands, J.; and Thompson, S. Y. 1962. The composition of the milk of reindeer. *J. Dairy Res.* 29:324–28.

Banfield, A. W. F. 1954. Preliminary investigation of the barren ground caribou. Part 2. Life history, ecology and utilization. Canada Department of Northern Affairs and Natural Resources, National Parks Branch, Canadian Wildlife Service, Ottawa. *Wildl. Mgmt. Bull.,* ser. 1, no. 10B:1–112.

Banwell, D. B. 1968. *The highland stags of Otago.* Wellington: Reed.

Barette, C. 1977. Some aspects of the behaviour of muntjacs in Wilpattu National Park. *Mammalia* 41:1–29.

Bateman, A. J. 1948. Intra-sexual selection in *Drosophila. Heredity* 2:349–68.

Bateson, P. P. G. 1977. Testing an observer's ability to identify individual animals. *Anim. Behav.* 25:247–48.

Bateson, P. P. G.; Lotwick, W.; and Scott, D. K. 1980. Similarities between the faces of parents and offspring in Bewick's swan and the differences between mates. *J. Zool.* 191:61–74.

Beddington, J. R., and May, R. M. 1980. A possible model for the effect of adult sex ratio and density on the fecundity of sperm whales. *Rep. Int. Whal. Comm.,* Special Issue, 2:75–76.

Beddington, J. R., and Taylor, D. B. 1973. Optimum age specific harvesting of a population. *Biometrics* 29:801–9.

Bell, R. H. V. 1970. The use of the herb layer by grazing ungulates in the Serengeti. In *Animal populations in relation to their food resources,* ed. A. Watson, pp. 111–24. Oxford: Blackwell.

———. 1971. A grazing ecosystem in the Serengeti. *Sci. Am.,* 225:86–93.

Belovsky, G. E. 1981. Optimal activity times and habitat choice of moose. *Oecologia* 48:22–30.

Belovsky, G. E., and Jordan, P. A. 1978. The time energy budget of moose. *Theor. Pop. Biol.* 14:76–104.

Benedict, F. G. 1938. *Vital energetics: A study in comparative basal metabolism.* Washington, D.C.: Carnegie Institution.

Beninde, J. 1937. *Zur Natur geschichte des Rothirsches.* Monographie der Wildsäugetiere, vol. 4. Leipzig: P. Schöps.

Berger, J. 1978. Group size, foraging, and antipredator ploys: An analysis of Bighorn sheep decisions. *Behav. Ecol. Sociobiol.* 4:91–99.

———. 1979. Weaning conflict in desert and mountain bighorn sheep *(Ovis canadensis):* An ecological interpretation. *Z. Tierpsychol.* 50:188–200.

Bergerud, A. T. 1971. The population dynamics of Newfoundland caribou. *Wildl. Monogr.* 25:1–55.

———. 1973. Movement and rutting behaviour of caribou *(Rangifer tarandus)* at Mount Alberta, Quebec. *Can. Field Nat.* 87:357–69.

———. 1974a. Rutting behaviour of Newfoundland caribou. In *The behaviour of ungulates and its relation to management,* eds. V. Geist and F. Walther, pp. 395–435. Morges: IUCN.

———. 1974b. Decline of caribou in North America following settlement. *J. Wildl. Mgmt.* 38:757–70.

Bertram, B. C. R. 1976. Kin selection in lions and in evolution. In *Growing points in ethology,* eds. P. P. G. Bateson and R. A. Hinde, pp. 281–301. Cambridge: Cambridge University Press.

———. 1978. Living in groups: Predators and prey. In *Behavioral ecology: An evolutionary approach,* eds. J. R. Krebs and N. B. Davies, pp. 64–96. Oxford: Blackwell.

Bischof, N. 1975. Comparative ethology of incest avoidance. In *Biosocial anthropology,* ed. R. Fox, pp. 37–67. London: Malaby.

Black, W. A. P. 1955. Seaweeds and their constituents in foods for man and animal. *Chem. Indus.,* 1955, pp. 1640–45.

Blaxter, K. L. 1961. Energy utilization in the ruminant. In *Digestive physiology and nutrition of ruminants,* ed. D. Lewis, pp. 183–99. London: Butterworth.

———. 1962. *The energy metabolism of ruminants.* London: Hutchinson.

———. 1975. Energy-protein relationships in ruminants. *Proc. Ninth Int. Congr. Nutr.,* Mexico, 1972, pp. 122–27.

Blaxter, K. L., and Hamilton, W. J. 1980. Reproduction in farmed red deer. 2. Calf growth and mortality. *J. Agric. Sci. Camb.* 95:275–84.

Blaxter, K. L.; Kay, R. N. B.; Sharman, G. A. M.; Cunningham, J. M. M.; and Hamilton, W. J. 1974. *Farming the red deer.* London: HMSO.

Blouch, R. A., and Atmosoedirdjo, S. 1978. Preliminary report on the status of the Bawean deer *(Axis kuhli).* In *Threatened deer,* pp. 49–55. Morges: IUCN.

Boonsong, L., and McNeely, D. 1977. *Mammals of Thailand.* Bangkok: Association for the Conservation of Wildlife.

Boonstra, R., and Krebs, C. J. 1979. Viability of large- and small-sized adults in fluctuating vole populations. *Ecology* 60:567–73.

Bouissou, M. F. 1972. Influence of body weight and presence of horns on social rank in domestic cattle. *Anim. Behav.* 20:474–77.

Boyd, I. L. 1981. Population changes and distribution of a herd of feral goats *(Capra* spp.) on Rhum, Inner Hebrides, 1960–78. *J. Zool. Lond.* 193:287–304.

Bramley, P. S. 1970. Territoriality and reproductive behaviour of roe deer. *J. Reprod. Fert.,* suppl., 11:43–70.

Brantas, G. C. 1968. On the dominance order in Friesian-Dutch dairy cows. *Z. Tierzüchtung Züchtungsbiol.* 84:127–51.

Brody, S.; Procter, R. C.; and Ashworth, U. S. 1934. Growth and development with special reference to domestic animals. Part 34. *Res. Bull. Missouri Agric. Exp. Stn.* 220:32.

Brown, J. L. 1975. *The evolution of behavior.* New York: Norton.

Bubenik, A. B. 1965. Beitrag zur Geburtskunde und zu den Mutterkind-Beziehungen des Reh- *(Capreolus capreolus* L.) und Rotwildes *(Cervus elaphus* L.). *Z. Säugetierk.* 30:65–228.

———. 1968. The significance of antlers in the social life of the Cervidae. *Deer* 1:208–14.

Bubenik, A. B., and Bubenikova, J. M. 1967. Twenty-four-hour periodicity in red deer *(Cervus elaphus* L.). *Proc. Int. Congr. Game Biol.* 7:343–49.

Buechner, H. K. 1961. Territorial behavior in Uganda kob. *Science* 133:698–99.

Bulmer, M. G. 1973. Inbreeding in the great tit. *Heredity* 30:313–25.

Burckhardt, D. 1958. Observations sur la vie sociale du cerf *(Cervus elaphus)* au Parc National Suisse. *Mammalia* 22:226–44.

Burfening, P. J. 1972. Prenatal and postnatal competition among twin lambs. *Anim. Prod.* 15:61–66.

Bützler, W. 1974. Kampf und Paarungsverhalten, sozial Rangordnung und Activitatsperiodik beim Rothirsch *(Cervus elaphus* L.). *Z. Tierpsychol.,* suppl., 16:1–80.

Cameron, A. G. 1923. *The wild red deer of Scotland.* Edinburgh: Blackwood.

Caughley, G. 1970. Eruption of ungulate populations, with emphasis on Himalayan thar in New Zealand. *Ecology* 51:53–72.

———. 1977. *Analysis of vertebrate populations.* London: Wiley.

Challies, C. N. 1973. The effect of the commercial venison industry on deer populations. In *Assessment and management of introduced animals in New Zealand forests.* New Zealand Forest Service Research Institute Symposium no. 14. Rangiora, N.Z.: Forest Service.

———. 1974. Trends in red deer *(Cervus elaphus)* populations in Westland forests. *Proc. New Zealand Ecol. Soc.* 21:45–50.

Chalmers, N. 1979. *Social behaviour in primates.* London: Arnold.

Chaplin, R. E., and White, R. W. G. 1972. The influence of age and season on the activity of the testes and epidiymides of the fallow deer, *Dama dama. J. Reprod. Fert.* 30:361–69.

Chapman, D. I., and Chapman, N. 1975. *Fallow deer: Their history, distribution and biology.* Lavenham, Eng.: Dalton.

Charcon, E., and Stobbs, T. H. 1976. Influence of progressive defoliation of a grass sward on the eating behavior of cattle. *Austr. J. Agric. Res.* 27:709–27.

Charles, W. N.; McCowan, D.; and East, K. 1977. Selection of upland swards by red deer *(Cervus elaphus* L.) on Rhum. *J. Appl. Ecol.* 14:55–64.

Charles-Dominique, P., and Martin, R. D. 1972. Behaviour and ecology of nocturnal prosimians. *Z. Tierpsychol.,* suppl., 9:1–89.

Charlesworth, B. 1980. *Evolution in age-structured populations.* Cambridge: Cambridge University Press.

Charlesworth, B., and Léon, J. A. 1976. The relation of reproductive effort to age. *Am. Nat.* 110:449–59.

Child, G. 1972. Observations on a wildebeest die-off in Botswana. *Arnoldia* 31:1–13.

Child, G.; Robbel, H.; and Hepburn, C. P. 1972. Observations on the biology of tsessebe *Damaliscus lunatus lunatus* in northern Botswana. *Mammalia* 36:342–88.

Clark, A. B. 1978. Sex ratio and local resource competition in a prosimian primate. *Science* 201:163–65.

Clegg, M. T. 1959. Factors affecting gestation length and parturition. In *Reproduction in domestic animals,* ed. H. H. Cole and P. T. Cupps, pp. 509–33. London: Academic Press.

Clutton-Brock, T. H. 1974. Primate social organisation and ecology. *Nature* 250:539–42.

Clutton-Brock, T. H. 1977. Some aspects of intraspecific variation in feeding and ranging behaviour in primates. In *Primate ecology: Studies of feeding and ranging behaviour in lemurs, monkeys and apes,* ed. T. H. Clutton-Brock, pp. 539–56. London: Academic Press.

———. 1982. The functions of antlers. *Behaviour.* In press.

Clutton-Brock, T. H., and Albon, S. D. 1979. The roaring of red deer and the evolution of honest advertisement. *Behaviour* 69:145–70.

———. 1982. Parental investment and the sex ratio of progeny in mammals. In *Current problems in sociobiology,* ed. King's College Sociobiology Group, pp. 223–47. Cambridge: Cambridge University Press. In press.

Clutton-Brock, T. H.; Albon, S. D.; Gibson, R. M.; and Guinness, F. E.

1979. The logical stag: Adaptive aspects of fighting in red deer (*Cervus elaphus* L.). *Anim. Behav.* 27:211–25.

Clutton-Brock, T. H.; Albon, S. D.; and Guinness, F. E. 1981. Parental investment in male and female offspring in polygynous mammals. *Nature* 289:487–89.

Clutton-Brock, T. H.; Albon, S. D.; and Harvey, P. H. 1980. Antlers, body size and breeding group size in the Cervidae. *Nature* 285:565–67.

Clutton-Brock, T. H.; Greenwood, P. J.; and Powell, R. P. 1976. Ranks and relationships in Highland ponies and Highland cows. *Z. Tierpsychol.* 41:202–16.

Clutton-Brock, T. H., and Guinness, F. E. 1975. Behaviour of red deer (*Cervus elaphus* L.) at calving time. *Behaviour* 55:287–300.

Clutton-Brock, T. H., and Harvey, P. H. 1976. Evolutionary rules and primate societies. In *Growing points in ethology,* ed. P. P. G. Bateson and R. A. Hinde, pp. 195–237. Cambridge: Cambridge University Press.

———. 1977a. Species differences in feeding and ranging behaviour in primates. In *Primate ecology: Studies of feeding and ranging behaviour in primates,* ed. T. H. Clutton-Brock, pp. 557–84. London: Academic Press.

———. 1977b. Primate ecology and social organisation. *J. Zool.* 183:1–39.

———. 1978. Mammals, resources and reproductive strategies. *Nature* 273:191–95.

———. 1982. The functional significance of variation in body size among mammals. In *Symposium of the National Zoological Park.* Washington, D.C.: Smithsonian Institution. In press.

Clutton-Brock, T. H.; Harvey, P. H.; and Rudder, B. 1977. Sexual dimorphism, socionomic sex ratio and body weight in primates. *Nature* 269:797–99.

Cockerill, R. A. n.d. Functional aspects of mother-offspring relationships in red deer. Ph.D. diss., Univ. of Cambridge. In preparation.

Cody, M. L. 1971. Finch flocks in the Mohave Desert. *Theor. Pop. Biol.* 2:142–48.

Cole, L. C. 1949. The measurement of interspecific association. *Ecology* 30:411–24.

Colquhoun, I. R. 1971. The grazing ecology of red deer and blackface sheep in Perthshire, Scotland. Ph.D. diss., University of Edinburgh.

Cooper, A. B. 1969. Golden eagle kills red deer calf. *J. Zool. Lond.* 158:215–16.

Corbet, G. B. 1978. *The mammals of the Palearctic region: A taxonomic review.* London: British Museum.

Corfield, T. 1973. Elephant mortality in Tsavo National Park, Kenya. *E. Afr. Wildl. J.* 11:339–68.

Cory, V. L. 1927. Activities of livestock on the range. *Bull. Texas Agric. Exp. Stn.* 367:1–47.

Coulson, J. C., and Hickling, G. 1961. Variation in the secondary sex-ratio of the grey seal *Halichoerus grypus* during the breeding season. *Nature* 190:281.

Cowan, I. McT. 1950. Some vital statistics of big game on over-stocked mountain range. *Trans. N. Am. Wildl. Conf.* 15:581–88.

———. 1956. What and where are the mule and black-tailed deer? In *The deer of North America,* ed. W. P. Taylor, pp. 334–59. Harrisburg: Stackpole.

Cowan, I. McT., and Wood, A. J. 1955. The growth rate of black-tailed deer. *J. Wildl. Mgmt.* 19:331–36.

Cox, C. R., and LeBoeuf, B. J. 1977. Female incitation of male competition: A mechanism of mate selection. *Am. Nat.* 111:317–35.

Craighead, J. J.; Craighead, F. C.; Ruff, R. L.; and O'Gara, B. W. 1973. Home ranges and activity patterns of non-migratory elk of the Madison Drainage herd as determined by biotelemetry. *Wildl. Monogr.,* no. 33.

Craighead, J. J., and Shoesmith, M. 1966. An intensive study of elk behaviour on the Muror Plateau summer range, Yellowstone National Park. *Quart. Rpt. Montana Coop. Wildl. Res. Unit* 17:6–8.

Crew, F. A. E. 1937. The sex ratio. *Am. Nat.* 71:529–59.

Crook, J. H. 1970. The socio-ecology of primates. In *Social behaviour in birds and mammals: Essays on the social ethology of animals and man,* ed. J. H. Crook, pp. 103–66. New York: Academic Press.

———. 1972. Sexual selection, dimorphism, and social organisation in the primates. In *Sexual selection and the descent of man,* ed. B. G. Campbell, pp. 231–81. Chicago: Aldine.

———. 1973. Darwinism and the sexual politics of primates. *Accad. Naz. Lincei* 182:199–217.

Daly, M. 1979. Why don't male mammals lactate? *J. Theor. Biol.* 78:325–45.

Dansie, O. 1970. *Muntjac.* British Deer Society Publication no. 2. Welwyn Garden City: Broadwater Press.

Darling, F. F. 1937. *A herd of red deer.* London: Oxford University Press.

Darwin, C. 1859. *The origin of species by means of natural selection.* London: Murray.

———. 1871. *The descent of man and selection in relation to sex.* 1888 ed. London: Murray.

———. 1876. Sexual selection in relation to monkeys. *Nature* 15:18–19.

Dasmann, R. F., and Taber, R. D. 1956. Behaviour of Columbian black-tailed deer with reference to population ecology. *J. Mammal.* 37:143–64.

Dasmann, W. P., and Blaisdell, J. A. 1954. Deer and forage relationship on the Lassen-Washoe interstate winter deer range. *Calif. Fish Game* 40:215–34.

Davies, G. 1931. An unusual sex ratio in red deer. *Nature* 127:94.

Davies, N. B., and Halliday, T. R. 1977. Optimal mate selection in the toad, *Bufo bufo. Nature* 269:56–58.

———. 1978. Deep croaks and fighting assessment in toads, *Bufo bufo. Nature* 274:683–85.

Dawkins, R. 1976. *The selfish gene.* Oxford: Oxford University Press.

———. 1978. Replicator selection and the extended phenotype. *Z. Tierpsychol.* 47:61–76.

————. 1982. Vehicles and replicators. In *Current problems in sociobiology*, ed. King's College Sociobiology Group, pp. 45–64. Cambridge: Cambridge University Press. In press.

Defries, J. C.; Touchberry, R. W.; and Hays, R. L. 1959. Heritability of the length of the gestation period in dairy cattle. *J. Dairy Sci.* 42:598–606.

Delpech, F., and Suire, C. 1974. La faune mésolithique et post-mésolithique du gisement de Rouffignac. *Publ. Institut. d'Art Préhistorique de l'Université de Toulouse-le-Mirail. Meon-Institut Art Préhistorique II Rouffignac,* fasc. 2.

Demment, M. W. 1982. Feeding ecology and the evolution of body size of baboons. *Afr. J. Ecol.* In press.

Dhillon, J. S.; Acharya, R. M; Tiwana, M. S.; and Aggarwal, S. C. 1970. Factors affecting the interval between calving and conception in Hariana cattle. *Anim. Prod.* 12:81–87.

Dittus, W. P. J. 1977. The social regulation of population density and age-sex distribution in the toque monkey. *Behaviour* 63: 281–322.

————. 1979. The evolution of behaviors regulating population density and age-specific sex ratios in a primate population. *Behaviour* 69: 265–301.

Douglas-Hamilton, I. 1972. On the ecology and behaviour of the African elephant: The elephants of Lake Manyara. Ph.D. diss., University of Oxford.

Drickamer, L. C. 1974. A ten-year summary of reproductive data for free-ranging *Macaca mulatta. Folia Primat.* 21:61–80.

Dubost, G. 1978. Un aperçu sur l'écologie du chevrotain africain *Hyemoschus aquaticus* Ogilby, Artiodactyle Tragulidé. *Mammalia* 42: 1–62.

Dunbar, R. I. M., and Dunbar, E. P. 1977. Dominance and reproductive success among female gelada baboons. *Nature* 266:351–52.

Duncan, P. 1975. Topi and their food supply. Ph.D. diss., University of Nairobi.

————. 1980. Time-budgets of Camargue horses. II. Time-budgets of adult horses and weaned sub-adults. *Behaviour* 72:26–49.

Duncan, P., and Vigne, N. 1979. The effect of group size on the rate of attacks by blood-sucking flies. *Anim. Behav.* 27:623–25.

Dzieciolowski, R. 1969. *The quantity, quality and seasonal variation of food resources available to red deer in various environmental conditions of forest management.* Warsaw: Polish Academy of Sciences, Forest Research Institute.

————. 1970a. Foods of the red deer determined by rumen content analysis. *Acta Theriol.* 15:89–110.

————. 1970b. Food selectivity in the red deer towards twigs of trees, shrubs and dwarf shrubs. *Acta Theriol.* 15:361–65.

————. 1979. Structure and spatial organisation of deer populations. *Acta Theriol.* 24:3–21.

Eaton, G. G. 1978. Longitudinal studies of sexual behaviour in the Oregon troop of Japanese macaques. In *Sex and behaviour,* ed. T. E. McGill, D. A. Dewsbury, and B. D. Sachs, pp. 35–59. New York: Plenum.

Eggeling, W. J. 1964. A nature reserve management plan for the Island of Rhum, Inner Hebrides. *J. Appl. Ecol.* 1:405–19.

Eisenberg, J. F. 1966. The social organization of mammals. *Handb. Zool.,* Band 8, Lieferung 39; 10:1–92.

Ellerman, J. R., and Morrison-Scott, T. C. S. 1951. Check list of Palearctic and Indian mammals, 1758–1946. London: British Museum.

Emlen, S. T., and Oring, L. W. 1977. Ecology, sexual selection and the evolution of mating systems. *Science* 197:215–23.

England, G. J. 1954. Observations on the grazing of different breeds of sheep at Pantyrhuad Farm, Carmarthenshire. *Brit. J. Anim. Behav.* 2:56–60.

Erlinge, S. 1972. Interspecific relations between otter, *Lutra lutra* and mink, *Mustela vison,* in Sweden. *Oikos* 23:327–35.

Eshel, I. 1975. Selection on sex-ratio and the evolution of sex-determination. *Heredity* 34:351–61.

Espmark, Y. 1964. Studies in dominance-subordination relationship in a group of semi-domestic reindeer (*Rangifer tarandus* L.). *Anim. Behav.* 12:420–25.

Estes, R. D. 1974. Social organization of the African Bovidae. In *The behaviour of ungulates and its relation to management,* ed. V. Geist and F. Walther, pp. 166–205. Morges: IUCN.

Estes, R. D., and Goddard, J. 1967. Prey selection and hunting behavior of the African wild dog. *J. Wildl. Mgmt.* 31:52–70.

Evans, H. 1890. *Some account of Jura red deer.* Derby: Carter. Private publication.

Everitt, B. 1974. *Cluster analysis.* London: Heinemann Educational Books.

Fagen, R. M. 1972. An optimal life history in which reproductive effort decreases with age. *Am. Nat.* 106:258–61.

Ferrar, A. M., and Kerr, M. A. 1971. A population crash of the reedbuck *Redunca arundinum* (Boddaert) in Kyle National Park, Rhodesia. *Arnoldia* 16:1–19.

Ferreira, R. E. C. 1970. *Vegetation map of the Isle of Rhum.* London: Nature Conservancy.

Festing, M. F. W. 1972. Hamsters. In *Handbook on the care and management of laboratory animals,* 4th ed., pp. 242–56. London: Universities Federation for Animal Welfare.

Fisher, R. A. 1930. *The genetical theory of natural selection.* Oxford: Oxford University Press.

Fitzgerald, B. M. 1977. Weasel predation on a cyclic population of the montane vole (*Microtus montanus*) in California. *J. Anim. Ecol.* 46:367–97.

Flerov, K. K. 1952. *Fauna of the U.S.S.R.* Vol.1. *Mammals: Musk deer and deer.* Moscow: USSR Academy of Sciences. Translated by the Israel Program of Scientific Translation.

Fletcher, T. J. 1974. The timing of reproduction in red deer *(Cervus elaphus)* in relation to latitude. *J. Zool.* 172:363–67.

Fletcher, T. J., and Short, R. V. 1974. Restoration of libido in castrated red deer stag *(Cervus elaphus)* with oestradiol 17β. *Nature* 248: 616–18.

Flook, D. R. 1970. *A study of sex differential in the survival of wapiti.* Canadian Wildlife Service Report Series no. 11. Ottawa: Department of Indian Affairs and Northern Development.

Ford, E. B. 1964. *Ecological genetics.* London: Chapman and Hall.

———. 1975. *Ecological genetics.* 4th ed. London: Chapman and Hall.

Frame, L. H.; Malcolm, J. R.; Frame, G. W.; and van Lawick, H. 1979. Social organisation of African wild dogs *(Lycaon pictus)* in the Serengeti Plains. *Z. Tierpsychol.* 50:225–49.

Franklin, W. L., and Lieb, J. W. 1979. The social organisation of a sedentary population of North American elk: A model for understanding other populations. In *North American elk: Ecology, behavior and management,* ed. M. S. Boyce and L. D. Hayden-Wing, pp. 185–98. Laramie: University of Wyoming.

Franklin, W. L.; Mossman, A. S.; and Dole, M. 1975. Social organization and home range of Roosevelt elk. *J. Mammal.* 56:102–18.

Fraser, D., and Morley Jones, R. 1975. The "teat order" of suckling pigs. I. Relation to birth weight and subsequent growth. *J. Agric. Sci. Camb.* 84:387–91.

Gadgil, M., and Bossert, W. H. 1970. Life history consequences of natural selection. *Am. Nat.* 104:1–24.

Garnett, M. C. 1981. Body size, its heritability and influence on juvenile survival among great tits, *Parus major. Ibis* 123:31–41.

Gartlan, J. S. 1968. Structure and function in primate society. *Folia Primat.* 8:89–120.

Gasaway, W. C., and Coady, J. W. 1974. Review of energy requirements and rumen fermentation in moose and other ruminants. *Natur. Can.* 101:227–62.

Gates, C., and Hudson, R. J. 1978. Energy costs of locomotion in wapiti. *Acta Theriol.* 23:365–70.

Gautier-Hion, A. 1975. Dimorphisme sexuel et organisation sociale chez les cercopithecinés forestiers africains. *Mammalia* 39:365–74.

———. 1980. Seasonal variations of diet related to species and sex in a community of *Cercopithecus* monkeys. *J. Anim. Ecol.* 49:237–69.

Geisler, P. A., and Fenlon, J. S. 1979. The effects of body weight and its components on lambing performance in some commercial flocks in Britain. *Anim. Prod.* 28:245–55.

Geist, V. 1963. On the behaviour of the North American moose *(Alces alces andersoni* Peterson 1950) in British Columbia. *Behaviour* 20:377–416.

———. 1966. The evolution of horn-like organs. *Behaviour* 27:175–214.

———. 1971*a.* The relation of social evolution and dispersal in ungulates during the Pleistocene, with emphasis on the Old World deer and the genus *Bison. Quart. Res.* 1:283–315.

———. 1971*b. Mountain sheep: A study in behavior and evolution.* Chicago: University of Chicago Press.

———. 1974*a.* On the relationship of social evolution and ecology in ungulates. *Am. Zool.* 14:205–20.

———. 1974*b.* On the evolution of reproductive potential in moose. *Nat. Can.* 101:527–37.

———. 1974*c.* On fighting strategies in animal combat. *Nature* 250:354.

———. 1978. On weapons, combat and ecology. In *Aggression, dominance and individual spacing,* ed. L. Kramer, P. Pliner, and T. Alloway, pp. 1–30. New York: Plenum.

———. 1982. Adaptive aspects in the behaviour of elk. In press.

Geist, V., and Bromley, P. T. 1978. Why deer shed antlers. *Z. Säugetierk.* 43:223–231.

Georgii, von B. 1980. Home range patterns of female red deer (*Cervus elaphus* L.) in the Alps. *Oecologia* 47:278–85.

Georgii, von B., and Schröder, W. 1978. Radiotelemetrisch gemessene Aktivatät weiblichen Rotwildes (*Cervus elaphus* L.). *Z. Jagdwiss.* 24:9–23.

Gibson, R. M. 1978. Behavioural factors affecting reproduction in red deer stags. Ph.D. diss., University of Sussex.

Gibson, R. M., and Guinness, F. E. 1980*a.* Differential reproductive success in red deer stags. *J. Anim. Ecol.* 49:199–208.

———. 1980*b.* Behavioural factors affecting male reproductive success in red deer (*Cervus elaphus* L.). *Anim. Behav.* 28:1163–74.

Giles, R. H., Jr. 1978. *Wildlife management.* San Francisco: Freeman.

Gill, E. 1871. The eared seals. *Am. Nat.* 4:675–84.

Glucksman, A. 1974. Sexual dimorphism in mammals. *Biol. Rev.* 49:423–75.

———. 1978. *Sex determination and sexual dimorphism in mammals.* London: Wykeham.

Goodall, J. 1968. The behaviour of free-living chimpanzees in the Gombe Stream Reserv. *Anim. Behav. Monogr.* 1:165–311.

Gosling, L. M. 1974. The social behaviour of Coke's hartebeest *Alcephalus buselaphus cokei.* In *The behavior of ungulates and its relation to management,* ed. V. Geist and F. Walther, pp. 488–511. Morges: IUCN.

Gosling, L. M., and Petrie, M. 1982. The economics of social organisation. In *Physiological ecology: An evolutionary approach,* ed. P. Calow and C. R. Townsend. Oxford: Blackwell Scientific Publications. In press.

Goss, R. J. 1963. The deciduous nature of deer antlers. In *Mechanisms of hard tissue destruction,* pp. 339–69. Publication no. 75. Washington, D.C.: American Association for the Advancement of Science.

———. 1968. Inhibition of growth and shedding of antlers by sex hormones. *Nature* 220:83–85.

———. 1969*a.* Photoperiodic control of antler cycles in deer. I. Phase shift and frequency changes. *J. Exp. Zool.* 170:311–24.

———. 1969*b.* Photoperiodic control of antler cycles in deer. II. Alteration in amplitude. *J. Exp. Zool.* 171:223–34.

Goss, R. J., and Rosen, J. K. 1973. The effect of latitude and photo-period on the growth of antlers. *J. Reprod. Fert.,* suppl., 19:111–18.

Gossow, H. 1971. Allgemeine Forst und Jagdzeitung soziologische und Rangordnungs aspekte bei einer Alpinen Rotwild Populations. *Allg. Forst- J.-Ztg.* 142:169–73.

———. 1973. Natural mortality pattern in the Spitsbergen reindeer. *Proc. Int. Congr. Game Biol.* 10:91–95.

———. 1974. On some environmental effects on sociology and be-haviour in Alpine red deer. *Proc. Int. Congr. Game Biol.* 11:99–101.

Gossow, H., and Schürholz, G. 1974. Social aspects of wallowing be-haviour in red deer herds. *Z. Tierpsychol.* 34:329–36.

Gottschlich, H. J. 1968. Brunftkämpfe beim Rotwild. *Unsere Jagd.* Ber-lin: VEB Deutscher Landwirtschaftsverlag.

Gould, S. J. 1974. The origin and function of "bizarre" structures: Ant-ler size and skull size in the "Irish elk," *Megaloceros giganteus. Evolution* 28:191–220.

Grace, J., and Easterbee, N. 1979. The natural shelter for red deer *(Cervus elaphus)* in a Scottish glen. *J. Appl. Ecol.* 16:37–48.

Grace, J., and Woolhouse, H. W. 1970. A physiological and mathemat-ical study of growth and productivity of a *Calluna-Sphagnum* commu-nity. *J. Appl. Ecol.* 7:363–81.

Graf, W. 1956. Territorialism in deer. *J. Mammal.* 37:165–70.

Grant, S. A., and Hunter, R. F. 1962. Ecotypic differentiation of *Calluna vulgaris* (L) in relation to altitude. *New Phyt.* 61:44–45.

———. 1966. The effects of frequency and season of clipping on the morphology, productivity and chemical composition of *Calluna vul-garis* L. Hull. *New Phyt.* 65:125–33.

Grant, S. A.; Lamb, W. I. C.; Kerr, C. D.; and Bolton, G. R. 1976. The utilization of blanket bog vegetation by grazing sheep. *J. Appl. Ecol.* 13:857–69.

Greenwood, P. J. 1978. Functions of dispersal in birds and mammals. Ph.D. diss., University of Sussex.

———. 1980. Mating systems, philopatry and dispersal in birds and mammals. *Anim. Behav.* 28:1140–62.

Greenwood, P. J.; Harvey, P. H.; and Perrins, C. M. 1978. Inbreeding and dispersal in the great tit. *Nature* 271:52–54.

Grimble, A. 1901. *Deer-stalking and the deer forests of Scotland.* London: Kegan, Paul, Trench, Trubner.

Grubb, P. 1974. Population dynamics of the Soay sheep. In *Island sur-vivors: The ecology of Soay sheep on St. Kilda,* ed. P. A. Jewell, C. Milner, and J. M. Boyd, pp. 242–72. London: Athlone Press.

Guinness, F. E. 1979. Cailleach. *Deer* 4:526.

Guinness, F. E.; Albon, S. D.; and Clutton-Brock, T. H. 1978. Factors affecting reproduction in red deer (*Cervus elaphus* L.). *J. Reprod. Fert.* 54:325–34.

Guinness, F. E., Clutton-Brock, T. H., and Albon, S. D. 1978. Factors affecting calf mortality in red deer. *J. Anim. Ecol.* 47:817–32.

Guinness, F. E.; Gibson, R. M.; and Clutton-Brock, T. H. 1978. Calving times of red deer (*Cervus elaphus* L.) on Rhum. *J. Zool.* 185:105–14.

Guinness, F. E.; Hall, M. J.; and Cockerill, R. A. 1979. Mother-offspring association in red deer. *Anim. Behav.* 27:536–44.

Guinness, F. E.; Lincoln, G. A.; and Short, R. V. 1971. The reproductive cycle of the female red deer (*Cervus elaphus* L.). *J. Reprod. Fert.* 27:427–38.

Gunn, R. G. 1964a. Levels of first winter feeding in relation to performance of Cheviot hill ewes. I. Body growth and development during treatment period. *J. Agric. Sci. Camb.* 62:99–122.

———. 1964b. Levels of first winter feeding in relation to performance of Cheviot hill ewes. II. Body growth and development during the summer after treatment, 12–18 months. *J. Agric. Sci. Camb.* 62:123–49.

———. 1965. Levels of first winter feeding in relation to performance of Cheviot hill ewes. III. Tissue and joint development to 12–18 months of age. *J. Agric. Sci. Camb.* 64:311–21.

———. 1970. A note on the effect of broken mouth on the performance of Scottish blackface hill ewes. *Anim. Prod.* 12:517–20.

Gunvalson, V. E.; Erickson, A. B.; and Burcalow, D. W. 1952. Hunting season statistics as an index to range conditions and deer population fluctuations in Minnesota. *J. Wildl. Mgmt.* 16:121–31.

Hall, M. 1978. Mother-offspring relationships in red deer (*Cervus elaphus* L.) and the social organization of an enclosed group. Ph.D. diss., University of Sussex.

Halliday, T. R. 1978. Sexual selection and mate choice. In *Behavioural ecology: An evolutionary approach,* ed. J. R. Krebs and N. B. Davies, pp. 180–213. Oxford: Blackwell.

Hamilton, W. D. 1964a. The genetical evolution of social behavior. I. *J. Theor. Biol.* 7:1–16.

———. 1964b. The genetical evolution of social behaviour. II. *J. Theor. Biol.* 7:1–16.

———. 1967. Extraordinary sex ratios. *Science* 156:477–88.

———. 1971. Selection of selfish and altruistic behavior in some extreme models. In *Man and beast: Comparative social behavior,* ed. J. F. Eisenberg, and W. S. Dillon, pp. 55–91. Washington, D.C.: Smithsonian Institution Press.

Hamilton, W. J., and Blaxter, K. L. 1980. Reproduction in farmed red deer. 1. Hind and stag fertility. *J. Agric. Sci. Camb.* 95:261–73.

Hancock, J. 1953. Grazing behaviour of cattle. *Anim. Breed. Abstr.* 21:1–13.

Harcourt, A. H. 1978. Strategies of emigration and transfer by primates with particular reference to gorillas. *Z. Tierpsychol.* 48:401–20.

Harcourt, A. H.; Harvey, P. H.; Larson, S. G.p and Short, R. V. 1981. Testis weight, body weight and breeding system in primates. *Nature* 293:55–57.

Harper, J. A. 1964. Movement and associated behavior of Roosevelt elk in southwestern Oregon. *Western Assoc. State Game Fish Comm., Ann. Conf. Proc.* 44:139–41.

Harper, J. A.; Harn, J. H.; Bentley, W. W.; and Yocom, C. F. 1967. The status and ecology of the Roosevelt elk in California. *Wildl. Monogr.* 16:1–49.

Harvey, P. H.; Kavanagh, M. J.; and Clutton-Brock, T. H. 1978. Sexual dimorphism in primate teeth. *J. Zool.* 186:475–86.

Haukioja, E., and Salovaara, R. 1978. Summer weight of reindeer *(Rangifer tarandus)* calves and its importance for their future survival. *Rep. Kevo. Subarctic Res. Stat.* 14:1–4.

Healy, W. B., and Ludwig, T. G. 1965. Wear of sheep's teeth. I. The role of ingested soil. *New Zealand J. Agric. Res.* 8:737–52.

Heape, W. 1913. *Sex antagonism.* London.

Henshaw, J. 1968. A theory for the occurrence of antlers in females of the genus *Rangifer. Deer* 1:222–26.

———. 1969. Antlers: The bones of contention. *Nature* 224:1036–37.

———. 1970. Consequences of travel in the rutting of reindeer and caribou. *Anim. Behav.* 18:256–58.

———. 1971. Antlers: The unbrittle bones of contention. *Nature* 231:469.

Heptner, W. A.; Nasimovitsch, A. A.; and Bannikov, A. G. 1961. *Mammals of the Soviet Union.* Jena: Fischer-Verlag.

Hesselton, W. T.; Severinghaus, C. W.; and Tanck, J. E. 1965. Population dynamics of deer at the Seneca Army Depot. *N.Y. Fish Game J.* 12:17–30.

Hirth, D. H. 1977. Social behavior of white-tailed deer in relation to habitat. *Wildlife Monogr.* 53:1–55.

Hirth, D. H., and McCullough, D. R. 1977. Evolution of alarm signals in ungulates with special reference to white-tailed deer. *Am. Nat.* 111:31–42.

Hodgson, J. 1977. Factors limiting herbage intake by the grazing animal. *Proc. Int. Meet. Anim. Prod. Temperate Grassland,* 70075. Dublin.

Hodgson, J., and Milne, J. A. 1978. The influence of weight of herbage per unit acre and per animal upon the grazing behaviour of sheep. *Seventh Gen. Meeting Europ. Grassld. Fed.* Gent.

Hofmann, R. R. 1973. *The ruminant stomach: Stomach structure and feeding habits of East African game ruminants.* Nairobi: East African Game Bureau.

———. 1976. Zur adaptieven Differenzierung der Wiederkäuer: Untersuchsergebriesse auf der Basis der vergleichenden funktionellen Anatomie des Verauungstrakts. *Prakt. Tierartz* 6:351–58.

Hofmann, R. R.; Geiger, G.; and König, R. 1976. Vergleichend-anatomische Untersuchungen an der Vormagenschleimhaut von Rehwild *(Capreolus capreolus)* und Rotwild *(Cervus elaphus).* Z. *Säugetierk.* 41:167–93.

Hofmann, R. R., and Stewart, D. R. M. 1972. Grazer or browser: A classification based on the stomach structure and feeding habits of East African ruminants. *Mammalia* 36:226–40.

Holter, J. B.; Urban, W. E.; Hayes, H. H.; and Silver, H. 1976. Predicting metabolic rate from telemetered heart rate in white tailed deer. *J. Wildl. Mgmt.* 40:626–29.

Hoppe, P. P. 1977. How to survive heat and aridity: Ecophysiology of the dik dik antelope. *Vet. Med. Rev.* 1:77–86.

Horwood, M. T., and Masters, E. H. 1970. *Sika deer.* British Deer Society Publication no. 3. Welwyn Garden City: Broadwater Press.

Houston, D. B. 1974. Aspects of the social organisation of moose. In *The behaviour of ungulates and its relation to management,* ed. V. Geist and W. Walther, pp. 690–96. Morges: IUCN.

Howard, R. D. 1979. Estimating reproductive success in natural populations. *Am. Nat.* 114:221–31.

Hrdy, S. B. 1977. *The langurs of Abu.* Cambridge: Harvard University Press.

———. 1979. Infanticide among animals: A review, classification and examination of the implications for the reproductive strategies of females. *Ethol. Sociobiol.* 1:13–40.

Hughes, G. P., and Reid, D. 1951. Studies on the behaviour of cattle and sheep in relation to the utilization of grass. *J. Agric. Sci. Camb.* 41:350–66.

Hungate, R. E. 1966. *The rumen and its microbes.* New York: Academic Press.

Hungate, R. E.; Phillips, G. D.; McGregor, A.; Hungate, D. P.; and Buechner, H. K. 1959. Microbial fermentation in certain mammals. *Science* 130:1192–94.

Hunter, R. F., and Davies, G. E. 1963. The effect of method of rearing on the social behaviour of Scottish blackface hoggets. *Anim. Prod.* 5:183–94.

Hunter, R. F., and Milner, C. 1963. The behaviour of individual, related and groups of south country Cheviot hill sheep. *Anim. Behav.* 11:507–13.

Hutchinson, G. E. 1978. *An introduction to population biology.* New Haven: Yale University Press.

Huxley, J. S. 1926. The annual increment of the antlers of the red deer (*Cervus elaphus*). *Proc. Zool. Soc.* 58:1021–36.

———. 1931. The relative size of antlers in deer. *Proc. Zool. Soc. Lond.* 72:819–64.

———. 1932. *Problems of relative growth.* London: Methuen.

Hyvarinen, H.; Kay, R. N. B.; and Hamilton, W. J. 1977. Variation in the weight, specific gravity, and composition of the antlers of red deer (*Cervus elaphus* L.). *Brit. J. Nutr.* 38:301–11.

Jackes, A. D. 1973. The use of wintering grounds by red deer in Ross-shire, Scotland. Ph.M. diss., University of Edinburgh.

Jaczewski, Z. 1954. The effect of changes in length of daylight on the growth of antlers in deer (*Cervus elaphus* L.). *Folia Biol.* 2:133–43.

Janis, C. 1976. The evolutionary strategy of the Equidae and the origins of rumen and caecal digestion. *Evolution* 30:757–74.

Jardine, N., and Sibson, R. 1968. The construction of hierarchic and nonhierarchic classification. *Comput. J.* 11:177–84.

Jarman, M. V. 1979. *Impala social behaviour: Territory, hierarchy, mating and the use of space.* Advances in Ethology, no. 21. Berlin and Hamburg: Paul Parey Verlag.

Jarman, P. J. 1974. The social organisation of antelope in relation to their ecology. *Behaviour* 48:215–67.

———. 1982. Prospects for interspecific comparison in sociobiology. In *Current problems in sociobiology,* ed. King's College Sociobiology Group, pp. 323–42. Cambridge: Cambridge University Press. In press.

Jensen, P. V. 1968. Food selection of the Danish red deer (*Cervus elaphus* L.) as determined by examination of the rumen contents. *Danish Rev. Game Biol.* 51:1–44.

Jeter, L. K., and Marchinton, R. L. 1964. Preliminary report on telemetric study of deer movements and behavior on the Elgin Field reservation in Northwestern Florida. *Proc. Nineteenth Ann. Conf. S.E. Assoc. Game Fish Comm.,* pp. 140–52.

Jewell, P. A.; Milner, C.; and Morton-Boyd, J., eds. 1974. *Island survivors: The ecology of the Soay sheep of St. Kilda.* London: Athlone Press.

Johnson, E. D. 1951. Biology of the elk calf *(Cervus canadensis nelsoni). J. Wildl. Mgmt.* 15:396–410.

Julander, O.; Robinette, L. W.; and Jones, D. A. 1961. Relation of summer range condition to mule deer herd productivity. *J. Wildl. Mgmt.* 25:54–60.

Kay, R. N. B. 1978. Seasonal changes of appetite in deer and sheep. *A. R. C. Res. Rev.* 5:13–15.

Kay, R. N. B.; Engelhardt, W. V.; and White, R. G. 1979. The digestive physiology of wild ruminants. In *Digestive physiology and metabolism in ruminants,* ed. Y. Ruckenbusch and P. Thivend, pp. 743–61. Lancaster: MTP Press.

Kay, R. N. B., and Goodall, E. C. 1976. The intake, digestibility and retention time of roughage diets by red deer *(Cervus elaphus)* and sheep. *Proc. Nutr. Soc.* 35:98A.

Kay, R. N. B., and Hobson, P. N. 1963. Reviews of the progress of dairy science. I. The physiology of the rumen. *J. Dairy Res.* 30:261–313.

Kay, R. N. B., and Staines, B. W. 1981. The nutrition of red deer *(Cervus elaphus). Nutr. Abstr. Rev.* 51:601–22.

King, C. M., and Moors, P. J. 1979. On co-existence, foraging strategy and the biogeography of weasels and stoats *(Mustela nivalis* and *M. erminea)* in Britain. *Oecologia* 39:129–50.

Kinnaird, J. W. 1974. Effect of site conditions on the regeneration of birch (*Betula pendula* Roth and *B. pubescus* Ehrh). *J. Ecol.* 62:467–72.

Kirkwood, T. B. L., and Holliday, R. 1979. The evolution of ageing and longevity. *Proc. Roy. Soc.,* ser. B., 205:97–112.

Kitchen, D. W. 1974. Social behaviour and ecology of the pronghorn. *Wildl. Monogr.* 38:1–96.

Kitts, W. D.; Cowan, I. McT.; Bandy, P. J.; and Wood, A. J. 1956. The immediate post-natal growth in Columbian black-tailed deer in relation to the composition of the milk of the doe. *J. Wildl. Mgmt.* 20:212–14.

Kleiber, M. 1961. *The fire of life: An introduction to animal energetics.* New York: Wiley.

Klein, D. R. 1964. Range related differences in growth of deer reflected in skeletal ratios. *J. Mammal.* 45:226–35.

———. 1968. The introduction, increase and crash of reindeer on St. Matthew Island. *J. Wildl. Mgmt.* 32:350–67.

———. 1970. Food selection by North American deer and their response to over-utilisation of preferred plant species. In *Animal populations in relation to their food resources,* ed. A. Watson, pp. 25–46. Oxford: Blackwell.

Klein, D. R., and Olson, S. T. 1960. Natural mortality patterns of deer in southeast Alaska. *J. Wildl. Mgmt.* 24:80–88.

Klein, D. R., and Strandgaard, H. 1972. Factors affecting growth and body size of roe deer. *J. Wildl. Mgmt.* 36:64–79.

Knight, R. R. 1970. The Sun River elk herd. *Wildl. Monogr.* 23:1–66.

Krebs, C. J., and Cowan, I. McT. 1962. Growth studies of reindeer fawns. *Can. J. Zool.* 40:863–69.

Krebs, C. J., and Myers, J. H. 1974. Population cycles in small mammals. *Adv. Ecol. Res.* 8:267–399.

Krebs, J. R., and Davies, N. B., eds. 1978. *Behavioural ecology: An evolutionary approach.* Oxford: Blackwell.

Krebs, J. R.; MacRoberts, M. H.; and Cullen, J. M. 1972. Flocking and feeding in the great tit *Parus major:* An experimental study. *Ibis* 114:507–30.

Kruuk, H. 1972. *The spotted hyena.* Chicago: University of Chicago Press.

Kurland, J. A. 1977. Kin selection in the Japanese monkey. *Contrib. Primat.* 12:1–145.

Kurt, F. 1978. Socio-ecological organisation and aspects of management in South Asian deer. In *Threatened deer,* pp. 219–39. Morges: IUCN.

Lack, D. 1954. *The natural regulation of animal numbers.* Oxford: Clarendon Press.

———. 1968. *Ecological adaptations for breeding in birds.* London: Methuen.

Latham, R. M. 1947. Differential ability of male and female game birds to withstand starvation and climatic extremes. *J. Wildl. Mgmt.* 11:139–49.

Lawton, J. H., and Hassell, M. P. 1981. Asymmetrical competition in insects. *Nature* 289:793–96.

Leader-Williams, N. 1979. Age-related changes in the testicular and antler cycles of reindeer, *Rangifer tarandus. J. Reprod. Fert.* 57:117–26.

———. 1980. Population dynamics and mortality of reindeer introduced into South Georgia. *J. Wildl. Mgmt.* 44:640–57.

Leader-Williams, N., and Ricketts, C. 1982. Seasonal and sexual patterns of growth and condition in introduced reindeer on South Georgia. *Oikos* 38:27–39.

LeBoeuf, B. J. 1974. Male-male competition and reproductive success in elephant seals. *Am. Zool.* 14:163–76.

LeBoeuf, B. J., and Briggs, K. T. 1977. The cost of living in a seal harem. *Mammalia* 41:167–95.

Lee, R. 1979. A study into the population performance of red deer on Rhum. Dip. Stat. thesis, University of Cambridge.

Leigh, E. G. 1970. Sex ratio and differential mortality between the sexes. *Am. Nat.* 104:205–10.

Lent, P. C. 1965. Rutting behaviour in a barren-ground caribou population. *Anim. Behav.* 13:259–64.

———. 1974. Mother-infant relationships in ungulates. In *The behaviour of ungulates and its relation to management,* ed. V. Geist and F. Walther, pp. 14–55. Morges: IUCN.

Leutenegger, W., and Kelly, J. T. 1977. Relationship of sexual dimorphism in canine size and body size to social behavioral and ecological correlates in anthropoid primates. *Primates* 18:117–36.

Leuthold, W. 1977. *African ungulates: A comparative review of their ethology and behavioural ecology.* Zoophysiology and Ecology 8. Berlin: Springer-Verlag.

Lewontin, R. C. 1978. Adaptation. *Sci. Am.* 239:156–69.

Lincoln, G. A. 1971*a*. The seasonal reproductive changes in the red deer stag *(Cervus elaphus). J. Zool. Lond.* 163:105–23.

———. 1971*b*. Puberty in a seasonally breeding male, the red deer stag *(Cervus elaphus L.). J. Reprod. Fert.* 25:41–54.

———. 1972. The role of antlers in the behaviour of red deer. *J. Exp. Zool.* 182:233–50.

Lincoln, G. A., and Guinness, F. E. 1973. The sexual significance of the rut in red deer. *J. Reprod. Fert.,* suppl., 19:475–89.

———. 1977. Sexual selection in an herd of red deer. In *Reproduction and evolution,* ed. J. H. Calaby and C. H. Tyndale-Biscoe, pp. 33–38. Canberra: Australian Academy of Sciences.

Lincoln, G. A.; Guinness, F. E.; and Short, R. V. 1972. The way in which testosterone controls the social and sexual behaviour of the red deer stag *(Cervus elaphus). Horm. Behav.* 3:375–96.

Lincoln, G. A.; Youngson, R. W.; and Short, R. V. 1970. The social and sexual behaviour of the red deer stag. *J. Reprod. Fert.,* suppl., 11:71–103.

Linzell, J. L. 1972. Milk yield, energy loss in milk and mammary gland weight in different species. *Dairy Sci. Abstr.* 34:357–60.

Lockie, J. D. 1966. Territory in small carnivores. *Symp. Zool. Soc. Lond.* 18:143–66.

Longhurst, W. M., and Douglas, J. R. 1953. Parasite interrelationships of domestic sheep and Columbian black-tailed deer. *Trans. N. Am. Wildl. Conf.* 18:168–88.

Lott, D. F. 1979. Dominance relations and breeding rate in mature male American bison. *Z. Tierpsychol.* 49:418–32.

Lowe, V. P. W. 1961. A discussion of the history, present status and future conservation of red deer *(Cervus elaphus* L.) in Scotland. *Terre et Vie* 1:9–40.

———. 1966. Observations on the dispersal of red deer on Rhum. In *Play, exploration and territory in Mammals,* ed. P. A. Jewell and C. Loizos, pp. 211–28. London: Academic Press.

———. 1969. Population dynamics of the red deer *(Cervus elaphus* L.) on Rhum. *J. Anim. Ecol.* 38:425–57.

————. 1971. Some effects of a change in estate management on a deer population. In *The Scientific management of plant and animal communities,* ed. E. Duffey and A. S. Watt, pp. 437–56. Oxford: Blackwell.

————. 1977. Sika deer, *Cervus nippon.* In *The handbook of British mammals,* 2d ed., ed G. B. Corbet and H. N. Southern, pp. 423–28. Oxford: Blackwell.

MacArthur, R. H. 1972. *Geographical ecology: Patterns in the distribution of species.* New York: Harper and Row.

McCabe, R. A. 1949. Notes on live-trapping mink. *J. Mammal.* 30:416–23.

McClure, P. A. 1981. Sex biased litter reduction in food restricted wood rats *(Neotoma floridana). Science* 211:1058–60.

MacCulloch, J. 1819. *A description of the Western Isles of Scotland.* London.

McCullough, D. R. 1969. *The Tule elk: Its history, behavior and ecology.* Berkeley: University of California Press.

————. 1974. *Status of larger mammals in Taiwan.* Taipai: Taiwan Tourism Bureau.

————. 1979. *The George River deer herd: Population ecology of a k-selected species.* Ann Arbor: University of Michigan Press.

McDonald, J. W., and Warner, A. C. I., eds. 1975. *Digestion and metabolism in the ruminant.* Armidale, N.S.W.: University of New England Publishing Unit.

McDougall, E. I., and Lowe, V. P. W. 1968. Transferrin polymorphism and serum proteins of some British deer. *J. Zool. Lond.* 155:131–40.

Mace, G. M. 1979. The evolutionary ecology of small mammals. Ph.D. diss., University of Sussex.

McEwan, E. H. 1968. Growth and development of barren ground caribou. II. Postnatal growth rates. *Can. J. Zool.* 46:1023–29.

McEwan, E. H., and Whitehead, P. E. 1970. Seasonal changes in energy and nitrogen intake in reindeer and caribou. *Can. J. Zool.* 48:905–13.

————. 1971. Measurement of the milk intake of reindeer and caribou calves using tritiated water. *Can. J. Zool.* 49:443–47.

————. 1972. Reproduction in female reindeer and caribou. *Can. J. Zool.* 50:43–46.

McEwan, E. H., and Wood, A. J. 1966. Growth and development of the barren-ground caribou. I. Heart girth, hindfoot length and body weight relationships. *Can. J. Zool.* 44:401–11.

McHugh, T. 1958. Social behavior of the American buffalo *(Bison bison bison). Zoologica* 43:1–40.

McNaughton, S. J. 1976. Serengeti migratory wildebeest: Facilitation of energy flow by grazing. *Science* 191:92–94.

McVean, D. N., and Ratcliffe, D. A. 1962. *Plant communities of the Scottish Highlands.* Nature Conservancy Monograph no. 1. London: HMSO.

Magruder, N. D.; French, C. E.; McEwan, L. C.; and Swift, R. W. 1957. Nutritional requirements of white-tailed deer for growth and antler development. II. *Pa. State Univ., Agric. Expt. Stat. Bull.,* no. 628.

Mann, K. H. 1973. Growth strategy of seaweeds. *Science* 182:975–81.

Marburger, R. G., and Thomas, J. W. 1965. A die-off in white-tailed deer of the central mineral region of Texas. *J. Wildl. Mgmt.* 29:706–16.

Markgren, G. 1969. Reproduction of moose in Sweden. *Viltrevy* 6:127–285.

Martin, M. 1703. *A description of the Western Isles of Scotland.* New ed. 1934. Stirling.

Martin, R. 1980. Body temperature, activity and energy costs. *Nature* 283:335–36.

Mathews, M. O. 1972. Red deer, heather, sugar percent, altitude. B.Sc. thesis, Department of Forestry and Natural Resources, Edinburgh.

Mautz, W. W., and Petrides, G. A. 1971. Food passage rate in the white tailed deer. *J. Wildl. Mgmt.* 35:723–31.

Maynard Smith, J. 1964. Group selection and kin selection: A rejoinder. *Nature* 201:1145–47.

———. 1974. *Models in ecology.* Cambridge: Cambridge University Press.

———. 1976. Group selection. *Quart. Rev. Biol.* 51:277–83.

———. 1978. *The evolution of sex.* Cambridge: Cambridge University Press.

———. 1980. A new theory of sexual investment. *Behav. Ecol. Sociobiol.* 7:247–51.

Maynard Smith, J., and Price, G. R. 1973. The logic of animal conflict. *Nature* 246:15–18.

Mayr, E. 1963. *Animal species and evolution.* Harvard: Belknap Press.

Medawar, P. B. 1979. *Advice to a young scientist.* New York: Harper and Row.

Michael, E. D. 1970. Activity patterns of white tailed deer in south Texas. *Texas J. Sci.* 21:417–28.

Miers, K. H. 1962. Herd composition and effective reproduction of wapiti *(Cervus canadensis)* of Eastern Fiordland. *Proc. New Zealand Ecol. Soc.* 9:31–33.

Miller, G. R. 1971. Grazing and regeneration of shrubs and trees: First progress report. In *Range ecology research,* pp. 27–40. Edinburgh: Nature Conservancy.

Miller, I. 1979. The red deer problem as perceived by the Red Deer Commission. *Deer* 4:408–9.

Miller, R., and Denniston, R. H., II. 1979. Interband dominance in feral horses. *Z. Tierpsychol* 51:41–47.

Miller, R. S. 1967. Pattern and process in competition. *Adv. Ecol. Res.* 4:1–74.

Miller, W. C. 1932. A preliminary note upon the sex ratio of Scottish red deer. *Proc. Roy. Phys. Soc.* 22:99–101.

Milne, J. A.; Macrae, J. C.; Spence, A. M.; and Wilson, S. 1976. Intake and digestion of hill-land vegetation by the red deer and sheep. *Nature* 263:763–64.

Mirarchi, R. E.; Scanlon, P. F.; and Kirkpatrick, R. L. 1977. Annual changes in spermatozoan production and associated organs of white-tailed deer. *J. Wildl. Mgmt.* 41:92–99.

Mitchell, B. 1967. Research on the Scottish mainland. In *Red deer research in Scotland.* Progress report no. 1. Edinburgh: Nature Conservancy.

————. 1973. The reproductive performance of wild Scottish red deer *Cervus elaphus. J. Reprod. Fert.,* suppl., 19:271–85.

Mitchell, B., and Brown, D. 1974. The effects of age and body size on fertility in female red deer (*Cervus elaphus* L.). *Proc. Int. Congr. Game Biol.* 11:89–98.

Mitchell, B., and Crisp, J. M. 1981. Some properties of red deer (*Cervus elaphus*) at exceptionally high population-density in Scotland. *J. Zool.* 193:157–69.

Mitchell, B., and Lincoln, G. A. 1973. Conception dates in relation to age and condition in two populations of red deer in Scotland. *J. Zool.* 171:141–52.

Mitchell, B.; McCowan, D.; and Nicholson, I. A. 1976. Annual cycles of body weight and condition in Scottish red deer. *J. Zool. Lond.* 180:107–27.

Mitchell, B.; McCowan, D.; and Parish, T. 1973. Some characteristics of natural mortality among wild Scottish red deer (*Cervus-elaphus* L.). *Proc. Int. Congr. Game Biol.* 10:437–50.

Mitchell, B.; Staines, B. W.; and Welch, D. 1977. *Ecology of red deer: A research review relevant to their management.* Cambridge: Institute of Terrestrial Ecology.

Moen, A. N. 1973. *Wildlife ecology.* San Francisco: Freeman.

————. 1978. Seasonal changes in heart rates, activity, metabolism and forage intake of white-tailed deer. *J. Wildl. Mgmt.* 42:715–38.

Monro, D. 1549. *A description of the Western Isle of Scotland called Hybrides.* New ed. 1934. Stirling.

Montgomery, G. G. 1963. Nocturnal movements and activity rhythms of white-tailed deer. *J. Wildl. Mgmt.* 27:422–27.

Moors, P. J. 1980. Sexual dimorphism in the body size of mustelids (Carnivora): The roles of food habits and breeding systems. *Oikos* 34:147–58.

Morgan, B. J. T.; Simpson, M. J. A.; Hanby, J. P.; and Hall-Craggs, J. 1974. Visualizing interaction and sequential data in animal behaviour: Theory and application of cluster-analysis methods. *Behaviour* 56:1–43.

Morgan, J. T. 1951. *Observations on the behaviour and grazing habits of dairy cattle at Pantyrhuad Farm, Carmarthenshire.* London: Association for the Study of Animal Behaviour.

Morrison, J. A. 1960. Characteristics of estrus in captive elk. *Behaviour* 26:84–92.

Morton, E. S. 1977. On the occurrence and significance of motivation-structural rules in some bird and mammal sounds. *Am. Nat.* 111:855–69.

Müller-Using, D., and Schloeth, R. 1967. Das Verhaltan der Hirshe, Kükenthal. *Handb. Zool.* 10(28):1–60.

Murie, O. J. 1956. *The elk of North America.* Harrisburg: Stackpole.

Mutch, W. E. S.; Lockie, J. D.; and Cooper, A. N. 1976. *The red deer in*

South Ross: A report on wildlife management in the Scottish Highlands. Edinburgh: Department of Forestry and Natural Resources, University of Edinburgh.

Myers, J. H., and Krebs, C. J. 1971. Sex ratios in open and enclosed vole populations: Demographic implications. *Am. Nat.* 105:325–44.

Nahlik, A. J. de. 1974. *Deer management: Improved herds for greater profit.* Newton Abbot: David and Charles.

Nellis, C. H. 1968. Productivity of mule deer on the National Bison Range, Montana. *J. Wildl. Mgmt.* 32:344–49.

Nesbit-Evans, E. M. 1970. The reaction of a group of Rothschild's giraffe to a new environment. *E. Afr. Wildl. J.* 8:53–62.

Newsome, A. E. 1977. Imbalance in the sex-ratio and age structure of the red kangaroo in central Australia. In *Biology and environment,* vol. 2, *The biology of marsupials,* ed. D. Gilmore and B. Stonehouse, pp. 221–33. London: Macmillan.

Newton, I. 1979. *Population ecology of raptors.* Berkhamstead: Poyser.

Nicholson, I. A. 1974. Red deer range and problems of carrying capacity in the Scottish Highlands. *Mammal Rev.* 4:103–18.

Nicholson, I. A., and Robertson, R. A. 1958. Some observations on the ecology of an upland grazing in north east Scotland with special reference to *Callunetum. J. Ecol.* 46:239–70.

Nie, N. H.; Hull, C. H.; Jenkins, J. G.; Steinbrenner, K.; and Bent, D. H. 1975. *SPSS: Statistical Package for the Social Sciences.* 2d ed. New York: McGraw-Hill.

Nietzsche, F. 1954 [1889]. *Twilight of the idols: The portable Nietzsche.* Reprinted in translation by W. Kaufmann. New York: Viking Press.

Nievergelt, B. 1966. *Der Alpensteinbock.* Mammalia Depicta, 1. Berlin: Parey Verlag.

Nordan, H. C.; Cowan, I. McT.; and Wood, A. J. 1970. The feed intake and heat production of the young black-tailed deer *(Odocoileus hemionus columbianus). Can. J. Zool.* 48:275–82.

Nowosad, R. F. 1975. Reindeer survival in the Mackenzie Delta herd, birth to four months. In *Proceedings of the First International Reindeer and Caribou Symposium,* pp. 199–208. Biological Papers of the University of Alaska, Special Report no. 1. Fairbanks: University of Alaska.

Odum, E. P. 1971. *Fundamentals of ecology.* 3d ed. Philadelphia: Saunders.

Ørskov, E. R. 1975. Manipulation of rumen fermentation for maximum food utilization. *World Rev. Nutr. Diet.* 22:152–82.

Osborne, B. C. 1980. The grazing ecology and management of red deer and hill sheep on Ardtornish Estate, Argyll. Ph.M. thesis, University of Sussex.

Otter, W. 1824. *The life and remains of the late Edward Daniel Clerke.* London.

Owen-Smith, N. 1977. On territoriality in ungulates and an evolutionary model. *Quart. Rev. Biol.* 52:1–38.

————. n.d. Factors influencing the transfer of plant products into large herbivore populations. In preparation.

Ozoga, J. J. 1972. Aggressive behaviour of white-tailed deer at winter cuttings. *J. Wildl. Mgmt.* 36:861–68.

Ozoga, J. J., and Verme, L. J. 1975. Activity patterns of white-tailed deer during estrus. *J. Wildl. Mgmt.* 39:679–83.

Packer, C. 1977. Inter-troop transfer and inbreeding avoidance in *Papio anubis* in Tanzania. Ph.D. diss., University of Sussex.

———. 1979*a*. Inter-troup transfer and inbreeding avoidance in *Papio anubis. Anim. Behav.* 27:1–36.

———. 1979*b*. Male dominance and reproductive activity in *Papio anubis. Anim. Behav.* 27:37–45.

Panwar, H. S. 1978. Decline and restoration success of the central Indian Barasingha *(Cervus duvanceli branderi).* In *Threatened deer,* pp. 143–58. Morges: IUCN.

Parker, G. A. 1974. Assessment strategy and the evolution of fighting behaviour. *J. Theor. Biol.* 47:223–43.

———. 1978. Selfish genes, evolutionary games and the adaptiveness of behaviour. *Nature* 274:849–55.

Partridge, L. 1980. Mate choice increases a component of offspring fitness in fruit flies. *Nature* 283:290–91.

Patterson, I. J. 1965. Timing and spacing of broods in the black-headed gull, *Larus ridibundus. Ibis* 107:433–59.

Payne, R. N. 1979. Sexual selection and intersexual differences in variance of breeding success. *Am. Nat.* 114:447–52.

Peart, J. N. 1968. Some effects of live weight and body condition on the milk production of Blackface ewes. *J. Agric. Sci. Camb.* 70:331–38.

Peek, J. M. 1962. Studies of moose in the Gravelly and Snowcrest Mountains. *J. Wildl. Mgmt.* 26:360–65.

Peek, J. M.; Lovaas, A. L.; and Rouse, R. A. 1967. Population changes within the Gallatin elk herd, 1932–1965. *J. Wildl. Mgmt.* 31:304–15.

Pennant, T. 1774. *A tour in Scotland and voyage to the Hebrides.* Chester.

Peterson, R. L. 1955. *North American moose.* Toronto: University of Toronto Press.

Phillipson, A. T., ed. 1970. *Physiology of digestion and metabolism in the ruminant.* Newcastle-upon-Tyne: Oriel Press.

Pianka, E. R. 1973. The structure of lizard communities. *Ann. Rev. Ecol. Syst.* 4:53–74.

———. 1976. Natural selection of optimal reproductive tactics. *Am. Zool.* 16:775–84.

———. 1978. *Evolutionary ecology.* 2d ed. New York: Harper and Row.

Pianka, E. R., and Parker, W. S. 1975. Age-specific reproductive tactics. *Am. Nat.* 109:453–64.

Picton, H. D. 1960. Migration patterns of the Sun River elk herd, Montana. *J. Wildl. Mgmt.* 24:279–90.

Pielowski, Z. 1969. Die Wiedereinbürgerung des Elches *Alces alces* (L.) im kampinos–Nationalpark in Polen. *Z. Jagdwirsenschaft* 15:6–17.

Pohlig, H. 1892. Die Cerviden des Turingischen Diluvialtravertinos. *Paleographica* 39:215–63.

Pollock, J. I. 1977. The ecology and sociology of feeding in *Indri indri.*

In *Primate ecology: Studies of feeding and ranging behaviour in lemurs, monkeys and apes,* ed. T. H. Clutton-Brock, pp. 38–69. London: Academic Press.

Pond, C. M. 1977. The significance of lactation in the evolution of mammals. *Evolution* 31:177–99.

Pond, M. 1978. Morphological aspects and the ecological and mechanical consequences of fat deposition in wild vertebrates. *Ann. Rev. Ecol. Syst.* 9:519–70.

Poore, M. E. D., and McVean, D. N. 1957. A new approach to Scottish mountain vegetation. *J. Ecol.* 45:401–39.

Post, D. G. 1978. Feeding and ranging behaviour of the Yellow Baboon *(Papio cynocephalus).* Ph.D. diss., Yale University.

Povilitis, A. 1978. The Chilean huemal *(Hippocamelus bisulcus)* project: A case history (1975–76). In *Threatened deer,* pp. 109–28. Morges: IUCN.

Pratt, D. M., and Anderson, V. A. 1979. Giraffe cow-calf relationships and social development of the calf in the Serengeti. *Z. Tierpsychol.* 51:233–51.

Prins, R. A., and Geelen, M. J. H. 1971. Rumen characteristics of red deer, fallow deer and roe deer. *J. Wildl. Mgmt.* 35:673–80.

Prior, R. 1968. *The roe deer of Cranborne Chase: An ecological survey.* London: Oxford University Press.

Ralls, K. 1976. Mammals in which the female is larger than the male. *Quart. Rev. Biol.* 51:245–75.

Ralls, K.; Brownwell, R. L.; and Ballou, J. 1980. Differential mortality by sex and age in mammals, with specific reference to the sperm whale. *Rep. Int. Whal. Comm.,* special issue, 2:223–43.

Ralls, K.; Brugger, K.; and Ballou, J. 1979. Inbreeding and juvenile mortality in small populations of ungulates. *Science* 206:1101–3.

Ransom, B. A. 1967. Reproductive biology of white-tailed deer in Manitoba. *J. Wildl. Mgmt.* 31:114–23.

Red Deer Commission. 1961–80. *Annual reports,* 1960–79. London: HMSO.

Redfield, J. A.; Taitt, M. J.; and Krebs, C. J. 1978a. Experimental alteration of sex ratios in populations of *Microtus townsendii,* a field vole. *J. Can. Zool.* 56:17–27.

————. 1978b. Experimental alterations of sex-ratios in populations of *Microtus oregoni,* the creeping vole. *J. Anim. Ecol.* 47:55–69.

Reimers, E. 1972. Growth in domestic and wild reindeer in Norway. *J. Wildl. Mgmt.* 36:612–19.

Reiter, J.; Stinson, N. L.; and LeBoeuf, B. J. 1978. Northern elephant seal development: The transition from weaning to nutritional development. *Behav. Ecol. Sociobiol.* 3:337–67.

Richards, S. M. 1974. The concept of dominance and methods of assessment. *Anim. Behav.* 22:914–30.

Richie, W. F. 1970. Regional differences in weight and antler measurements of Illinois deer. *Trans. Ill. State Acad. Soc.* 63:189–97.

Ritchie, J. 1920. *The influence of man on animal life in Scotland.* Cambridge: Cambridge University Press.

Rivers, J. P. W., and Crawford, M. A. 1974. Maternal nutrition and the sex ratio at birth. *Nature* 252:297–98.

Robbins, C. T., and Moen, A. N. M. 1975. Milk consumption and weight gain of white-tailed deer. *J. Wildl. Mgmt.* 39:355–60.

Robbins, C. T.; Prior, R. L.; Moen, A. N.; and Visek, W. J. 1974. Nitrogen metabolism of white-tailed deer. *J. Anim. Sci.* 38:186–91.

Robbins, C. T., and Robbins, B. L. 1979. Fetal and neonatal growth patterns and maternal reproductive effort in ungulates and subungulates. *Am. Nat.* 114:101–16.

Robinette, W. L. 1966. Mule deer home range and dispersal in Utah. *J. Wildl. Mgmt.* 30:334–49.

Robinette, W. L.; Gashwiler, J. S.; Jones, D. A.; and Crane, H. S. 1955. Fertility of mule deer in Utah. *J. Wildl. Mgmt.* 19:115–36.

Robinette, W. L.; Gashwiler, J. S.; Low, J. B.; and Jones, D. A. 1957. Differential mortality by sex and age among mule deer. *J. Wildl. Mgmt.* 21:1–16.

Roe, N. A. 1975. Mimeographed report to *Red Data Book*. Cambridge: International Union for the Conservation of Nature.

Roe, N. A., and Rees, W. B. 1976. Preliminary observations of the Taruca (*Hippocamelus antisensis:* Cervidae) in southern Peru. *J. Mammal.* 57:722–30.

Roseberry, J. L., and Klimstra, W. D. 1974. Differential vulnerability during a controlled deer harvest. *J. Wildl. Mgmt.* 38:499–507.

Rowell, T. E. 1974. The concept of social dominance. *Behav. Biol.* 11:131–54.

Russell, W. S. 1976. Effect of twin birth on growth of cattle. *Anim. Prod.* 22:167–73.

Ryder, M. L. 1977. Seasonal coat changes in grazing red deer (*Cervus elaphus*). *J. Zool. Lond.* 181:137–43.

Sade, D. S. 1972. Sociometrics of *Macaca mulatta*. 1. Linkages and cliques in grooming matrices. *Folia Primat.* 18:196–223.

Sadleir, R. M. F. S. 1969. *The ecology of reproduction in wild and domestic mammals.* London: Methuen.

———. 1980. Energy and protein intake in relation to growth of suckling black-tailed deer fawns. *Can. J. Zool.* 58:1347–54.

Schaffer, A. L.; Young, B. A.; and Chimnano, A. M. 1978. Ration digestion and retention times of digesta in domestic cattle (*Bos taurus*), American bison (*Bison bison*) and Tibetan yak (*Bos grunniens*). *Can. J. Zool.* 56:2355–58.

Schaller, G. B. 1967. *The deer and the tiger.* Chicago: University of Chicago Press.

———. 1972. *The Serengeti lion: A study of predator-prey relations.* Chicago: University of Chicago Press.

———. 1979. *Mountain monarchs: Wild sheep and goats of the Himalaya.* Chicago: University of Chicago Press.

Schaller, G. B., and Hamer, A. 1978. Rutting behaviour of Père David's deer, *Elaphurus davidianus. Zool. Garten,* n.s. (Jena) 48:1–15.

Schein, M. W., and Fohrman, M. H. 1955. Social dominance relationships in a herd of dairy cattle. *Brit. J. Anim. Behav.* 3:45–55.

Schinkel, P. G., and Short, B. F. 1961. The influence of nutritional level during pre-natal and early post-natal life on adult fleece and body characteristics. *Aust. J. Agric. Res.* 12:176–202.

Schloeth, V. R. 1961. Das sozialleben des Camargue-Rindes: Qualitative und quantitative Untersuchungen über die sozialen Berziehungen— insbesondere die soziale Rangordung—des halbarilden französischen Kampfriendes. *Z. Tierpsychol.* 18:547–627.

Schmidt-Nielsen, K. 1975. Scaling in biology: The consequences of size. *J. Exp. Zool.* 194:287–307.

Scott, D. K. 1978. Social behaviour of wintering Bewick's swans. Ph.D. diss., University of Cambridge.

———. 1979. Indentification of individual Bewick's swans by bill patterns. In *Animal marking*, ed. B. Stonehouse, pp. 160–74. London: Macmillan.

Scrimshaw, N. S.; Taylor, G. E.; and Gordon, J. E. 1968. *Interactions of nutrition and infection.* Publication no. 57. Geneva: World Health Organization.

Scrope, W. 1897. *The art of deer stalking.* New ed. London: Arnold.

Searcy, W. A. 1979. Sexual selection and body size in male red-winged blackbirds. *Evolution* 33:649–61.

Sekulig, R., and Estes, R. D. 1977. A note on bone chewing in the sable antelope in Kenya. *Mammalia* 41:550–52.

Selander, R. K. 1965. On mating systems and sexual selection *Am. Nat.* 99:129–41.

———. 1972. Sexual selection and dimorphism in birds. In *Sexual selection and the descent of man*, ed. B. Campbell, pp. 180–230. Chicago: Aldine.

Severinghaus, C. N., and Cheatum, E. L. 1956. Life and times of the white-tailed deer. In *Deer of North America*, ed. W. P. Taylor, pp. 57–186. Harrisburg: Stackpole.

Shaposhnikov, F. D. 1956. Ecology of the sable in North-eastern Altai (1956). In *Studies on mammals in government reserves*, ed. P. B. Yurgenson, pp. 18–20. Translated by Israel Program for Scientific Translation, 1961.

Sherman, P. W. 1976. Natural selection among some group-living organisms. Ph.D. diss., University of Michigan.

———. 1977. Nepotism and the evolution of alarm calls. *Science* 197: 1246–53.

Short, C. 1970. Mophological development and aging of mule and white-tailed deer fetuses. *J. Wildl. Mgmt.* 34:383–88.

Short, H. L. 1975. Nutrition of southern deer in different seasons. *J. Wildl. Mgmt.* 39:321–29.

Short, R. V. 1979. Sexual selection and its component parts, somatic and genital selection, as illustrated by man and the great apes. *Adv. Study Behav.* 9:131–58.

Short, R. V., and Mann, T. 1966. The sexual cycle of a seasonally breeding mammal, the roebuck (*Capreolus capreolus*). *J. Reprod. Fert.* 12:337–51.

Siegel, S. 1956. *Non-parametric statistics for the behavioral sciences.* Tokyo: McGraw-Hill Kogakusha.

Silver, H.; Colvos, N. F.; Holter, J. B.; and Hayes, H. H. 1969. Fasting metabolism of white-tailed deer *J. Wildl. Mgmt.* 33:490–98.

Simpson, A. M. 1976. A study of the energy metabolism and seasonal cycles of captive red deer. Ph.D. diss., University of Aberdeen.

Sinclair, A. R. E. 1977. *The African buffalo: A study of resource limitation of populations.* Chicago: University of Chicago Press.

Singh, O. N.; Singh, R. N.; and Srivastava, R. R. P. 1965. Study in post-partum interval to first service in Tharparkar cattle. *Indian J. Vet. Sci.* 35:245–48.

Sinh, Rajit. 1975. Keibul Lamjao Sanctuary and the brow-antlered deer—1972 with notes on a visit in 1975. *J. Bombay Nat. Hist. Soc.* 72:243–55.

Slater, P. J. B. 1974. The temporal pattern of feeding in the zebra finch. *Anim. Behav.* 22:506–15.

Slee, J. 1970. Resistance to body cooling in male and female sheep, and the effects of previous exposure to chronic cold, acute cold and repeated cold shocks. *Anim. Prod.* 12:13–21.

———. 1972. Habituation and acclimatization of sheep to cold following exposures of varying length and severity. *J. Physiol.* 227:51–70.

Smith, R. H. 1978. Fallow deer. *Reading Nat.* 30:12–14.

———. 1979. On selection for inbreeding in polygynous animals. *Heredity* 43:205–11.

Snedecor, G. W., and Cochran, W. G. 1967. *Statistical methods.* 6th ed. Ames: Iowa State University Press.

Sobanskii, G. G. 1979. Selective elimination in the Siberian stag population in the Altais as a result of the early winter of 1976/77. *Soviet J. Ecol.* 10:78–80. Translated from Russian, Consultants Bureau, New York.

Soest, P. J. van. 1965. Voluntary intake in relation to chemical composition and digestibility. *J. Anim. Sci.* 24:834–43.

Sokal, R. R., and Rohlf, F. J. 1969. *Biometry.* San Francisco: Freeman.

Southern, H. N. 1979. Population processes in small mammals. In *Ecology of small mammals,* ed. D. M. Stoddart, pp. 63–101. London: Chapman and Hall.

Sparks, J. 1967. Allogrooming in primates: A review. In *Primate ethology: Essays in the socio-sexual behaviour of apes and monkeys,* ed. D. Morris, pp. 148–75. London: Weidenfeld and Nicolson.

Spassov, N. B. 1979. Sexual selection and evolution of intraspecific display means in baboons (Primates, Cercopithecidae). *C.R. Acad. Bulgare Sci.* 32:225–28.

Spinage, C. A. 1970. Giraffid horns. *Nature* 227:735–36.

Staines, B. W. 1969. Our knowledge of deer behaviour and its possible effects on deer husbandry. In *The husbanding of red deer,* ed. M. M.

Bannerman and K. L. Blaxter, pp. 29–31. Aberdeen: Aberdeen University Press.

————. 1970. The management and dispersion of a red deer population in Glen Dye, Kincardineshire. Ph.D. diss., University of Aberdeen.

————. 1974. A review of factors affecting deer dispersion and their relevance to management. *Mammal. Rev.* 4:79–91.

————. 1976. The use of natural shelter by red deer *(Cervus elaphus)* in relation to weather in north-east Scotland. *J. Zool.* 180:1–8.

————. 1977. Factors affecting the seasonal distribution of red deer *(Cervus elaphus* L.) in Glen Dye, N.E. Scotland. *Ann. Appl. Biol.* 87:495–512.

————. 1978. The dynamics and performance of a declining population of red deer *(Cervus elaphus)*. *J. Zool.* 184:403–19.

Staines, B. W., and Crisp, J. M. 1978. Observations on food quality in Scottish red deer *(Cervus elaphus)* as determined by chemical analysis of the rumen contents. *J. Zool.* 185:253–59.

Stewart, L. K. 1976. The Scottish red deer census. *Deer* 3:529–32.

Stirling, I. 1971. Variation in sex ratio of newborn Weddell seals during the pupping season. *J. Mammal.* 52:842–44.

————. 1975. Factors affecting the evolution of social behaviour in the pinnipedia. *P. Réun. Cons. Int. Explor. Mer* 169:205–12.

Stobbs, T. H. 1973a. The effect of plant structure on the intake of tropical pastures. I. Variation in the bite size of grazing cattle. *Aust. J. Agric. Res.* 24:809–19.

————. 1973b. The effect of plant structure on the intake of tropical pastures. II. Differences in sward structure, nutritive value and bite size of animals grazing *Setaria anceps* and *Chloris gayana* at various stages of growth. *Austr. J. Agric. Res.* 24:821–29.

Stonehouse, B. 1968. Thermoregulatory function of growing antlers. *Nature* 218:870–72.

Struhsaker, T. T. 1967. Behavior of elk *(Cervus canadensis)* during the rut. *Z. Tierpsychol.* 24:80–114.

Suttie, J. M. 1979. The effect of antler removal on dominance and fighting behaviour in farmed red deer stags. *J. Zool.* 190:217–24.

————. 1980. Influence of nutrition on growth and sexual maturation of captive red deer stags. In *Proceedings of the Second International Reindeer/Caribou Symposium.* (Røros, Norway), ed. E. Reimers, E. Gaare, and S. Skjenneberg. Trondheim: Direktoratet for vilt og ferskvannsfisk.

Syme, G. J. 1974. Competitive orders as measures of social diminance. *Anim. Behav.* 22:931–40.

Taber, R. D., and Dasmann, R. F. 1954. A sex difference in mortality in young Columbian black-tailed deer. *J. Wildl. Mgmt.* 18:309–15.

Talbot, L. M., and Talbot, M. H. 1963. The wildebeest in western Masailand, E. Africa. *Wildl. Monogr.* 12:1–88.

Teer, J. G.; Thomas, J. W.; and Walker, E. A. 1968. Ecology and management of white tailed deer in the Llano Basin of Texas. *Wildl. Monogr.* 15:1–62.

Tener, J. S. 1954. A preliminary study of the musk-oxen of Fosheim Peninsula, Ellesmere Island, N.W.T. Canada Dept. Natl. Parks Board. Canadian Wildl. Service, Ottawa. *Wildl. Mgmt. Bull.*, ser. 1, 9:1–34.

Thompson, W. A.; Vertinsky, I.; and Krebs, J. R. 1974. The survival value of flocking in birds: A simulation model. *J. Anim. Ecol.* 43:785–820.

Thomson, A. M., and Thomson, W. 1953. Effect of diet on milk yield of the ewe and growth of her lamb. *Brit. J. Nutr.* 2:290–305.

Thomson, B. R. 1971. *Wild reindeer activity.* IBP Report, Grazing Project of the Norwegian IBP Committee. Trondheim: IBP.

Tinkle, D. W., and Hadley, N. F. 1975. Lizard reproductive effort: Caloric estimates and comments on its evolution. *Ecology* 56:427–34.

Topinski, P. 1974. The role of antlers in establishment of the red deer herd hierarchy. *Acta Theriol.* 19:509–14.

Treisman, M. 1975a. Predation and the evolution of gregariousness. I. Models for concealment and evasion. *Anim. Behav.* 23:779–800.

———. 1975b. Predation and the evolution of gregariousness. II. An economic model for predator-prey interaction. *Anim. Behav.* 23:801–25.

Tribe, D. E. 1949. Some seasonal observations on the grazing habits of sheep. *Emp. J. Exp. Agric.* 27:106–15.

Trivers, R. L. 1972. Parental investment and sexual selection. In *Sexual selection and the descent of man,* ed. B. Campbell, pp. 136–79. Chicago: Aldine.

———. 1974. Parent-offspring contact. *Am. Zool.* 249–65.

Trivers, R. L., and Willard, D. E. 1973. Natural selection of parental ability to vary the sex ratio of offspring. *Science* 179:90–92.

Tuomi, J. 1980. Mammalian reproductive strategies: A generalised relation of litter size to body size. *Oecologia* 45:39–44.

Ullrey, D. E.; Youatt, W. G.; Johnson, H. E.; Fay, L. D.; and Bradley, B. L. 1967. Protein requirement of white-tail deer fawns. *J. Wildl. Mgmt.* 31:679–85.

Underwood, R. 1981. Companion preference in an eland herd. *Afr. J. Ecol.* 19:341–54.

Veen, H. E. van de. 1979. Food selection and habitat use in the red deer (*Cervus elaphus* L.). Ph.D. diss., University of Groningen.

Verheyen, R. 1951. *Contribution à l'étude éthologique des mammifères du parc National de l'Upemba.* Brussels.

Verme, L. J. 1963. Effect of nutrition on growth of white-tailed deer fawns. *Trans. N. Am. Wildl. Conf.* 20:431–43.

———. 1965. Reproduction studies on penned white-tailed deer. *J. Wildl. Mgmt.* 29:74–79.

———. 1969. Reproductive patterns of white-tailed deer related to nutritional plane. *J. Wildl. Mgmt.* 33:881–87.

Vine, I. 1971. Risk of visual detection and pursuit by a predator and selective advantage of flocking behaviour. *J. Theor. Biol.* 30:405–22.

Viret, J. 1961. Artiodactyla. In *Traité de paléontologie VI,* ed. J. Piveteau, pp. 1001–21. Paris: Masson.

Vos, A. de.; Brokx, P.; and Geist, V. 1967. A review of social behavior of the North American cervids during the reproductive period. *Amer. Midl. Nat.* 77:390–417.

Walker, E. O. ed. 1975. *Mammals of the world.* Vols. 1 and 2. 3d ed. Baltimore: Johns Hopkins University Press.

Walther, F. 1965. Verhaltensstudien an der Grantgazelle (*Gazella granti* Brooke, 1872) im Ngorongoro krater. *Z. Tierpsychol.* 22:167–208.

———. 1968. *Verhalten der Gazellen.* Nene Brehun-Bücherei no. 373. W. Hemberg Lutherstadt: Ziemsen.

———. 1969. Flight behaviour and avoidance of predators in Thomson's gazelle (*Gazella thomsoni* Gunther, 1884). *Behaviour* 34:184–221.

Walvius, M. R. 1961. A discussion of the size of recent red deer (*Cervus elaphus* L.) compared with prehistoric specimens. *Beaufortia* 97:75–82.

Waser, P. 1977. Feeding, ranging and group size in the Mangabey *Cercocebus albigena.* In *Primate ecology: Studies of feeding and ranging behaviour in lemurs, monkeys and apes,* ed. T. H. Clutton-Brock, pp. 183–222. London: Academic Press.

Watson, A., and Staines, B. W. 1978. Differences in the quality of wintering areas used by male and female red deer (*Cervus elaphus*) in Aberdeenshire. *J. Zool.* 286:544–50.

Wegge, P. 1975. Reproduction and early calf mortality in Norwegian red deer. *J. Wildl. Mgmt.* 39:92–99.

Weiner, J. 1977. Energy metabolism of roe deer. *Acta Theriol.* 22:3–24.

Wesley, D. E.; Knox, K. L.; and Nagy, J. G. 1973. Energy metabolism of pronghorn antelope. *J. Wildl. Mgmt.* 37:563–73.

Western, D. 1971. Giraffe chewing a Grant's gazelle carcass. *E. Afr. Wildl. J.* 9:156–57.

White, F. N. 1914. Variation in the sex ratio of *Mus rattus* associated with an unusual mortality of adult females. *Proc. Roy. Soc. Lond.* 87:335–44.

Whitehead, G. K. 1972. *Deer of the world.* London: Constance.

Widdowson, E. 1976. The response of the sexes to nutritional stress. *Proc. Nutr. Soc.* 35:1175–80.

Widdowson, E. M., and McCance, R. A. 1956. The effects of chronic undernutrition and of total starvation on growing and adult rats. *Brit. J. Nutr.* 10:363–73.

Widdowson, E. M.; Mavor, W. O.; and McCance, R. A. 1964. The effect of undernutrition and rehabilitation on the development of the reproductive organs: Rats. *J. Endocr.* 29:119–26.

Wiens, J. A. 1976. Population responses to patchy environments. *Ann. Rev. Ecol. Syst.* 7:81–120.

Wiersema, G. J. 1974. Observations on the supplementary feeding of red deer on an estate in the Central Highlands of Scotland. M.Sc. thesis, Agricultural University of Wageningen, Netherlands.

Williams, G. C. 1966. *Adaptation and natural selection: A critique of some current evolutionary thought.* Princeton: Princeton University Press.

———. 1971. *Group selection.* Chicago: Aldine-Atherton.

———. 1975. *Sex and evolution.* Princeton: Princeton University Press.

Williams, J. P. G.; Tanner, J. M.; and Collins, J. P. 1974. Catch-up growth in male rats after growth retardation during the suckling period. *Paed. Res.* 8:149–56.

Willis, M. B., and Wilson, A. 1974. Factors affecting birth weight of Santa Gertrudis calves. *Anim. Prod.* 18:231–36.

Wilson, E. O. 1971. Competitive and aggressive behavior. In *Man and beast: Comparative social behavior,* ed. J. F. Eisenberg and W. S. Dillon, pp. 182–217. Washington, D.C.: Smithsonian Institution Press.

———. 1975. *Sociobiology: The new synthesis.* Cambridge, Mass.: Harvard University Press.

Wilson, E. O., and Bossert, W. H. 1971. *A primer of population biology.* Sunderland, Mass.: Sinauer.

Wood, A. J.; Cowan, I. McT.; and Nordan, H. C. 1962. Periodicity of growth in ungulates as shown by deer of the genus *Odocoileus. Can. J. Zool.* 40:593–603.

Woodgerd, W. 1964. Population dynamics of bighorn sheep on Wildhorse Island. *J. Wildl. Mgmt.* 28:381–91.

Woodward, R. H., and Goldsmith, P. L. 1964. *Cumulative sum tests: Theory and practice.* Imperial Chemical Industries Monograph no. 3. Edinburgh: Oliver and Boyd.

Woolf, A., and Harder, J. D. 1979. Population dynamics of a captive white-tailed deer herd with emphasis on reproduction and mortality. *Wildl. Monogr.* 67:1–53.

Wrangham, R. W. 1975. The behavioral ecology of chimpanzees in Gombe National Park, Tanzania. Ph.D. diss., Cambridge University.

———. 1977. Feeding behaviour of chimpanzees in Gombe Stream National Park, Tanzania. In *Primate ecology: Studies of feeding and ranging behaviour in lemurs, monkeys and apes,* ed. T. H. Clutton-Brock, pp. 504–38. London: Academic Press.

———. 1979. On the evolution of ape social systems. *Soc. Sci. Inf.* 18:335–68.

———. 1980. An ecological model of female-bonded primate groups. *Behaviour* 75:262–300.

———. 1982. Mutualism, kinship and social evolution. In *Current problems in sociobiology,* ed. King's College Sociobiology Group, pp. 269–89. Cambridge: Cambridge University Press. In press.

Wyatt, J. R. 1971. Osteophagia in Masai giraffe. *E. Afr. Wildl. J.* 9:157.

Wynne-Edwards, V. C. 1962. *Animal dispersion in relation to social behaviour.* Edinburgh: Oliver and Boyd.

Yanushko, P. A. 1957. The way of life of the Crimean deer and their influence on the natural cycle. *Trans. Moscow Soc. Nat.* 35:39–52.

Zahavi, A. 1975. Mate selection: A selection for a handicap. *J. Theor. Biol.* 53:205–14.

Author Index

Ahlén, I., 278
Albon, S. D., 5, 46, 76, 86, 88, 89, 105, 107, 126, 128, 129, 132, 133, 134, 135, 137, 138, 150, 151, 164, 169, 170, 172, 291
Alexander, G., 62, 168
Alexander, R. D., 6, 192
Allden, W. G., 151, 222, 227, 231, 267
Altmann, M., 65, 177, 178, 301
Altmann, S. A., 300
Anderson, A. E., 293
Anderson, C. C., 279
Anderson, J., 302
Anderson, J. E. M., 220, 221
Anderson, V. A., 182
Appleby, M. C., 191, 206, 212, 213
Aristotle, 286
Arman, P., 62, 70, 72, 220, 326
Arnold, G. W., 168, 227
Aschaffenburg, R., 69
Atmosoedirdjo, S., 303

Ballou, J., 196, 294
Bandy, P. J., 267
Banfield, A. W. F., 279
Bannikov, A. G., 133, 279
Banwell, D. B., 275
Barette, C., 302
Bateman, A. J., 4
Bateson, P. P. G., 34
Bell, R. H. V., 239, 249
Belovsky, G. E., 224
Benedict, F. G., 165
Beninde, J., 10, 11, 136
Bent, D. H., 41
Berger, J., 93, 152, 193
Bergerud, A. T., 134, 151, 278, 292, 301, 302
Bertram, B. C. R., 34, 188, 300
Bischof, N., 196

Blaisdell, J. A., 275
Blaxter, K. L., 62, 74, 87, 88, 90, 156, 239
Blouch, R. A., 303
Bolton, G. R., 233
Boonsong, L., 303
Boonstra, R., 7, 259, 279
Bossert, W. H., 3, 93
Bouissou, M. F., 151
Bowdon, D. C., 293
Boyd, I., 20
Bradley, B. L., 276
Bramley, P. S., 293, 302
Briggs, K. T., 169
Brokx, P., 302
Bromley, P. T., 193
Brown, D., 83, 93
Brown, J. L., 6, 196
Brownwell, R. L., 294
Brugger, K., 196
Bubenik, A. B., 65, 69, 72, 136, 223, 224
Bubenikova, J. M., 223, 224
Buechner, H. K., 150
Bulmer, M. G., 196
Burckhardt, D., 203, 301
Bützler, W., 107, 202, 203, 204

Cameron, A. G., 11
Caughley, G., 275, 297, 315, 316
Challies, C. N., 275, 276
Chalmers, N., 202
Chaplin, R. E., 293
Chapman, D. I., 302
Chapman, N., 302
Charcon, E., 227
Charles, W. N., 24, 25, 233, 234, 239
Charles-Dominique, P., 296
Charlesworth, B., 3, 93
Cheatum, E. L., 233

Child, G., 278, 279,295
Clark, A. B., 84, 171, 196, 296
Clegg, M. T., 62
Clutton-Brock, T. H., 5, 6, 20, 46, 64,
    65, 66, 67, 68, 76, 86, 88, 89, 105,
    107, 126, 128, 129, 132, 133, 134,
    135, 137, 138, 150, 151, 164, 169,
    170, 172, 181, 194, 212, 213, 214,
    239, 245, 289, 291, 292, 300, 317,
    323
Coady, J. W., 220
Cochran, W. G., 49
Cockerill, R. A., 47, 72, 73, 182, 183,
    327
Cody, M. L., 192
Cole, L. C., 177, 189
Collins, J. P., 151
Colovos, N. F., 220
Colquhoun, I. R., 228, 233
Cooper, A. B., 85
Cooper, A. N., 177, 268
Corbet, G. B., 10
Corfield, T., 296
Coulson, J. C., 164
Cowan, I. McT., 165, 267, 293, 302
Cox, C. R., 60, 61
Craighead, F. C., 134, 177, 224, 242,
    243
Craighead, J. J., 134, 177, 224, 242,
    243
Crane, H. S., 62
Crisp, J. M., 17, 48, 239, 240
Crook, J. H., 6, 47
Cullen, J. M., 192
Cunningham, J. M. M., 62, 74, 90,
    239

Daly, M., 220
Darling, F. F., 13, 164, 176, 177, 201,
    214, 242, 243, 245, 278, 302
Darwin, C., 3, 4, 9, 80, 143, 219, 258
Dasmann, R. F., 278
Dasmann, W. P., 275, 302
Davies, G., 164
Davies, G. E., 186
Davies, N. B., 138, 151
Davis, H. L., 168
Dawkins, R., 3
Defries, J. C., 165
Delpech, F., 11
Denniston, R. H., 193
Dittus, W. P. J., 90, 165, 296

Dole, M., 177, 212
Douglas, J. R., 278
Douglas-Hamilton, I., 34
Dubost, G., 294, 296
Dudzinski, M. L., 168
Dunbar, E. P., 212
Dunbar, R. I. M., 212
Duncan, P., 192
Dzieciolowski, R., 181, 233

East, K., 233, 239
Eaton, G. G., 60
Eggeling, W. J., 20
Ellerman, J. R., 10
Emlen, S. T., 299, 300
England, G. J., 222
Espmark, Y., 137, 151, 291, 301, 317
Estes, R. D., 134, 193, 234, 300
Evans, H., 12, 298
Everitt, B., 47

Fay, L. D., 276
Fenlon, J. S., 90
Ferrar, A. M., 295
Ferreira, R. E. C., 21, 27, 28, 323
Festing, M. F. W., 295
Fisher, R. A., 4, 94, 162, 168, 280, 316
Flerov, K. K., 10
Fletcher, T. J., 18, 54, 106
Flook, D. R., 164, 278, 279, 293, 302
Fohrman, M. H., 151
Ford, E. B., 3
Franklin, W. L., 177, 207, 212
Fraser, D., 151
French, C. E., 123

Gadgil, M., 3, 93
Garnett, M. C., 152
Gartlan, J. S., 202
Gasaway, W. C., 220
Gashwiler, J. S., 62, 278, 279
Gautier-Hion, A., 6
Geiger, G., 12
Geisler, P. A., 90
Geist, V., 4, 5, 65, 70, 107, 134, 136,
    147, 193, 203, 213, 267, 278, 290,
    302
Georgii, von, B., 224, 227, 243
Gibson, R. M., 5, 60, 88, 118, 128,
    129, 132, 133, 134, 135, 146, 149,
    150, 151
Gill, E., 6

Glucksman, A., 6, 156, 165, 249
Goddard, J., 134
Goldsmith, P. L., 50
Goodall, E. D., 70, 72, 220, 326
Goodall, J., 34
Gordon, J. E., 90, 279
Gosling, L. M., 7, 149, 170
Goss, R. J., 105, 106
Gossow, H., 118, 177, 178, 191, 202, 207, 278
Gottschlich, H. J., 120
Gould, S. J., 299
Grace, J., 25
Graf, W., 177
Grant, S. A., 25, 26, 233
Greenwood, P. J., 20, 61, 192, 196, 212, 214, 294, 296, 317
Gregory, M. E., 69
Grimble, A., 275
Grubb, P., 259, 267, 275, 278, 279
Guinness, F. E., 5, 27, 46, 47, 52, 56, 60, 64, 65, 66, 67, 68, 76, 86, 88, 89, 105, 106, 107, 128, 129, 132, 133, 134, 135, 146, 149, 150, 151, 164, 169, 172, 182, 183, 204, 249, 315, 323
Gunn, R. G., 151

Hadley, N. F., 94
Hall, M. J., 47, 62, 63, 70, 72, 182, 183, 212
Hall-Craggs, J., 187
Halliday, T. R., 52, 60, 138, 151
Hamer, A., 301
Hamilton, W. D., 2, 3, 171, 196
Hamilton, W. J., 62, 74, 87, 88, 90, 151, 239, 270
Hanby, J. P., 187
Hancock, J., 227
Harcourt, A. H., 6, 292
Harder, J. D., 278
Hare, H., 170
Harper, J. A., 177
Harvey, P. H., 6, 61, 181, 194, 196, 239, 245, 289, 291, 292, 300
Hassell, M. P., 297
Haukioja, E., 88
Hayes, H. H., 220
Hays, R. L., 165
Healy, W. B., 233
Heape, W., 51
Henshaw, J., 136, 291, 302

Hepburn, C. P., 279
Heptner, W. A., 133, 279
Hesselton, W. T., 278
Hickling, G., 164
Hirth, D. H., 181, 214, 301, 302
Hodgson, J., 227
Hofmann, R. R., 12
Holter, J. B., 220
Hoogland, J. L., 6
Horwood, M. T., 301, 302
Houston, D. B., 302
Howard, R. D., 4, 6, 153
Hrdy, S. B., 2
Hughes, G. P., 224
Hull, C. H., 41
Hunter, R. F., 25, 26, 186
Huxley, J. S., 11, 12, 151, 275, 289
Hyvarinen, H., 151, 270

Ineson, 239

Jackes, A. D., 178, 224, 239, 247, 248, 249
Jaczewski, Z., 106
Jardine, N., 187
Jarman, M. V., 137, 150
Jarman, P. J., 149, 178, 181, 193, 249, 300
Jenkins, J. G., 41
Jensen, P. V., 233
Johnson, E. D., 293
Johnson, H. E., 276
Jones, D. A., 62, 276, 278, 279
Jordan, P. A., 224
Julander, O., 276

Kavanagh, M. J., 6, 289, 292
Kay, R. N. B., 25, 62, 70, 72, 74, 90, 123, 151, 220, 224, 230, 239, 270, 312, 326
Kelly, J. T., 6
Kerr, C. D., 233
Kerr, M. A., 295
Kirkpatrick, R. L., 293
Kitchen, D. W., 149
Kitts, W. D., 267
Klein, D. R., 259, 275, 278, 279, 280, 295, 302
Klimstra, W. D., 275
Knight, R. R., 177, 181, 188, 189, 245
Kon, S. K., 69
König, R., 12

Krebs, C. J., 7, 165, 259, 279
Krebs, J. R., 192
Kruuk, H., 134, 192, 295
Kurland, J. A., 194, 202
Kurt, F., 303

Lack, D., 192, 196, 279
Lamb, W. I. C., 233
Larson, S. G. P., 6, 292
Latham, R. M., 279, 295
Lawton, J. H., 297
Leader-Williams, N., 154, 276, 279, 293
LeBoeuf, B. J., 4, 5, 60, 61, 87, 150, 162, 166, 167, 169
Leigh, E. G., 169
Lent, P. C., 65, 301
Léon, J. A., 93
Leutenegger, W., 6
Lieb, J. W., 177, 207, 212
Lincoln, G. A., 27, 52, 54, 56, 60, 76, 88, 93, 105, 106, 107, 137, 202, 245
Lockie, J. D., 177, 268
Longhurst, W. M., 278
Lott, D. F., 151
Lotwick, W., 34
Loudon, A., 166
Lovaas, A. L., 279
Low, J. B., 278, 279
Lowe, V. P. W., 11, 12, 15, 16, 18, 19, 76, 83, 239, 246, 299, 302
Ludwig, T. G., 233

McCance, R. A., 276, 277
McClure, P. A., 170
McCowan, D., 46, 75, 90, 124, 164, 233, 234, 239, 278, 293
McCullough, D. R., 64, 112, 120, 123. 177, 214, 275, 276, 293, 302, 303
McDougall, E. I., 16
McEwan, E. H., 62, 165, 167, 220, 276
McEwan, L. C., 123
McHugh, T., 317
McNaughton, S. J., 26
McNeely, D., 303
MacRoberts, M. H., 192
McVean, D. N., 21, 23
Magruder, N. D., 123
Mann, T., 293
Marburger, R. G., 275
Markgren, G., 62

Martin, M., 13
Martin, R., 299
Martin, R. D., 296
Masters, E. H., 301, 302
Mathews, M. O., 25
Mavor, W. O., 277
Maynard Smith, J., 2, 3, 6, 135, 162, 170
Medawar, P. B., 33
Medin, D. E., 293
Michael, E. D., 224
Miers, K. H., 275
Miller, G. R., 23
Miller, I., 13
Miller, R., 193
Miller, R. S., 5, 15
Miller, W. C., 164
Milne, J. A., 227
Mirarchi, R. E., 293
Mitchell, B., 17, 18, 25, 46, 48, 54, 62, 74, 75, 76, 83, 88, 90, 93, 98, 118, 120, 124, 156, 164, 165, 178, 181, 191, 228, 230, 275, 278, 293, 302
Moen, A. N., 165, 220, 242, 293
Mohun, 239
Monro, D., 13
Morgan, B. J. T., 187
Morgan, J. T., 222
Morley-Jones, R., 151
Morrison, J. A., 54, 112
Morrison-Scott, T. C. S., 10
Morton, E. S., 137
Mossman, A. S., 177, 212
Müller-Using, D., 133
Murie, O. J., 243
Mutch, W. E. S., 177, 268
Myers, J. H., 259, 279

Nahlik, A. J. de, 298
Nasimovitsch, A. A., 133, 279
Nature Conservancy Council, 15, 20, 39, 261
Nellis, C. H., 62
Nesbit-Evans, E. M., 234
Newsome, A. E., 279
Newton, I., 294
Nicholson, I. A., 23, 25, 46, 75, 90, 124, 164, 293
Nie, N. H., 41
Nietzsche, F., 104
Nievergelt, B., 278

Noonan, M., 6
Nordan, H. C., 165
Nowosad, R. F., 87

O'Gara, B. W., 134, 177, 224, 242, 243
Old Statistical Accounts (1796), 13, 14
Olson, S. T., 278, 279, 280
Oring, L. W., 299, 300
Osborne, B. C., 249, 297, 298
Otter, W., 14
Owen-Smith, N. Q., 231, 300
Ozoga, J. J., 54, 206

Packer, C., 4, 5, 61, 196
Parish, T., 278
Parker, G. A., 2, 105, 135
Parker, W. S., 93, 94
Partridge, L., 52
Patterson, I. J., 192
Payne, R. N., 4, 153
Peart, J. N., 77
Peek, J. M., 279, 293
Pennant, T., 13, 14
Perrins, C. M., 61, 196
Peterson, R. L., 302
Petrie, M., 170
Pianka, E. R., 93, 94, 154, 316
Picton, H. D., 181
Pielowski, Z., 134
Pollock, J. I., 196, 296
Pond, C. M., 5, 138, 156
Poore, M. E. D., 23
Povilitis, A., 303
Powell, R. P., 20, 212, 214, 317
Pratt, D. M., 182
Price, G. R., 135

Ralls, K., 196, 294
Ransom, B. A., 62
Ratcliffe, D. A., 21, 23
Red Deer Commission Annual Reports, 13, 190
Redfield, J. A., 259
Rees, W. B., 303
Reid, D., 224
Reimers, E., 276
Reiter, J., 87, 162, 166, 167
Richards, S. M., 317
Richie, W. F., 275
Ricketts, C., 154, 276

Ritchie, J., 12
Robbel, H., 279
Robbins, C. T., 165
Robertson, R. A., 23, 25
Robinette, W. L., 62, 276, 278, 279
Roe, N. A., 303
Rohlf, F. J., 49
Roseberry, J. L., 275
Rosen, J. K., 105, 106
Rouse, R. A., 279
Rowell, T. E., 202
Rowlands, J., 69
Rudder, B., 6
Ruff, R. L., 134, 177, 224, 242, 243
Russell, W. S., 151
Ryder, M. L., 267

Sade, D. S., 317
Sadleir, R. M. F. S., 261
Salovaara, R., 88
Scanlon, P. F., 293
Schaller, G. B., 134, 279, 301, 303
Schein, M. W., 151
Schinkel, P. G., 151
Schloeth, R., 133
Schloeth, V. R., 177
Schröder, W., 224, 227
Schürholz, G., 207
Scott, D. K., 34, 139, 194
Scrimshaw, N. S., 90, 279
Scrope, W., 12
Searcy, W. A., 7
Seger, 170
Sekulig, R., 234
Selander, R. K., 7
Severinghaus, C. W., 233, 278
Sharman, G. A. M., 62, 70, 72, 74, 90, 220, 239, 326
Sherman, P. W., 4, 6, 153, 214
Shoesmith, M., 177
Short, B. F., 151
Short, C., 165
Short, H. L., 220
Short, R. V., 6, 27, 52, 56, 88, 105, 106, 202, 245, 292, 293
Sibson, R., 187
Siegel, S., 49
Silver, H., 220
Simpson, A. M., 123, 220
Simpson, M. J. A., 187
Sinclair, A. R. E., 230, 275, 279

Sinh, R., 303
Slater, P. J. B., 187, 323, 325
Slee, J., 249
Smith, R. H., 196
Snedecor, G. W., 49
Sobanskii, G. G., 278
Sokal, R. R., 49
Southern, H. N., 259
Sparks, J., 214
Spassov, N. B., 6
Staines, B. W., 17, 18, 25, 48, 62, 76,
    83, 118, 120, 156, 164, 165, 178,
    181, 191, 193, 220, 224, 228, 230,
    239, 240, 243, 245, 246, 247, 275,
    278, 302, 312
Steinbrenner, K., 41
Stewart, L. K., 13
Stinson, N. L., 87, 162, 166, 167
Stirling, I., 164, 294
Stobbs, T. H., 227
Stonehouse, B., 291
Strandgaard, H., 302
Struhsaker, T. T., 107, 123, 301
Suire, C., 11
Suttie, J. M., 151, 213
Swift, R. W., 123
Syme, G. J., 317

Taber, R. D., 278, 302
Taitt, M. J., 259
Talbot, L. M., 279
Talbot, M. H., 279
Tanck, J. E., 278
Tanner, J. M., 151
Taylor, G. E., 90, 279
Teer, J. G., 62
Tener, J. S., 279
Thomas, J. W., 62, 275
Thompson, S. Y., 69
Thompson, W. A., 192
Thomson, A. M., 267
Thomson, B. R., 224
Thomson, W., 267
Tinkle, D. W., 94
Topinski, P., 136, 213
Touchberry, R. W., 165
Treisman, M., 192
Tribe, D. E., 224
Trivers, R. L., 1, 4, 5, 60, 93, 162,
    168, 170

Ullrey, D. E., 276
Underwood, R., 191
Urban, W. E., 220

Veen, H. E. van de, 12, 224, 230, 233
Verme, L. J., 54, 62, 220, 267
Vertinsky, I., 192
Vigne, N., 192
Vine, I., 192
Viret, J., 10
Vos, A. de, 302

Walker, E. A., 62
Walther, F., 65
Walvius, M. R., 11
Watson, A., 193, 239, 240
Wegge, P., 276
Welch, D., 17, 18, 25, 48, 62, 76, 83,
    118, 120, 156, 164, 165, 178, 181,
    191, 228, 230, 275, 278, 302
Western, D., 234, 296
White, F. N., 279
White, R. W. G., 293
Whitehead, G. K., 11, 301, 303
Whitehead, P. E., 62, 165, 167, 220
Whittaker, I. A. MacD., 222, 227, 231
Widdowson, E. M., 276, 277, 279
Wiens, J. A., 181
Wiersema, G. J., 202
Willard, D. E., 162
Williams, G. C., 2, 161
Williams, J. P. G., 151
Willis, M. B., 61, 196
Wilson, A., 61, 196
Wilson, E. O., 5, 192
Wood, A. J., 165, 267, 293
Woodgerd, W., 278
Woodward, R. H., 50
Woolf, A., 278
Woolhouse, H. W., 25
Wrangham, R. W., 192, 193, 196, 216,
    300
Wyatt, J. R., 234
Wynne Edwards, V. C., 3

Yanushko, P. A., 178, 245
Youatt, W. G., 276
Youngson, R. W., 27, 105, 202, 245

Subject Index

Aberdeenshire, 240
Activities, defined, 46, 48
Activity budgets: area differences in, 269–70; of estrous hinds, 54; of harem holders, 121–23, 274; sex differences in, 224–27, 247
Activity patterns, 221, 222–27; measurement of, 42–44; of orphans, 195; of rutting stags, 107, 121–28
Affiliative behavior, 214
Age of hinds: and association with relatives, 186–87; and dominance, 210–12, 214, 216; at first breeding, 18, 144; and range size, 184–86
Age of stags: and association, 178, 191, 193–94; and body condition, 120–21; and dominance, 204, 213, 214, 216; and fighting success, 131–32, 135, 150; and reproductive success, 144, 146, 150, 153; and reproductive value, 154, 155; and rutting activities, 117–21; and weight, 120, 213
Agelaius phoenicius, 7
Agrostis canina, 23
Agrostis/Festuca grassland, 22–23, 27–28, 30, 230, 233, 249, 311, 312, 314
Agrostis spp., 22, 25, 233
Agrostis tenuis, 22–23, 312
Alces, taxonomy of, 307
Alces alces, 11, 291, 293
Altitude: use of, 221, 227–29, 242, 243, 247; and vegetation, 25, 314
Ammophila spp., 22
Antelopes, 249
Anthoxanthum odoratum, 22–23
Anthyllus vulneraria, 22
Antlers: chewing of, 234; and fighting

ability, 287; growth of, 221, 270, 275, 287; growth of, and population size, 270, 287; interspecific comparisons of, 289–92
Antler size: and assessment of opponents, 136–37; and dispersal, 190; and dominance, 213; factors affecting, 151, 152, 213, 275–76; interspecific comparisons of, 291; and reproductive success, 136–37, 151
Appetite, 123, 220
Approaches: effects of, on harem holder, 124–26, 127–28, 129, 130, 135; by rutting stags, 113
Area differences: in adult sex ratio, 273; in antler growth, 270; in feeding behavior, 28, 241, 269–70; in food availability, 27–28, 241; in mortality, 90, 264, 270–72; in mother/offspring association, 270; in movements, 270, 273; in population density, 28, 241, 259–60; in reproductive performance, 84, 88, 90, 93, 238, 262, 263, 264, 275; in threat rates, 270
Artificial feeding, 178, 202
Assessment of potential opponents, 105, 135–39, 155
Association: between adult female relatives, 186–87, 192, 194–96, 216, 275, 287; defined, 47; long-term, 177, 186, 187–90; between males and females, 193–94; sex differences in, 202, 214, 216; among stags, 190–91, 212
Axis, 10; taxonomy of, 307
Axis axis, 291
Axis kuhlii, 291
Axis porcinus, 291

Barasingh, 301
Bawean deer, 300
Bellis perennis, 22
Betula spp., 312
Bewick swans, 139
Birth, 62–64
Birth date: and antler length, 152; and birth weight, 90; in calculating conception date, 146, 148; and calf mortality, 87–88, 89–90, 91, 167; determination of, 46; factors affecting, 88; and sex of previous offspring, 167; and stag dispersal, 190; and offspring sex ratio, 163
Birth weight: and antler length, 152; and calf mortality, 87, 88, 89, 90, 91, 92; factors affecting, 76, 88, 95, 98, 260–62; interspecific comparisons of, 293–94; of males and females, 164–65, 173; of male and female caribou, 167; and stag dispersal, 190; and suckling-bout duration, 72
Biting flies, 192, 193, 228
Black-backed gulls, 20
Blanket bog, 21
Blastocerus, taxonomy of, 307
Body condition of females: and coat change, 267; factors affecting, 68, 74–77, 81, 83–84, 95, 167; and offspring sex ratio, 162–64; and reproductive performance, 54, 68–69, 83–84, 85, 88, 89, 90, 92–93, 95, 98, 146, 152
Body condition of males: and age, 120; and dominance, 150, 213; and mortality, 279; in the rut, 120, 123, 137, 150
Body size: and antlers, 151, 213; of calves, and calf mortality, 89; and conception date, 93; costs of, 6–7, 277, 287; and dominance, 151, 213; and fecundity, 83–84, 93; and food quality, 249; and growth, 151, 156–57; and heat loss, 247–48; of males competing for females, 4, 278; of mothers and calves, 90, 152; and nutritional requirements, 239, 248, 277; and reproductive success of females, 93, 155, 156; and reproductive success of males, 150–51, 155, 156, 287; and sexual dimorph-

ism in, 6–7, 144, 155, 156–57, 247, 287; and stag age, 213
Bones, chewing of, 234
Brocket deer, 300
Bushbaby, 171

Calf catching, 36–39, 85
Calf growth: factors affecting, 68, 74, 90, 98, 151, 152; sex difference in, 165–67
Calf/hind ratios, 267
Calluna, 21–22, 27, 230, 233, 240, 241, 244–45, 311
Calluna vulgaris, 21, 312
Calluneatum, 229
Calving, timing of, 18, 76, 167, 173
Capreolus, taxonomy of, 307
Capreolus capreolus, 291, 293
Carex, 21, 22, 233, 312
Carex arenaria, 22
Carex nigra, 22
Caribou, 151, 167, 278, 279, 290, 292, 293, 301
Carnivores, 279
Casting date: defined, 49; and dominance, 212; factors affecting, 212, 270; interspecific comparisons of, 290
Cattle, 20, 151, 222
Cervidae, taxonomy of, 307–8
Cervus, 10–11; taxonomy of, 307
Cervus canadensis, 10, 11, 291, 293
Cervus duvauceli, 291
Cervus eldi, 291
Cervus nippon, 291
Chasing by stags, 117, 204
Chinese water deer, 301, 304
Chivying, 112, 126, 127, 328
Cirsium palustre, 22
Cleaning date, 49, 105, 270
Climate of Rhum, 20–21, 310
Cluster analysis, 182–88, 191, 329
Coat change: defined, 49; factors affecting, 74, 249, 267
Competition for mates, 1, 4, 105, 144, 155; and group living, 193; and population density, 277–78
Conception date: factors affecting, 88, 90, 93, 146, 262; measurement of, 46, 146, 148; and offspring sex ratio, 165; and stag reproductive

success, 123, 135, 146–48; synchrony of, 54, 61
Conception peak, 54, 61, 149
Copulations, 54–55; data collection, 44; involving close relatives, 61
Core area: area differences in size of, 270; defined, 48, 321–23; of hinds, 238; plant communities in, 238, 247; size of, 242–45; of stags, 191, 238; use of, by nonrelatives, 172–73. *See also* Range
Core area overlap, 47–48; in matrilineal groups, 188, 195–96; between mothers and daughters, 170, 184–86, 192. *See also* Range overlap
Costs of reproduction: body condition, 68, 74–76, 123; energetic costs of fighting, 135; energy expenditure by mothers, 162; harem disruption during fights, 134–35; injuries to rutting stags, 123, 132–34; and maternal age, 94; and matrilineal group size, 170; mortality, 77, 147; and offspring sex, 164, 167, 168–69, 170; sex difference in, 220; subsequent reproductive performance, 76–77
Coypu, 7
Crysops relicta, 228
CT. See *Calluna*
Cuckoo, 261
Cumulative sum testing, 45, 49–50

Dama, taxonomy of, 307
Dama dama, 291, 293
Data collection, 39–44
Date of death, determination of, 46
Day range length, 48, 53, 221, 242
Deer, phylogeny of, 10–12
Deschampsia, 21, 25
Deschampsia flexuosa, 312
Desertion of calves, by mothers, 39, 85
Dicrocerus, 10
Digestibility of food from different plant communities, 23–25, 230, 312
Disease, 279
Dispersal: sex differences in, 170, 192, 197, 296; by stags, 190–91, 192, 196–97, 272–73. *See also* Emigration
Displacing, 113, 202–3, 206, 213

Diurnal variation: in feeding behavior, 222–29; in ranging behavior, 242
Dominance among females: factors affecting, 210–12, 214–16; and mother/offspring association, 186, 194–95; and reproductive performance, 90, 212, 216
Dominance among males, 135–36, 204, 212–13; and dispersal, 191; factors affecting, 151, 213, 214, 216; and reproductive success, 150, 212, 216

Eagle, 20, 85
Early growth: and calf mortality, 89; and lifetime reproductive success, 151, 155–57, 171; and milk yield of mother, 207; sex differences in, 155–57, 173, 275–76
EC. *See* Blanket bog
Elaphodus, taxonomy of, 307
Elaphurus, taxonomy of, 307
Elaphurus davidianus, 291
Elephants, 296
Elephant seal, 61, 167, 169–70
Elk, 10, 11, 65, 120, 123, 177, 188–90, 224, 242, 245, 279, 288
Emigration and population density, 268–69, 272–73, 275, 276. *See also* Dispersal
Energy requirements, 220–21, 248
Equids, 214
Erica, 21
Erica cinerea, 22
Erica tetralix, 21
Eriophorum, 21, 24, 29, 233, 311, 314
Eriophorum angustifolium, 22
Eriophorum vaginatum, 21, 312
Estrus: behavior of hinds in, 54, 56, 59, 61, 112, 214; reaction of stags to, 54–55, 124, 127; failure to conceive at, 55
Etorphine, 39
Euphrasia officinalis, 22
Euphrasia spp., 22

Fallow deer, 196, 292, 304
Fat deposition, sex differences in, 156–57, 249, 279, 287
Fecundity: and calf/hind ratios, 267;

defined, 45; factors affecting, 76, 83–84, 93, 146, 238, 262–63; and lifetime reproductive success, 82–83, 155

Feeding behavior, 220–50; area differences in, 269–70; diurnal variation in, 181, 222–29; sex differences in, 220–50, 259, 287, 296

Female choice of mate, 5, 52, 56–61

Feral goats, 20

Festuca, 25, 233

Festuca rubra, 22, 312

Festuca vivipara, 22, 23

Fiber content of food, 25, 239

Fighting ability: assessment of, 105, 135– 39, 155; and body size, 151, 287; and reproductive success, 155

Fighting success: defined, 46, 317; estimation of, 317–18; factors affecting, 131– 32, 136–37, 150, 151; and reproductive success, 149–52; and roaring rates, 137; and rutting grounds, 150

Fights, 105, 113–55; assessment of potential opponents in, 135–39; benefits and duration of, 135; benefits and frequency of, 135; and conception peak, 135; course of events in, 128–31; effects of, on harem size, 149–50; energetic costs of, 135; frequency of, and population density, 273; harem disruption during, 134–35, 139; involving harem holders, 128–39; initiation of, 113, 129, 135; injuries in, 123, 132–34

Flehman, 110, 328

Flight distance, 65, 67

Food availability: and birth weight, 261; and calf mortality, 88; effects of, 259, 288; and growth, 267, 275–76, 288; and milk yield, 267; and plant community, 230–33, 248; and range size, 244; and reproductive success of females, 155, 171; and reproductive success of males, 150, 171, 277–78; seasonal changes in, 23–26, 220, 227; and segregation of the sexes, 194; and threat frequencies, 204

Food competition: and association with relatives, 194–96; among

females, 1, 5, 155; and female range overlap, 84, 194–96; and food dispersion, 181; and party size, 192, 193; sex differences in, 7, 208; and stag dominance, 213; and threats, 195

Food dispersion: and breeding systems, 299–304; for herbivores, 5; and party size, 181; and range size, 245; and social interactions, 202

Food intake, 220, 227, 248

Food quality: and body size, 249; factors affecting, 23–25, 27–28, 230–31, 248; and juvenile growth, 267; and milk yield, 267; sex differences in, 239–41, 248

Food selection, between different plant communities, 230–41

G1. See *Agrostis/Festuca* grassland; Herb-rich heath

G2. See *Agrostis/Festuca* grassland

Galaginae, 171

Galium verum, 22

Geology of Rhum, 20, 27–28, 311

Gestation length, 61–62; in calculating conception dates, 46, 146, 148; costs of gestation, 74, 220; and offspring sex, 62, 148, 165, 173

Glen Dye, 17

Glenfeshie (Invernesshire), 17, 76, 90, 93, 95, 96, 99, 156, 247

Glen Fiddich, 17

Glensaugh, 74

Gray fox, 294

Graze types, 52, 65–66. *See also* Plant communities

Grazing, 221; defined, 48; as proportion of adult's activity budgets, 222–27; by rutting stags, 121–22, 126

Grazing bout: definition of, 48, 324–25; length, variation in, 222–23, 227

Grazing pressure: area differences in, 28, 241; and plant growth, 25–26

Greens, and range size, 244–45

Grooming, 214, 216

Group living, advantages of, 192–93, 216, 300

Group size. *See* Party size

Growth: and population density, 275, 276; sex differences in, 156–57, 173, 276, 287, 293–94. *See also* Calf growth

Habitat use, 220–50; and population density, 259, 269–70
Haematopota crassicornis, 228
Hamster, 295
Harem: competition for, 105; defined, 44; disruption during fights, 134–35; duration of holding, 144–50; membership of, 56–61
Harem holders: activity budgets of, and population density, 274; behavior of, 117–35; fights involving, 128–39; reactions to approaches, 127–28, 135; reactions to young stags, 127–28, 184; activity budgets, 328
Harem size: defined, 46; effects of young stags on, 118; and fighting ability, 149, 150; and population density, 273; and roaring, 123–24; and stag age, 144; and stag dispersal, 191; and stag reproductive success, 149, 155
Harem stability, 53, 56–61; changes in, 56; defined, 45; and population density, 273
Heather moorland, 25–26, 233, 236, 238, 239, 246, 248, 314
Heat loss: and body size, 247–48; sex differences in, 249, 279, 287; and timing of feeding, 224
Herb-rich heath, 22; use of, 233
Herding, 53, 54, 110, 112, 126, 328
Heterogametic sex, mortality of, 279
Hiding response, 63–68
Highland ponies, 20, 85
Hippocamelus, taxonomy of, 307
Hippocamelus antisensis, 291
Hippocamelus bisulcus, 291
Hog deer, 301
Holcus, 25
Holcus lanatus, 22
Holcus molis, 312
Home range. *See* Core area; Range
Hooded crows, 20
Hormones, 105–6, 123, 139, 190, 293
Huemul, 301

Hydropotes, taxonomy of, 307
Hymenoptera, social, 3

Immigration by stags, 272–73
Impala, 150
Incest avoidance, 196
Ingestion rate: seasonal changes in, 227; sex differences in, 222, 248, 276
Injuries: from hinds, 195; to rutting stags, 123, 132–34
Invermark, 17
Interspecific comparisons of breeding systems, 299–304
Interspecific comparisons of sex differences, 288–96
Irish elk, 299

Juncus acutiflorus, 22, 311
Juncus articulatus, 233
Juncus effusus, 22
Juncus marsh, 22, 23, 27, 28, 29, 30, 108, 230, 233, 236, 240, 241, 243, 247, 311, 314
Juncus squarrosus, 233
Juniperus communis, 312

KFI. *See* Kidney fat index
Kidney fat index: defined, 46; of mothers and calves, 90, 95, 96, 98–99; of different sex and age categories, 156
Knob length. *See* Antlers
Koeleria cristata, 22

Lactation, 68–70; deficiencies in, 85; energetic costs of, 5, 74, 220; nutritional requirements in, 239; and offspring sex, 165–68
Larder weight, 16–18, 48
Life span of hinds and reproductive success, 81–83, 97, 155
Life span of stags and reproductive success, 146–47, 148–49, 155
Lifetime reproductive success of hinds: defined, 46; estimates of, 81–82, 315; factors affecting, 82–83, 91–93, 155; variation in, 155
Lifetime reproductive success of stags: defined, 47; estimates of, 318; factors affecting, 143–57; variation in, 155

Linum catharticum, 22

Long greens, use of, 229, 236, 240, 241, 243, 247. See also *Agrostis/ Festuca* grassland

Lotus corniculatus, 22

Macropods, 279

Mallard, 295

Management, 297–99

Marsh deer, 301, 304

Maternal investment: birth and early care, 62–68; gestation, 61–62; and early growth, 74; lactation, 68–73; and mother's age, 93–99; and offspring sex, 162–74, 287; after weaning, 170–71

Mate selection, 56–61

Mating, 52–56

Matrilineal group: defined, 47; and long-term associations, 177, 187–90; and postweaning investment, 170–73; and social interactions, 202; threat frequencies in, 210, 216

Matrilineal group size: and mortality of calves and yearlings, 90; and offspring sex ratio, 164; and reproductive success, 84, 93, 155, 170, 171–72

Mazama, taxonomy of, 307–8

Mazama americana, 291

Mazama gouazoubira, 291

Mazama rufina, 291

Megaceros gigantius, 299

Metabolic rate: seasonal changes in, 220; sex differences in, 155, 249, 276, 279, 287

MG. See *Molinia* grassland

Milk composition, 69, 326

Milk yield, 68–70; and calf mortality, 88, 264; and food avalability, 267; and food quality, 267; and juvenile growth, 267; and offspring sex, 167–68; seasonal decline in, 88; in sheep, 77

Molinia, 21, 22, 311

Molinia caerulea, 21, 312

Molinia grassland, 23–24, 27, 28, 30, 230, 233, 235, 236, 238, 240, 241, 242, 243, 246, 311, 314

Moor burning, 298–99

Moose, 220, 224, 289, 292–93, 300

Mortality: age specific, 18–19; of calves, 76, 84–98, 155, 167, 263–67, 275; of hinds, 77, 268–69; measurement of, 39; and population density, 90, 259, 263, 268–69, 275, 276, 278–80; sex differences in, 156–57, 270–72, 278–80, 287, 294–96; of stags, 272, 278–80; of yearlings, 90

Moschus, taxonomy of, 307

Mosses, 21

Mother/daughter range overlap and feeding competition, 84

Mother/offspring association, 42, 182–84; with adult daughters, 170–73, 184–86, 194–95; area differences in, 270; sex differences in, 183–84, 192; with sons, 190

Mother's age: and association with adult daughters, 186–87; and body condition, 81, 88, 95; defined, 45; and offspring sex ratio, 163; and reproductive performance, 72, 81, 83, 88–89, 91, 93–98, 146, 152; and reproductive value, 93–99, 154–55

Mother's home range area, 83–84, 90, 93, 152, 164, 190; defined, 45

Mother's previous reproductive status: and birth date, 88, 146, 152; and birth weight, 88, 152; and calf growth, 152; and calf mortality, 88, 90–91; defined, 45; and maternal age, 95–97; and offspring sex ratio, 163–64; and suckling bout duration, 72

Mother's subsequent reproductive status: and association with offspring, 182–83; defined, 45; and dispersal of sons, 190; and maternal age, 97; and sex of offspring, 167; and weaning, 73, 166

Mountain sheep, 147, 149, 267

MR. See *Juncus* marsh

Mule deer, 276, 278, 279, 292–93, 304

Muntiacus, taxonomy of, 307

Muntiacus muntjac, 291

Muntiacus reevesi, 291

Muntjac, 289, 292–93, 300

Musk deer, 10

Musk oxen, 279

Myocastor coypus, 7

Nardus heath, 21, 311

Natural selection, levels of, 2–3, 168, 280

Nature Conservancy Council, annual mortality counts by, 39

Nearest neighbor distance: defined, 47; and party size, 207; sex differences in, 208, 210, 214, 216; and threat rates, 207, 208

Nutritional requirements: of calves, 239; of hinds, 220, 224, 236, 239; of males, 220–21, 236; and management, 298; sex differences in, 221, 247, 248–49, 259, 276, 279, 287, 298

Odocoileus, 290, taxonomy of, 307

Odocoileus hemionus, 291, 293

Odocoileus virginianus, 291, 293

Olfactory cues, 52, 106–9, 134

Orphans, 195

Ozotoceros, taxonomy of, 307

Parallel walks, 44, 113, 129, 138

Parasites, 85

Parental investment: defined, 4; by males, 105; and offspring sex, 162–74; terminal investment, 93–99; and variance in offspring reproductive success, 162. *See also* Maternal investment

Partridge, 295

Party size: defined, 47, 319–20; functional considerations, 192–93; interspecific compari-

sons, 288–89, 292–93, 299–304; and rates of movement, 242; in the rut, 52, 55, 61; and threat frequency, 207; variation in, 178–82

Paternity, 59, 146–48

Père David's deer, 301

Pheasant, 295

Photoperiod, 54, 106, 220, 267

Phylogeny of the *Cervidae,* 10–11

Plantago lanceolata, 22

Plant communities: and grazing bout length, 227; and party size, 179, 181; and range size, 244–45; on Rhum, 21–26; sex differences in use of, 193–94, 197, 207–8, 220–50; in the study area, 27–28, 323; and threat rates, 207–10

Playback experiments, 137

Population density: and antler lengths, 275–76; area differences in, 28, 241, 259–60; and birth date, 88; and birth weight, 260–62; and calf/hind ratios, 267; and emigration by hinds, 268–69, 275, 276; and emigration by stags, 272–73, 276; and growth, 275, 276; and harem size, 144, 273; and habitat use, 259, 269–70; and immigration by stags, 272–73; and intermale competition, 277–78; compared with mainland, 16; and matrilineal group size, 172; and mortality, 90, 259, 263, 268–69, 275, 276, 278–80; and reproductive performance, 259, 275, 276; and the rut, 273–74; and sex ratio, 272–73, 276, 287; and sexual dimorphism, 275–78, 287; and stag dispersal, 272–73; and suckling, 264–67; interaction with weather, 259

Population dynamics, 259–80

Population size: 309, 313; and antlers, 270; and birth weight, 260; and conception date, 262; defined, 49; and fecundity, 262; and rutting behavior, 273–74; and stag mortality, 272

Potentilla, 21

Potentilla erecta, 22

Predation: on calves, 20, 64, 85; and group living, 192–93, 196, 216, 300; and sex differences in mortality, 279

Primates, 90, 202, 212, 214, 289, 292

Primula vulgaris, 22

Protein requirements, 220–21, 239

Prunella vulgaris, 22

Pudu, taxonomy of, 308

Quail, 295

Range: changes at calving, 64; changes in the rut, 52, 53, 105; defined, 48, 321–23; and male age, 190; and male fighting success, 150; and matrilineal groups, 188; sex differences in, 193; size, 184–86, 221, 242–45, 270

Range overlap: calculation of, 323; defined, 47–48; and feeding competition, 84; mother/offspring, 84, 170, 192, 238; with relatives, 171, 195–96, 238; between stags and hinds, 239

Rangifer, 292; taxonomy of, 307

Rangifer tarandus, 291, 293

Ranging behavior, 242–49

Ranunculus acris, 22

Ranunculus flammula, 22

Rats, 276–77

Redwing blackbirds, 7

Reedbuck, 295

Reindeer, 151, 154, 220, 224, 276, 279, 289, 290–92, 293, 295, 301

Rejections of attempts to suckle, 73, 166, 173

Reproductive performance: area differences in, 93; of hinds, changes in, 260–70; compared with mainland, 28; and population density, 259; and previous reproductive status, 76–77

Reproductive status of hinds: and activity budgets, 224–26; and body condition, 74–75; and coat change, 74, 267; and nutritional requirements, 224, 248; and use of different plant communities, 235–36, 238–39, 240, 248; and

range size, 244; and threats, 211. *See also* Mother's previous reproductive status; Mother's subsequent reproductive status

Reproductive success of hinds: area differences in, 275; and breeding sex ratio, 168–69; and calf survival, 52, 91– 93; defined, 46; and dominance, 212; and matrilineal group size, 170, 171–72; effects of non-relatives on, 172–73; effects of relatives on, 172–73; causes of variation in, 5, 82–83, 91–93, 152–53, 154–55, 156, 197, 287. *See also* Lifetime reproductive success of hinds

Reproductive success of stags, 143–57; and breeding sex ratio, 168–69; defined, 46; dominance and, 213; estimates of, 144, 146, 147–48, 318; factors affecting, 105, 136–37, 144–52, 154–55, 156, 197, 287. *See also* Lifetime reproductive success of stags

Reproductive success, variation in: among females, 5, 81–89; among males, 4, 144–56; and parental investment, 162; sex differences in 287

Reproductive value of hinds: and age, 93–98, 154–55; defined, 46, 316; estimation of, 316

Reproductive value of stags: and age, 154, 155; defined, 47; estimation of, 316

Rhachomitrum, 21, 23

Rhum: history of red deer on, 13–15; compared with mainland, 28–29

Rodents, 170, 220, 279

Roe deer, 289, 292–93, 300

Rumination, 48, 222, 224, 269–70

Rut, 52–61, 104–39, 143–57; age, and timing of, 120, 149; behavior of young stags in, 56–57; and female range size, 52, 53; and male range changes, 53; and population density, 273–74; reaction of stags to estrous hinds in, 54–55, 127; and sex differences in mortality, 279; timing of, 18, 52, 54, 105, 106, 123, 149, 273

Rutting grounds: and fighting success,

150; occupation of, 52–53; and reproductive success, 149
Rutting stags: age of, 117–21; energy
expenditure by, 287; reactions of,
to calves and yearlings, 117; reactions of, to young males, 53, 105,
117, 126, 127–28, 184; weights of,
117
Rutting stags, displays and interactions of, 107–17, 123, 155. *See also*
Assessment of potential opponents

Sambar, 301, 304
Scarba, 17
Scotland, deer forests in, 12–13
Seals, 164, 294
Seasons: defined, 48; and feeding,
220–50; and nutritional requirements, 220–21; and party
size, 178–81; and ranging behavior,
242–47; and threat rates, 204–6;
and appetite, food intake, metabolic
rate, and weight, 220
Seaweed, use of, 230, 240, 241, 247
Secondary sexual characteristics: and
food availability, 275, 278; in
monogamous and polygynous
species, 5–6; and reaction of harem
holders, 127; changes in, in the rut,
105
Segregation between the sexes, 178,
192, 193–94, 239, 247
Sex differences, 287–88; in association, 183–84, 192, 202, 214, 216; in
birth weight, 165; in body condition, 287; in calf mortality, 90; in
costs of reproduction, 220; in dispersal, 170, 192; in dominance
hierarchies, 212; and evolution,
299; in factors affecting reproductive success, 154–55, 197, 216; in fat
deposition, 156–57, 249, 279, 287;
in feeding behavior, 222–50, 259,
276, 287; in food quality, 239; in
function of dominance, 216; in
gestation length, 46, 62, 148, 165,
173; in growth, 155, 156–57, 173,
275–76, 287; in habitat use, 197,
239–41; in heat loss, 249, 279, 287;
in home range areas, 193; interspecific comparisons of, 288–96;
and management, 297–99; in

metabolic rate, 155, 249, 276, 279,
287; in mortality, 156, 270–72,
278–80, 287; in mother/offspring
distance 183; in nutritional requirements, 193, 197, 220, 221,
247, 259, 276, 279, 287, 298; in
parental investment received, 287;
and population dynamics, 296–97;
in ranging behavior, 242–49; in
suckling, 165–66; in threat frequency, 197, 207–9, 211, 214–16,
288; in yearling mortality, 90
Sex ratio: adaptive variation in,
162–64, 168–74; area differences
in, 273; breeding, 152; Fisher's
theory of, 168–69; and population
density, 272–73, 276, 287
Sexual dimorphism: constraints on,
6–7; interspecific comparisons of,
289; and maintenance re-
quirements, 220, 247; and party
size, 289; and population density,
275–78; and reproductive success,
144, 155
Sexual selection, 4, 6, 153
Sheep, 77, 186, 222, 224, 231–33, 249
Sheep, management of: in relation to
deer, 12, 13; on Rhum, 15, 18
Shelter, 48, 181, 221–22, 245, 246–47,
249
Shoenus fen, 21, 311
Shoenus nigricans, 21
Short greens: grazing pressure on, 26;
party size on, 181; plant production
on, 23; on rutting grounds, 52, 53,
149, 150; in stag ranges, 150; use
of, 229, 230, 231, 236, 238, 239,
240–41, 246, 247, 248. See also
Agrostis/Festuca grassland; Herb-rich
heath
Siberia, 278
Sielingia decumbens, 22
Sika, 293, 301
Silver-backed jackal, 294
Soay sheep, 278
Social groups, structure of, 177–98
Social interactions, 202–17; recorded
in day watches, 43, 44; functional
considerations of, 214–16; between
hinds and stags, 107, 126–27, 328
Soil characteristics of plant associations, 311

Sparring, 46, 204
Sparrowhawk, 294
Sperm competition, 292
Sphagnum, 21
Sphagnum recurvum, 22
Spotted hyena, 294
Stag groups, 191–92, 197; dominance in, 150; fragmentation before the rut, 105
Statistical analysis, 49–50
Study area: geology of, 27–28; location of, 26–27; plant communities in, 27–28
Succisa pratensis, 22
Suckling bout duration, 72–73; and antler growth, 270; defined, 46; and maternal age, 95; and offspring age, 165, 327; and offspring sex, 165, 173; and population density, 264–67
Suckling frequency, 70–72; and maternal age, 95; and offspring sex, 165– 66, 173
Sutherland, 85

Tabanus bisignatus, 228
Tabanus montanus, 228
Terminal investment, 93–99
Testes: interspecific comparisons of, 292–93; seasonal changes in, 105–6
Threats: costs of, to recipient, 195; frequency of, 204–7, 212, 214–16, 288; between hinds, 210–12; to male relatives, 190; to orphans, 195; between stags, 209, 212–13; types of, 202–4
Thymus drucei, 22
Toads, 151
Tooth wear, 76, 231–33
Topography of Rhum, 20, 181, 245
Toque macaques, 296
Trichophorum, 21, 24, 311
Trichophorum caespitosus, 312
Tsessebe, 279
Turkey, 295
Turning point, defined, 45, 50

Udder size, 36, 62, 46

Vaccinium, 21, 23
Vigilance: by mothers, 65–66; and party size, 193
Viola riviniana, 22
Vocalizations: of hinds, 62, 65, 214; of stags, 107, 117, 118, 123–26, 129, 137, 139, 214

Wallowing, 110, 126, 207, 328
Wapiti, 278, 293, 301
Weaning, 73, 97, 166, 169–70
Weather: and antler growth, 270; and birth weight, 88, 261; and calf/hind ratios, 267; and calf mortality, 88, 263; and conception date, 262; data collection on, 39; and fecundity, 262; and hind emigration, 268–69; and hind mortality, 268–69; interaction of, with population density, 259–60; and party size, 181; and plant communities, use of, 235–36; and ranging behavior, 246–47
Weights: of adults caught for marking, 39; and age, 81, 120; birth, 16, 39, 46, 72, 76, 87, 88, 89, 90, 91, 95; carcass, 156; and fecundity, 83; and fighting success, 131, 151; and male reproductive success, 150–51; maternal, and calf growth, 152; of mothers and calves, 88, 90, 152; of rutting stags, 117, 120, 123; and stag dominance, 151, 213; winter decline in, 220. *See also* Birth weight; Larder weight
Wet heath, 21
White-tailed deer, 220, 224, 276, 278, 292–94, 301, 304
Wildebeest, 278, 279, 295

Yearling males, antler length of, 152